Fire Inspection and Code Enforcement

FIFTH EDITION

VALIDATED BY

THE INTERNATIONAL FIRE SERVICE TRAINING ASSOCIATION

PUBLISHED BY

FIRE PROTECTION PUBLICATIONS OKLAHOMA STATE UNIVERSITY

COVER PHOTOS:
Fort Worth, Texas Fire Department
Fullerton, California Fire Department.

Dedication

*This manual is dedicated to the members of that unselfish organization of men and women who hold devotion to duty above personal risk, who count sincerity of service above personal comfort and convenience, who strive unceasingly to find better ways of protecting the lives, homes and property of their fellow citizens from the ravages of fire and other disasters ... **The Firefighters of All Nations.***

Dear Firefighter:

The International Fire Service Training Association (IFSTA) is an organization that exists for the purpose of serving firefighters' training needs. IFSTA is a member of the Joint Council of National Fire Organizations. Fire Protection Publications is the publisher of IFSTA materials. Fire Protection Publications staff members participate in the National Fire Protection Association and the International Society of Fire Service Instructors.

If you need additional information concerning our organization or assistance with manual orders, contact:

For assistance with training materials, recommended material for inclusion in a manual or questions on manual content, contact:

Customer Services
Fire Protection Publications
Oklahoma State University
Stillwater, OK 74078-0118
1-(800) 654-4055

Technical Services
Fire Protection Publications
Oklahoma State University
Stillwater, OK 74078-0118

First Printing, October 1987
Second Printing, October 1988

Third Printing, December 1990

TABLE OF CONTENTS

LIST OF TABLES

THE INTERNATIONAL FIRE SERVICE TRAINING ASSOCIATION

The International Fire Service Training Association is an educational alliance organized to develop training material for the fire service. The annual meeting of its membership consists of a workshop conference which has several objectives —

> . . . to develop training material for publication
> . . . to validate training material for publication
> . . . to check proposed rough drafts for errors
> . . . to add new techniques and developments
> . . . to delete obsolete and outmoded methods
> . . . to upgrade the fire service through training

This training association was formed in November 1934, when the Western Actuarial Bureau sponsored a conference in Kansas City, Missouri, to determine how all agencies that were interested in publishing fire service training material could coordinate their efforts. Four states were represented at this conference and it was decided that, since the representatives from Oklahoma had done some pioneering in fire training manual development, other interested states should join forces with them. This merger made it possible to develop nationally recognized training material which was broader in scope than material published by an individual state agency. This merger further made possible a reduction in publication costs, since it enabled each state to benefit from the economy of relatively large printing orders. These savings would not be possible if each individual state developed and published its own training material.

From the original four states, the adoption list has grown to forty-four American States; six Canadian Provinces; the British Territory of Bermuda; the Australian State of Queensland; the International Civil Aviation Organization Training Centre in Beirut, Lebanon; the Department of National Defence of Canada; the Department of the Army of the United States; the Department of the Navy of the United States; the United States Air Force; the United States Bureau of Indian Affairs; The United States General Services Administration; and the National Aeronautics and Space Administration (NASA). Representatives from the various adopting agencies serve as a voluntary group of individuals who govern policies, recommend procedures, and validate material before it is published. Most of the representatives are members of other international fire protection organizations and this meeting brings together individuals from several related and allied fields, such as:

> . . . key fire department executives and drillmasters,
> . . . educators from colleges and universities,
> . . . representatives from governmental agencies,
> . . . delegates of firefighter associations and organizations, and
> . . . engineers from the fire insurance industry.

This unique feature provides a close relationship between the International Fire Service Training Association and other fire protection agencies, which helps to correlate the efforts of all concerned.

The publications of the International Fire Service Training Association are compatible with the National Fire Protection Association's Standard 1001, "Fire Fighter Professional Qualifications (1981)," and the International Association of Fire Fighters/International Association of Fire Chiefs "National Apprenticeship and Training Standards for the Fire Fighter." The standards are an effort to attain professional status through progressive training. The NFPA and IAFF/IAFC Standards were prepared in cooperation with the Joint Council of National Fire Service Organizations of which IFSTA is a member.

The International Fire Service Training Association meets each July at Oklahoma State University, Stillwater, Oklahoma. Fire Protection Publications at Oklahoma State University publishes all IFSTA training manuals and texts. This department is responsible to the executive board of the association. While most of the IFSTA training manuals can be used for self-instruction, they are best suited to group work under a qualified instructor.

PREFACE

This is the Fifth Edition of **Fire Inspection and Code Enforcement**. The greatly expanded text and illustrations detail all aspects of inspection responsibilities and procedures, plan review, hazardous materials and their associated storage and handling, building construction, exiting systems, and general fire safety practices.

Acknowledgement and special thanks are extended to the members of the validating committee who contributed their time, wisdom, and knowledge to this manual.

Chairman
Howard Boyd
Consultant
Nashville, Tennessee

Vice Chairman
Douglas Forsman, Fire Chief
Champaign Fire Department
Champaign, Illinois

Other persons serving on the committee during its tenure were:

Charlotte Badgett
Lewisville, (TX) Fire Department

William Cooper
Huntington Beach, (CA) Fire Department

Hal Courtney
Banning, (CA) Fire Department

Lonnie Jackson
Illinois Fire Inspectors Association
Arlington Heights, IL

William Jenaway
INA Loss Control Services
King of Prussia, PA

Thomas Makey
Deputy Fire Commissioner
Edmonton, Alberta, Canada

Donald Meaney
New York State Office of
 Fire Prevention and Control
Montour Falls, NY

George Oster
Iowa Fire Service Training
Ames, IA

William Peterson
Fire Chief
Plano, (TX) Fire Department

Ron Williamson
Fire Marshal
Edmond, (OK) Fire Department

Bill Vandevort
California State Fire Service Training
Sacramento, CA

A book of this scope would be impossible to publish were it not for the assistance of many persons and organizations. To the following and others listed in the captions, we owe a great debt as they gave freely of time, advise, personnel, and/or equipment.

Bilco Company
Pat Brock, Associate Professor, Fire Protection and Safety Engineering Technology, OSU
Building Officials and Code Administrators International, Inc. (BOCA)
Fort Worth, Texas Fire Department
  Larry McMillen, Fire Chief
  Jeffrey M. Shapiro, P.E., Assistant to the Fire Marshal
Illinois Fire Inspectors Association
Institute of Makers of Explosives
International Conference of Building Officials (ICBO)
Justrite Manufacturing, Inc.

Mount Prospect, Illinois Fire Department
National Fire Protection Association
National Research Council of Canada
Plano, Texas Fire Department
   William Peterson, Fire Chief
Edward Prendergast, Chicago Fire Department
Southern Building Code Congress International, Inc. (SSBCC)

     Cover Photos courtesy of the Fort Worth, Texas Fire Department and the Fullerton, California Fire Department.

     Special acknowledgements and thanks are extended to Lonnie Jackson, Mount Prospect, Illinois Fire Department, Edward Prendergast of the Chicago, Illinois Fire Department, and Bill Jones of the Plano, Texas Fire Department for taking many of the photographs appearing in this manual.

     Gratitude is also extended to the following members of the Fire Protection Publications staff, whose contributions made the final publication of the manual possible.

Lynne Murnane, Senior Publications Editor
Michael A. Wieder, Associate Editor
Charles Donaldson, Associate Editor
Robert Fleischner, Publications Specialist/Photographer
Carol Smith, Publications Specialist
Cynthia Brakhage, Unit Assistant
Scott Stookey, Research Technician
Don Davis, Coordinator, Publications Production
Ann Moffat, Graphic Designer
Michael McDonald, Graphic Artist II
Karen Murphy, Phototypesetter Operator II
Desa Porter, Phototypesetter Operator II
Terri Jo Gaines, Senior Clerk Typist

Gene P. Carlson
Editor

GLOSSARY

A

ACCESSIBILITY — The ability of fire apparatus to get close enough to a building to conduct emergency operations.

ACUTE — Severe, rapid onset, usually of short duration.

ALARM-INITIATING DEVICE — A mechanical or electrical device that activates an alarm system. There are three basic types of alarm-initiating devices: manual, products of combustion detectors, and extinguishing system activation devices.

AMPERE — (1) The amount of current sent by one volt through one ohm of resistance; (2) Unit of measurement of electrical current.

ARSON — The willful and malicious burning of a property.

AUTHORITY — Relates to the empowered duties of an official, in this case the fire inspector. The level of an inspector's authority is commensurate with the enforcement obligations of the governing body.

AUTOMATIC SPRINKLER SYSTEM — A system of water pipes, discharge nozzles, and control valves designed to activate during fires by automatically discharging enough water to control or extinguish a fire.

B

BACKDRAFT — Instantaneous combustion that occurs when oxygen is introduced into a smoldering fire. The stalled combustion resumes with explosive force.

BASEMENT PLANS — Drawings showing the belowground view of a building. The thickness and external dimensions of the basement walls are given, as are floor joist locations, strip footings, and other attached foundations.

BEARING WALLS — Walls of a building that by design carry at least some part of the structural load of the building.

BLEVE (Boiling Liquid Expanding Vapor Explosion) — The failure of a closed container as a result of overpressurization caused by an external heat source.

BOARD OF APPEALS — A group of five to seven people with experience in fire prevention and code enforcement who arbitrate differences in opinion between fire inspectors and property owners or occupants.

BOILING POINT — The temperature at which the vapor pressure of a liquid is equal to the external pressure applied to it.

BONDING — The connection of two objects with a metal chain or strap in order to neutralize the static electrical charge between the two.

BOYLE'S LAW — This law states that the volume of a gas varies inversely with the applied pressure. The formula is: $P_1V_1 = P_2V_2$, where

P_1 = original pressure
V_1 = original volume
P_2 = final pressure
P_2 = final volume

BRANDS — (1) Large, flying, burning embers that are lifted by a fire's thermal column and carried away with the wind, (2) Small burning pieces of wood or charcoal used to test the fire resistance of roof coverings and roof deck assemblies.

BRITISH THERMAL UNIT (BTU) — The amount of heat required to raise the temperature of one pound (0.45 kg) of water one degree Fahrenheit.

BUREAU OF ALCOHOL, TOBACCO, AND FIRE-ARMS (BATF) — Division of the U.S. Department of Treasury; regulates the storage, handling, and transportation of explosives.

C

CALORIE — The amount of heat required to raise the temperature of one gram (0.035 oz.) of water one degree centigrade.

CARBON DIOXIDE (CO_2) — A gas heavier than air used to extinguish Class B or C fires by smothering or displacing the oxygen.

CENTIGRADE (Celsius) — A temperature scale where the boiling point of water is 100°C (212°F)

and the freezing point is 0°C (32°F) at normal atmospheric pressure.

CHARLES' LAW — Scientific law that says the increase or decrease of pressure in a constant volume of gas is directly proportional to corresponding increase or decrease of temperature. The formula is stated: $P_1T_2 = P_2T_1$ where

P_1 = original pressure
T_1 = original temperature
P_2 = final pressure
T_2 = final temperature

CHEMTREC — The Manufacturing Chemists Association's name for its Chemical Transportation Emergency Center. The center provides immediate information about handling hazardous material incidents. The toll-free phone number is 1-800-424-9300 (1-202-483-7616 in Washington, D.C., Alaska, and Hawaii).

CHRONIC — Of long duration (opposite of acute).

CIRCUIT — The complete path of an electric current.

CITATION — A legal reprimand for failure to comply with existing laws or regulations.

CODES — Rules or laws used to enforce requirements for fire protection, life safety, or building construction.

COMBUSTIBLE LIQUID — Any liquid having a flash point at or above 100°F (37.8°C) and below 200°F (93.3°C).

COMBUSTION — The self-sustaining process of rapid oxidation of a material that produces heat and light.

COMMON HAZARD — A condition likely to be found in almost all occupancies and generally not associated with a specific occupancy or activity.

COMPLAINT — An objection to existing conditions that is brought to the attention of the fire inspection bureau.

COMPLIANCE — Meeting the minimum standards set forth by applicable codes or regulations.

COMPRESSED GAS — Gas that, at normal temperature, exists solely as a gas when pressurized in a container.

CONDUCTION — The transfer of heat energy from one body to another through a solid medium.

CONDUCTOR — A substance or material that transmits electrical or heat energy.

CONSTRUCTION CLASSIFICATION — The rating given to a particular building based on the materials and methods used to construct and their ability to resist the effects of a fire situation.

CONVECTION — The transfer of heat energy by the movement of air or liquid.

CORROSIVES — Those materials that cause harm to living organisms by destroying body tissue.

CRYOGENICS — Gases that are converted into liquids by being cooled below -150°F (-101°C).

D

DETAILED VIEW — Additional, closeup information shown on a particular section of a larger drawing.

DIKES — Temporary or permanent barriers that prevent liquids from flowing into certain areas or that direct the flow as desired.

DIMENSIONING — A drawing that places a building on a site plan to clearly show its size and arrangement relative to existing conditions.

DRAFT CURTAINS — Dividers hung from the ceiling in large open areas that are designed to minimize the mushrooming effect of heat and smoke.

DRY CHEMICAL — Any one of a number of powdery extinguishing agents used to extinguish fires.

DRY STANDPIPE SYSTEM — A standpipe system that either has water supply valves closed or that has no fixed water supply to it.

DUCTS — Hollow pathways used to move air from one area to another in ventilation systems.

E

ELECTRIC SHOCK — Injury caused by electricity passing through the body. Severity of injury depends upon the path the current takes through the body, the amount of current, and the resistance of the skin.

ELECTRICAL SYSTEMS — Those wiring systems designed to distribute electricity throughout a building.

ELECTRON — A minute component of an atom that possesses a negative charge.

ELEVATION VIEW — An architectural drawing used to show the number of floors of a building, ceiling heights, and the grade of surrounding ground.

EMERGENCY RESPONSE GUIDEBOOK (E.R.G.) — A manual provided by the U.S. Department of Transportation that aids emergency response personnel in identifying hazardous materials placards. It also gives guidelines for initial actions to be taken at hazardous materials incidents.

EQUIVALENCY — Alternative practices that are acceptable for meeting a minimum level of fire protection.

EX — Rating symbol used on lift trucks that are safe for use in atmospheres containing flammable vapors or dusts.

EXHAUST SYSTEM — A ventilation system designed to remove stale air, smoke, vapors, or other airborne contaminants from an area.

EXIT — That portion of a means of egress that is separated from all other spaces of the building structure by construction or equipment and that provides a protected way of travel to the exit discharge.

EXIT ACCESS — The portion of a means of egress that leads to the exit. Hallways, corridors and aisles are examples of exit access.

EXIT CAPACITY — The maximum number of people who can discharge through a particular exit.

EXIT DISCHARGE — That portion of a means of egress that is between the exit and a public way.

EXIT STAIRS — Stairs that are used as part of a means of egress. The stairs may be part of either the exit access or the exit discharge when conforming to requirements in the *Life Safety Code®*.

EXPELLENT GAS — Any of a number of inert gases that are compressed and used to force extinguishing agents from a portable fire extinguisher. Nitrogen is the most commonly used expellent gas.

EXPLOSIVE — Any material or mixture that will undergo an extremely fast self-propagation reaction when subjected to some form of energy.

EXTINGUISHING AGENT — Any substance used for the purpose of controlling or extinguishing a fire.

F

FACTORY MUTUAL SYSTEM (FM) — Fire research and testing laboratory that provides loss control information for the Factory Mutual System and anyone else who may find it useful.

FAHRENHEIT — Temperature scale in which the boiling point of water is 212°F (100°C) and the freezing point is 32°F (0°C) at normal atmospheric pressure.

FIELD SKETCH — A rough drawing of an occupancy that is made during an inspection. The field sketch is used to make a final inspection drawing.

FIRE ALARM SYSTEM — (1) A system of alerting devices that takes a signal from fire detection or extinguishing equipment and alerts building occupants or proper authorities of a fire condition, (2) A system used to dispatch fire department personnel and apparatus to emergency incidents.

FIRE CAUSE — The combination of fuel supply, heat source, and a hazardous act that results in a fire.

FIRE CAUSE DETERMINATION — The process of establishing the cause of a fire incident through careful investigation and analysis of the available evidence.

FIRE DAMPER — A device that automatically interrupts air flow through all or part of an air handling system, thereby restricting the passage of heat and the spread of fire.

FIRE DEPARTMENT CONNECTION (FDC) — An inlet appliance that has two or more 2½-inch (65 mm) connections or one large-diameter (4-inch [100 mm] or larger) connection through which fire apparatus can boost the pressure or amount of water flowing through a sprinkler or standpipe system.

FIRE DETECTION SYSTEM — A system of detection devices, wiring, and supervisory equipment used for detecting fire or products of combustion and then signaling that these elements are present.

FIRE DOOR — A specially constructed, tested, and approved door installed to prevent fire spread.

FIRE DRILL — A training exercise to ensure that the occupants of a building can exit the building in a quick and orderly manner in case of fire.

FIRE EXTINGUISHER — A portable fire fighting device designed to combat incipient fires.

FIRE HAZARD — Any material, condition, or act that contributes to the start of a fire or that increases the extent or severity of fire.

FIRE LOAD — The maximum amount of heat that can be produced if all the combustible materials in a given area burn.

FIRE PARTITION — A fire barrier that extends from one floor to the bottom of the floor above or to the underside of a fire-rated ceiling assembly. A fire partition provides a lower level of protection than a fire wall.

FIRE POINT — The temperature at which a liquid fuel produces sufficient vapors to support combustion once the fuel is ignited. The fire point is usually a few degrees above the flash point.

FIRE PREVENTION CODE — A law enacted for the purpose of enforcing fire prevention and safety regulations.

FIRE RESISTANCE RATING — The amount of time a material or assembly of materials will resist a typical fire as measured on a standard time-temperature curve.

FIRE RETARDANT — A chemical applied to material or another substance that is designed to retard ignition or the spread of fire.

FIRE STOP — Materials used to prevent or limit the spread of fire in hollow walls or floors, above false ceilings, in penetrations for plumbing or electrical installations, or in cocklofts and crawl spaces.

FIRE WALL — A wall with a specified degree of fire resistance that is designed to prevent the spread of fire within a structure or between adjacent structures.

FLAMEOVER — Condition that occurs when a portion of the fire gases trapped at the upper level of a room ignite, spreading flame across the ceiling of the room.

FLAME SPREAD RATING — A numerical rating assigned to a material based on the speed and extent to which flame travels over its surface.

FLAME TEST — A test designed to determine the flame spread characteristics of structural components or interior finishes.

FLAMMABLE LIQUID — Any liquid having a flash point below 100°F (37.8°C) and having a vapor pressure not exceeding 40 psi absolute (276 kPa).

FLAMMABLE AND EXPLOSIVE LIMITS — The upper and lower concentrations of a vapor expressed in percent mixture with an oxidizer that will produce a flame at a given temperature and pressure.

FLASHOVER — The state of a fire at which all combustibles are heated to their ignition temperatures and the area becomes fully involved in fire.

FLASH POINT — The lowest temperature at which sufficient vapors are produced from a flammable liquid to form an ignitable mixture.

FLOOR PLAN — An architectural drawing showing the layout of a floor within a building as seen from above. It outlines where each room is and what the function of the room is.

FOAM — An extinguishing agent produced by mixing a foam-producing compound with water and aerating the solution for expansion. These agents are primarily used for extinguishing Class B fires, but in some cases may be used on Class A fires as well.

G

GAS — A compressible substance, with no specific volume, that tends to assume the shape of its container. Molecules move about most rapidly in this state.

GROUNDING — Reducing the difference in electrical potential between an object and the ground by the use of various conductors.

H

HALOGENATED AGENTS — Chemical compounds (halogenated hydrocarbons) that contain carbon plus one or more elements from the halogen series. Halon 1301 and Halon 1211 are most commonly used as extinguishing agents for Class B and C fires.

HAZARDOUS MATERIAL — Any material that possesses an unreasonable risk to the health and safety of persons and/or the environment if it is not properly controlled during handling, storage, manufacture, processing, packaging, use, disposal, or transportation.

HEAT — A form of energy that is proportional to molecular movement. To signify its intensity, it is measured in degrees of temperature.

HEAT TRANSFER — The flow of heat from a hot substance to a cold substance. This may be accomplished by convection, conduction, or radiation.

HOSELINE — A section of flexible conduit that is connected to a water supply source for the purpose of delivering water onto a fire.

HVAC SYSTEM (Heating, Ventilating, and Air Conditioning System) — The mechanical system used to provide environmental control within a structure.

HYDROSTATIC TEST — A testing method used to check the integrity of pressure vessels.

I

IGNITION SOURCE — A method (either wanted or unwanted) that provides a means for the initiation of self-sustained combustion.

IGNITION TEMPERATURE — The lowest temperature at which a fuel, when heated, will ignite in air and continue to burn; the temperature required to cause ignition of a substance.

IMMUNITY — Freedom from legal liability for an act or physical condition.

INCIPIENT PHASE — The first stage of the burning process where the substance being oxidized is producing some heat, but the heat has not spread to other substances nearby.

INSPECTION — A formal examination of an occupancy and its associated uses or processes to determine its compliance with the fire and life safety codes and standards.

IONIZATION — The process by which an object or substance gains or loses electrons, thus changing its electrical charge.

L

LEGEND — An explanatory list of symbols on a map or diagram.

LIABILITY — To be legally obligated or responsible for an act or physical condition.

LIFE SAFETY CODE® (NFPA 101®) — A building standard designed to protect lives in the event of a fire.

LIQUEFIED GAS — A confined gas that at normal temperatures exists in both liquid and gaseous states.

LIQUID — An incompressible substance that assumes the shape of its container. The molecules flow freely, but substantial cohesion prevents them from expanding as a gas would.

LIVE LOAD — The force placed upon a structure by the addition of people, objects, or weather.

M

MAGAZINE — A storage facility approved by the Bureau of Alcohol, Tobacco, and Firearms (BATF) for the storage of explosives.

MEANS OF EGRESS — A safe and continuous path of travel from any point in a structure leading to a public way. The means of egress is comprised of three parts: the exit access, the exit, and the exit discharge.

MECHANICAL SYSTEMS — Large equipment systems within a building that may include, but are not limited to, climate control systems; smoke, dust, and vapor removal systems; trash collection systems; and automated mail systems. These do not include general utility systems such as electric, gas, and water.

N

NATIONAL FIRE PROTECTION ASSOCIATION (NFPA) — A nonprofit educational and technical association dedicated to protecting life and property from fire by developing fire protection standards and educating the public.

NATIONAL RESPONSE CENTER — A federal organization charged with coordinating the response of numerous agencies to emergency incidents involving the release of significant amounts of hazardous materials.

NEUTRON — A part of the nucleus of an atom that has a neutral electrical charge.

NFPA 704 LABELING SYSTEM — A system for identifying hazardous materials in fixed facilities. The placard is divided into sections that identify the degree of hazard with respect to health, flammability, and reactivity, as well as special hazards.

NONCOMBUSTIBLE — Incapable of supporting combustion under normal circumstances.

NORMAL OPERATING PRESSURE — The normal amount of pressure that is expected to be available from a hydrant, prior to pumping.

O

OCCUPANCY CLASSIFICATION — The classifications given to structures for the National Building Code by the American Insurance Association or the authority having jurisdiction.

OCCUPATIONAL SAFETY AND HEALTH ADMINISTRATION (OSHA) — A federal agency that develops and enforces standards and regulations for occupational health and safety in the workplace.

OCCUPANT LOAD — The total number of people who may occupy a building or portion of a building at any given time.

Ohms — Units of measurement of electrical resistance.

ORDINANCE — A law set forth by a government agency, usually at the local municipal level.

OXIDIZER — Any material that provides oxygen for combustion.

P

PANIC HARDWARE — Hardware mounted on exit doors in public buildings that enables doors to be opened from the inside when pressure is applied.

PARAPET — A portion of a wall that extends above the level of the roof.

PITOT TUBE — An instrument containing a Bourdon Tube that is inserted into a stream of water to measure the velocity pressure of the stream. The gauge reads in units of pounds per square inch (psi [kPa]).

PLACARD — A diamond-shaped sign that is affixed to all sides of a vehicle transporting hazardous materials. The placard indicates the primary class of the material, and in some cases the exact material, being transported.

PLAN VIEW — A drawing containing the two-dimensional view of a building as seen from directly above the area.

PLANS REVIEW — The process of reviewing building plans and specifications to determine the safety characteristics of the proposed building. This is generally done before permission is granted to begin construction.

PLOT PLAN — An architectural drawing showing the layout of buildings and landscape features for a given plot of land. The view is from directly above.

POINT OF ORIGIN — The exact location at which a particular fire started.

POLICE POWER — The authority that may be given to an inspector to arrest, issue summons, or issue citations for fire code violations.

PREDISCHARGE ALARM — An alarm that sounds before a total flooding fire extinguishing system is about to discharge, thus giving occupants the opportunity to leave the area.

PRE-FIRE PLANNING — Advance planning of fire fighting operations at a particular location, taking into account all factors that will influence fire fighting tactics.

PRESSURE-REDUCING VALVE — A valve installed at standpipe connections that is designed to reduce the amount of water pressure at that discharge to a specific pressure, usually 100 psi (690 kPa).

PRESSURE RELIEF DEVICE — An automatic device designed to release excess pressure from a container.

PROTON — A part of an atom that possesses a positive charge.

PUBLIC WAY — A parcel of land such as a street or sidewalk that is essentially open to the outside and is used by the public for moving from one location to another.

PYROLYSIS — The chemical decomposition of a substance through the action of heat.

R

RADIATION — Heat transfer energy by electromagnetic waves.

RADIOACTIVE MATERIAL — A material whose atomic nucleus spontaneously decays or disintegrates, emitting radiation.

RADIOACTIVE PARTICLE — Particles emitted during the process of radioactive decay. There are three types of radioactive particles: alpha, beta, and gamma particles.

REACTIVE MATERIAL — A material that will react violently when combined with air or water.

REACTIVITY — The ability of two or more chemicals to react and release energy and the ease with which this reaction takes place.

REFRIGERANT — A substance used within a refrigeration system to provide the cooling action.

RESIDUAL PRESSURE — The pressure remaining at a given point in a water supply system while water is flowing.

RESPONSIBILITY — An act or duty for which someone is clearly accountable.

RIGHT OF ENTRY — The rights set forth by the administrative powers that allow the inspector to inspect buildings to ensure compliance with applicable codes.

RISER — A vertical water pipe used to carry water for fire protection systems above ground, such as a standpipe or sprinkler riser.

ROOF COVERING — The final outside cover that is placed on top of a roof deck assembly. Common roof coverings include composite or wood shake shingles, tile, slate, tin, or asphaltic tar paper.

ROOF DECK — The bottom components of the roof assembly that support the roof covering. The roof deck may be constructed of such components as plywood, wood studs (2 inches x 4 inches [51 mm x 102 mm]) or larger, lathe strips, and so on.

S

SALAMANDER — A portable heating device generally found on construction sites.

SANCTION — A notice or punishment attached to a violation for the purpose of enforcing a law or regulation.

SECTIONAL VIEW — A vertical view of a building as if it were cut into two parts. The purpose of a sectional view is to show the internal construction of each assembly.

SEMICONDUCTOR — Material that is neither a good conductor nor a good insulator, and therefore may be used as either in some applications.

SITE PLAN — A drawing that provides a view of the proposed construction in relation to existing conditions. It is generally the first sheet on a set of drawings.

SMOKE DAMPER — A device that restricts the flow of smoke through an air handling system.

SMOKEPROOF ENCLOSURES — Stairways that are designed to limit the penetration of smoke, heat, and toxic gases and that serve as part of a means of egress.

SOLID — A substance that has a definite shape and size. The molecules of a solid generally have very little mobility.

SPECIAL FIRE HAZARD — A fire hazard arising from the processes or operations that are peculiar to the individual occupancy.

SPECIFIC GRAVITY — The ratio of the weight of a liquid to an equal volume of water.

SPRINKLER — The waterflow discharge device in a sprinkler system. The sprinkler consists of a threaded intake nipple, a discharge orifice, and a deflector to create an effective fire stream pattern.

STANDARD — A document containing requirements and specifications outlining minimum levels of performance, protection, or construction.

STANDPIPE SYSTEM — Fire department hose outlets installed on different levels of a building to be used by firefighters and/or building occupants.

STATIC ELECTRICITY — An accumulation of electrical charges on opposing surfaces created by the separation of unlike materials or by the movement of surfaces.

STATIC PRESSURE — The pressure at a given point in a water system when no water is flowing.

STEINER TUNNEL — A horizontal furnace 25 feet (7.6 m) long, 17½ inches (445 mm) wide, and 12 inches (305 mm) high. A 5,000 BTU (5 270 Kj) flame is produced in the tunnel and the extent of flame travel across the surface of the test material is observed through ports in the side of the furnace.

STEINER TUNNEL TEST — A test used to determine the flame spread ratings of various materials.

STORAGE TANKS — Storage vessels that are larger than 60 gallons (227 L) and are located in a fixed location.

STRUCTURAL ABUSE — Using or changing a building beyond its original designed capabilities.

SUPERVISORY CIRCUIT — An electronic circuit within a fire protection system that monitors the system's readiness and transmits a signal when there is a problem with the system.

T

TEMPERATURE — The measurement of the intensity of the heating of a material.

THRESHOLD LIMIT VALUE (TLV) — The concentration of a given material that may be tolerated for an 8-hour exposure during a regular workweek without ill effects.

TITLE BLOCK — A small information section on the face of every plan drawing. The title block contains such information as name of project, title of the particular drawing, the scale used, and date of drawing and/or revisions.

TORT — A wrongful act (except for breach of contract) for which a civil action will lie.

TOXIC MATERIAL — Any material classified as a poison, asphyxiant, irritant, or anesthetic.

TOXICITY — The ability of a substance to do harm within the body.

TRAVEL DISTANCE — The distance from any given area in a structure to the nearest exit or to a fire extinguisher.

U

UNDERWRITERS LABORATORIES, INC. (UL) — An independent fire research and testing laboratory.

UNITED STATES DEPARTMENT OF TRANSPORTATION — The federal organization that regulates the transportation of hazardous materials.

UNSTABLE MATERIAL — Material that is capable of undergoing chemical changes or decomposition with or without a catalyst.

V

VAPOR DENSITY — The weight of a gas as compared to the weight of air.

VAPORIZATION — The passage from a liquid to a gaseous state. Rate of vaporization depends upon the substance involved, heat, and pressure.

VAPOR RECOVERY SYSTEM (VRS) — A system that recovers gasoline vapors emitted from a vehicle's gasoline tank during product dispensing.

VENTILATION — The systematic removal of heated air, smoke, and gases from a structure and replacing them with cooler air to reduce damage and facilitate fire fighting operations.

VIOLATION — An infringement of existing rules, codes, or laws.

VOLTAGE — Units of electric potential.

W

WET STANDPIPE SYSTEM — A standpipe system that has water supply valves open and maintains water in the system at all times.

INTRODUCTION

Fire suppression activities are not the only way to combat fires: a well-planned and executed fire prevention and inspection program is a less expensive and more effective way to accomplish the goal of the fire service. The goal is, of course, to minimize the risk of life and property loss from fire. By observing, making recommendations, and subsequently controlling or eliminating hazardous conditions, the inspector can make major strides toward accomplishing this goal before a fire occurs. The inspector also helps to educate occupants in ways to control hazards, in proper methods of evacuation, and in overall fire safety practices.

The fire prevention inspector may be a civilian fire prevention officer or a member of a fire department whose duties include conducting inspections. Throughout this manual the term *inspector* is used to refer to an individual who has the training and authority to inspect buildings, equipment, and processes for the purpose of effecting conditions that are conducive to fire and life safety.

Fire Inspection and Code Enforcement Practices, 5th edition, is designed to aid fire inspection officials in the performance of their duties. The manual addresses the administrative and technical areas that an inspector must be familiar with in order to ensure a minimum acceptable level of fire and life safety.

In recent years, several state courts have ruled that fire departments and/or individuals empowered to perform fire inspections may be held legally accountable for their performance. Chapter 1 addresses the issue of liability, as well as authority and responsibilities of the inspector. A discussion of police power, right of entry, organizational structure, and cooperation among other agencies is also included.

Chapter 2, Inspection Procedures, covers all aspects of conducting an inspection. These include preparation for the inspection, the inspection itself, photography, documentation, followup inspections, and taking cases to court.

The fire inspector must know how fires start and spread in order to recognize fire hazards and their potential consequences. Chapter 3, Principles of Combustion and Fire Cause Determination, discusses such fire basics as the combustion process, sources of heat energy, phases of fire, and methods of heat transfer. This information is intended as a review for the firefighter/inspector and as an overview of fire behavior for the civilian inspector. The last part of the chapter concentrates on the principles of fire cause determination and preserving physical evidence.

Chapter 4, General Fire Safety, addresses safety practices that are necessary in every occupancy. These include good housekeeping, regulating smoking, open burning, and the use of flammable decorations. Electrical safety is also critical, so basic electrical theory, electrical hazards, and the dangers of static electricity are covered. An additional aspect of fire safety is maintaining access for emergency apparatus if a fire does occur. Since the inspector's duties may also include performing safety standby at public events, guidelines for this duty are given.

In order to determine that a building meets the appropriate building and fire safety codes, the inspector must possess some knowledge of building construction and fire safety features. Chapter 5, Building Construction for Fire and Life Safety, describes construction classifications, testing of building components, and fire safety components in construction. The last part of the chapter focuses on evaluating buildings for structural safety.

Two vital areas in terms of occupant safety are covered in Chapter 6, Occupancy Classifications and Means of Egress. Occupancies must be properly identified because the regulations and safety requirements vary for different occupancies. Adequate and safe means of egress, of course, are essential for life safety. Much of the information about means of egress is referenced to NFPA 101, *Life Safety Code. (*NOTE: Life Safety Code® and 101® are Registered Trademarks of the National Fire Protection Association, Inc., Quincy, MA.)

Extinguishing equipment and fire protection systems are effective ways to increase occupant safety and to control fires during their early stages. Chapter 7 covers portable extinguishers, standpipe systems, fixed extinguishing systems, alarm systems, and water supply systems. Each type of equipment or system is described in depth, along with inspection and testing procedures.

Fire departments have come to understand that plans review is a necessary component of overall fire protection and prevention. Chapter 8 introduces the inspector to the many aspects of plans review, including developing a plans review system and evaluating architectural, structural, mechanical, and electrical drawings. Also included is information about evaluating fire protection features.

Because hazardous materials is such a complex subject, it is treated in two chapters. Chapter 9 discusses identification and properties of flammable and combustible liquids, compressed and liquefied gases, explosives, combustible metals and dusts, natural and synthetic fibers, and other hazardous chemicals. Chapter 10 discusses these same materials in terms of storage, handling, and transportation practices. The last part of the chapter discusses local, state, and federal response agencies that can provide assistance during hazardous materials incidents.

PURPOSE AND SCOPE

Fire Inspection and Code Enforcement Practices is designed to instruct the inspector in the principles and techniques of fire inspection and code enforcement. It is written to meet the professional standards set forth in NFPA 1031, *Standard for Professional Qualifications for Fire Inspector* levels I and II. Because today's fire inspector may be either a uniformed or civilian member of a fire department, this manual is designed for use by both types of individuals.

1

Authority, Responsibility, and Organization

This chapter provides information that addresses performance objectives described in NFPA 1031, *Standard for Professional Qualifications for Fire Inspector* (1987), particularly those referenced in the following sections:

Fire Inspector I

3-1 General.

3-1.7

3-1.9

3-1.10

3-20 Inspection and Code Enforcement Procedures.

3-20.2.2

10-90

Chapter 1
Authority, Responsibilities, and Organization

Fire prevention inspections are the single most important non-fire fighting activity performed by the fire service. This chapter presents the general concepts related to the fire inspector's responsibilities and legal authority. Because they differ widely from area to area, it is the inspector's duty to become thoroughly familiar with the statutes, codes, and regulations in his or her own area. Most of the information contained in this chapter is adapted from Fire Inspection Management Guidelines, a test developed by the National Fire Protection Association (NFPA) and the Fire Marshals Association of North America in cooperation with the United States Fire Administration.

The individual commanding a position of fire inspector must possess a great deal of knowledge regarding fire safety and building codes and a versatile personality. Although a college degree is not required to be a fire inspector, the individual must be at least 18 years of age and have graduated from high school or a state-recognized equivalent. The candidate should be proficient in written and oral communications. Through tests or interview, the candidate should be able to demonstrate the ability to interact with the public using tact and discretion without loss of authority. Furthermore, the candidate will have a thorough character investigation and evaluation completed before being accepted. As a final prerequisite in obtaining certification, the individual must also identify the organization of the fire department or entity for which the duties will be performed.

A fire inspector, fire marshal, or fire prevention specialist — whatever the title may be — must have a thorough knowledge and understanding of the following:

- The statutes that create the position
- The statutes that provide the legal basis and requirements for fire prevention activities
- The laws, codes, and ordinances that detail various fire safety requirements and establish a fire inspector's duties and responsibilities
- The statutes that set the limits of authority
- The ways in which the statutes or laws can be changed

These statutes may include federal regulations, state laws, and municipal ordinances. Statutes, laws, codes, and ordinances vary from jurisdiction to jurisdiction. Few are exactly alike, but most are similar in concept. Court decisions further affect authority and liability.

AUTHORITY

When fire inspectors reach the legal basis of and limits to their authority, they should consider certain key areas. These areas fall into two categories:

- Fire inspectors' status as members of the public sector, including the right, responsibilities, and liabilities inherent in being public officers or employees.
- The limits and scope of their authority as fire inspectors.

It is extremely important that fire inspectors contact the city or county attorney to obtain a clear understanding of the statutes that they are responsible for enforcing. The attorney can also clarify questions about the inspector's authority and liability (Figure 1.1). Assistance may also be obtained from legal research companies or legal libraries.

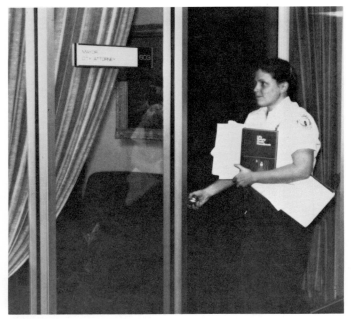

Figure 1.1 An attorney can clarify an inspector's authority and liability. *Courtesy of Des Plaines, Illinois Fire Department.*

Legal Status as Members of the Public Sector

As employees in the public sector, the legal status of fire inspectors is different from that of people employed in the private sector. In some jurisdictions, a distinction is made between a public employee and a public officer. This distinction can affect potential liability, compensation, and other benefits.

In general, inspectors should begin investigating legal status at the state level. State statutes often deal with the organization of fire services, the basis for the retirement system, and civil service regulations. These statutes also describe and define the organization and responsibilities of the state fire marshal, if there is one. Codes determining the liability of fire inspectors as public sector employees are also important. For example, fire inspectors should know if the state's motor vehicle code applies while they are performing their official duties.

Police Power

Fire inspectors must clearly understand any authority they have to arrest, to issue summons,

and to issue citations for fire code violations (Figure 1.2). Some jurisdictions give limited police power to fire inspectors. The statutes delegate authority to fire inspectors with specific words such as "to issue summons," "to write a ticket," "to issue a warrant," or "to arrest." When fire inspectors have this authority, they should have appropriate law enforcement training, particularly if they could be involved in prosecuting a fire code violator.

If the inspector does have limited police powers, the exact relationship between the inspector and the police department must be clearly defined to avoid conflicts if the fire inspector must exercise police powers that require police department support (Figure 1.3).

Figure 1.2 Fire inspectors must understand their authority to issue citations, summons, or to arrest violators. *Courtesy of Des Plaines, Illinois Fire Department.*

Figure 1.3 Sometimes police assistance is required during the performance of duty. *Courtesy of Des Plaines, Illinois Fire Department.*

Liability Incurred as a Result of Authority

In many jurisdictions, fire inspectors are considered to be public officers and therefore immune from certain kinds of liability. Immunity is not guaranteed; state or local statutes, or both, may impose liability. Generally, fire inspectors are not held liable for discretionary acts. Discretionary acts involve actions fire inspectors consider necessary to fulfill their responsibilities. However, fire inspectors can be held liable for acts that are considered to be ministerial. That is, they can be held liable for the manner in which they carry out or perform an act or policy.

Fire prevention codes typically contain language limiting the liability of the jurisdiction and its inspectors. If such language is to be effective, it must be adopted specifically as part of a statute, not just as part of a code adopted by reference. The BOCA Fire Prevention Code (1987) liability statement is an excellent example and can be found in Appendix A-2.

In recent years, courts have ruled against immunity provisions of codes. The courts have held that the immunity provisions conflict with statutes that establish an inspection authority and require the enforcement of codes and regulations. In other words, a community cannot be required to do something and, at the same time, be immune from liability if it or its officers (fire inspectors) do the job inadequately or negligently. Appendix B contains a summary of some recent court decisions on liability.

In general, fire inspectors are not immune from their actions even if the community is. Communities, however, normally indemnify their officers (that is, they assume the responsibilities for any claims in total) or provide liability insurance. The procedures for indemnification generally depend upon prevailing state law. It is most important for fire inspectors to determine whether they do or do not have indemnity or insurance against possible liability when they are performing their official duties. Such protection is important since courts today are awarding larger judgments in tort (wrongful act) cases. In addition, the cost of retaining counsel to defend against a lawsuit can financially ruin a fire inspector regardless of the outcome of the trial.

In 1976, the court held that fire inspectors, in conducting code enforcement inspections, had taken on a duty and must use reasonable care in the exercise of that duty. For example, if fire inspectors inspect a property and determine that there are violations of the fire code but fail to follow up to ensure that the violations are corrected, they can be held liable if a fire occurs. In addition, they can be held liable for the deaths or injuries resulting if they can be attributed foreseeably or in fact to the code violation.

Likewise, fire inspectors taking on a special duty to a person can be held liable. A special duty is one in which a person has moved from a position of safety to a position of danger because he or she relied on the inspector's expertise. For example, by issuing an occupancy permit, a fire inspector establishes the duty to ensure that the building complies with applicable codes and regulations (Figure 1.4 on pg. 8).

Most codes include a "duty to inspect" clause. These clauses are significant for a fire inspector because they normally do not allow selective enforcement; rather, they charge the inspector with total enforcement. Therefore, these clauses are also the basis for personal and professional liability.

Two examples of "duty to inspect" clauses from nationally recognized fire prevention codes are included in Appendix A-1. They represent the language used by most codes. From these two "duty to inspect" examples, it should be apparent that the fire inspector must know all provisions of the code and the various ways in which the codes can be interpreted. Knowing such information is no small task. Periodic training is necessary to keep abreast of changes.

Basically, it is better for fire inspectors to conduct fewer but more thorough inspections and to follow up on all violations than to perform many inspections in a haphazard, incomplete, or negligent manner. Failure to inspect a property does not impose a duty upon the inspector, unless the laws or statutes impose such a duty or there are known code violations present. Laws that single out a particular class occupancy for a predetermined number of inspections, however, establish a duty on the inspection staff. Fire inspectors need to re-

STILLWATER
CERTIFICATE OF OCCUPANCY

BUILDING PERMIT NO. _____

(MUST BE FILED 10 DAYS PRIOR TO OCCUPANCY)

ADDRESS OF PROPERTY REQUESTED FOR OCCUPANCY

NAME _____

MAILING ADDRESS _____

PROPOSED USE _____ ZONE _____

PREVIOUS USE _____ ZONE _____

OFF-SITE INSPECTION APPROVED:

OFF-SITE IMPROVEMENTS _____ _____
    (driveways, sidewalks, sewer & water)   Engineering Division   Date

OFF-SITE ELECTRICAL _____ _____
                         Electrical Department   Date

ZONING REQUIREMENT _____ _____
                     Code Enforcement Division   Date

ON-SITE INSPECTION APPROVED:

FIRE CODE REQUIREMENTS _____ _____
    (Exclusive of one & two family units)   Fire Department   Date

ELECTRIC _____ PLUMBING _____
      (City, part or total)-(Rural)   (City water) (City sewer) (Septic) (Gas)

HOUSING REHABILITATION (HCDA) _____
                                             Date

HEALTH DEPARTMENT _____
                                             Date

OTHER _____
                                             Date

CONDITIONS OF APPROVAL _____

_____

_____

FINAL APPROVAL _____ _____
                     Code Enforcement Division   Date

APPLICANT _____ _____
                                             Date

PROJECTED DATE OF OCCUPANCY _____ FEE _____

Figure 1.4 An occupancy permit is issued when applicable codes and regulations are met for a particular occupancy.

search these laws carefully so that they are fully aware of their duties.

Another liability area that fire inspectors need to be aware of is possible violations of civil rights laws. If inspectors perform inspections in a jurisdiction in a manner that discriminates against a certain group of people (by inadequately enforcing a code, for example), the jurisdiction might be held liable. Such cases can be extremely costly since the court can award damages plus attorney's fees.

Contracts with Others

While researching the extent of authority, fire inspectors should determine if there is a right to enter into contracts with outside agencies. The advice of other experts is particularly useful when inspectors must determine whether alternate materials or fire protection systems are consistent with code requirements. It may also be useful to hire private citizens to perform certain fire inspection activities.

Right of Entry

The right to enter property to inspect for code compliance is essential in order for fire inspectors to fulfill their duties. This part of a fire inspector's authority (the right to enter) is not a problem in most cases. However, the U.S. Supreme Court has ruled that property owners have the right to refuse to allow a fire inspection unless the inspector has obtained a warrant or has adequate reason to believe that the property contains fire hazards.

The U.S. Supreme Court has held that administrative entry, without consent, of the portions of commercial premises which are not open to the public may be compelled only through prosecution or physical force within the framework of a warrant procedure. See vs. City of Seattle 387 vl. 541, 87 S. Ct 1737.

The U.S. Supreme Court has also set forth guidelines for inspection agencies. The following recommendations are given to assist fire inspectors in operating within these guidelines:

- Inspectors must be adequately identified (Figure 1.5).
- Inspectors must state the reason for the inspection.

Figure 1.5 Fire inspectors should identify themselves upon arrival at the inspection site. *Courtesy of Mount Prospect, Illinois Fire Department.*

- Inspectors must request permission for the inspection.
- Inspectors should invite the person in charge to walk along during the inspection.
- Inspectors should carry and follow a written inspection procedure.
- Inspectors should request a search warrant if entry is denied.
- Inspectors may issue stop orders for extremely hazardous conditions, even if entry is denied, while search warrants are being prepared.
- Inspectors should develop a reliable record keeping system of inspections.
- Inspectors should have guidelines available that define conditions whereby they may stop operations without obtaining permission to enter or a search warrant.
- Inspectors should be sure that all licenses and permits indicate that compliance inspection can be made throughout the duration of the permit or license.
- Inspectors must be trained in fire hazard recognition and in applicable laws and ordinances (Figure 1.6 on pg. 10).

Many state and federal courts have handed down decisions that protect the "right of privacy" of owners of private dwellings where no known or

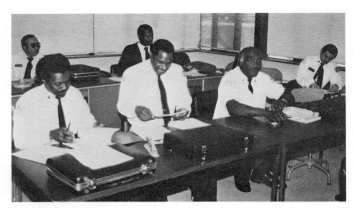

Figure 1.6 Inspectors must become familiar with applicable laws and ordinances. *Courtesy of Edward Prendergast.*

suspected fire hazards exist. Insistence on fire inspection under such conditions has been labeled an "unreasonable search."

Similar constitutional protection has been extended to owners of commercial property on the grounds that administrative entry without the owner's consent results in warrantless entry and, as such, violates the owner's rights under the Fourth and Fourteenth Amendments to the Constitution.

Challenges to the right to enter property may be rare, but fire inspectors should know the legal boundaries of their authority. Appendix A-3 shows the NFPA 1, *Fire Prevention Code* section on right of entry. Again, the language used is typical of such codes.

The state or local jurisdiction may already have developed a set of guidelines on right of entry by public personnel. If so, fire inspectors should review these guidelines to ensure that all provisions that affect fire prevention inspection are covered. It is important that they understand fully any limits to right of entry for inspection and code enforcement.

If fire inspectors must obtain a search warrant, they must understand the process required in granting the warrant so that they can provide the necessary supporting documentation. The form shown in Figure 1.7 is used by the Arvada Fire Protection District in Arvada, Colorado, when a property owner refuses entry for a routine inspection. Any form used should be reviewed by the legal counsel of the jurisdiction prior to its use.

Arvada Fire Protection District

FIRE MARSHALS' OFFICE
P. O. Box 3- D
ARVADA, COLORADO 80001
Telephone (303) 425-0850

LIFE IS FRAGILE · HANDLE WITH CARE ·

REFUSAL OF INSPECTION

I, ______ (BUSINESS REPRESENTATIVE), do hereby refuse ______ (INSPECTOR), authorized representative of the Arvada Fire Protection District, entry into my business, ______ (BUSINESS NAME), located at ______ (STREET ADDRESS), ______ (INCORPORATED/UNINCORPORATED AREA), to conduct a fire inspection.

______ (AUTHORIZED COMPANY OFFICIAL)

______ (TITLE)

______ (DATE)

The following conditions were observed in my attempt to secure entry:

1. ______
2. ______
3. ______

Past inspections have disclosed few/several fire code violations.

Date of last inspection: ______.

Occupancy Group ______ Occupant Load ______

Construction Type ______ Life Hazard ______

Distance to adjacent occupancies/structures ______

______ (ARVADA FIRE PROTECTION DISTRICT)

Figure 1.7 A refusal of entry should be documented on a form such as the one shown above. *Courtesy of Arvada Fire Protection District.*

Gaining Compliance with Codes

The inspector must understand fully the steps in the process of gaining compliance and in taking a person to court if necessary. In assessing the statutes of the local jurisdiction dealing with code compliance, the inspector should know the answers to the following questions:

- Is noncompliance with the code a violation of criminal or civil law?
- What have been the major methods of achieving compliance?
- What inducements are applied for voluntary action for compliance?
- What is the nature of penalties applied to ensure compliance?

Alternatives for Code Compliance

Fire inspectors will, from time to time, be asked to accept alternate materials, products, or systems as "equivalent" for code compliance. It is necessary, therefore, that the inspector be aware of what latitude, if any, the local code allows for such judgment. The inspector must also be aware of any requirements in the code that govern judgment in determining equivalency or the use of alternatives. NFPA 1, *Fire Prevention Code,* allows the use of alternatives. An example of the wording used in a statement on alternatives can be found in Appendix A-4.

Actions on Violations

One of the primary functions of fire inspectors is in gaining compliance with applicable codes. When fire inspectors find violations, they must know what actions they can take to ensure that the violations are corrected. When violations are discovered, normally the owner is notified and given a period of time to comply with the code. The inspector should also make the owner aware of any options that can be taken if he or she does not agree with the findings. Then, if the owner does not comply within the time frame established, fire inspectors should know the answers to several questions:

- Can the inspector issue a citation?
- Is it necessary to involve the jurisdiction's attorney?
- Is assistance from the police needed?

If fire inspectors find that the statutes in the local jurisdiction do not provide for adequate legal process to ensure code compliance, they should initiate action that is aimed at revising statutes to make them more effective. A model suggested by Appendix C of NFPA 1, *Fire Prevention Code,* may be found in Appendix A-5 of this manual.

Appeals

It is important for fire inspectors to know which individuals or agencies can override their authority and under what conditions. They should also know the avenues of appeal that are available to a property owner. Most codes establish an appeal procedure with a Board of Appeals or other body that is empowered to interpret the code and to issue a ruling (Figure 1.8). Fire inspectors need to understand the workings of such a board and the way in which a Board of Appeals can affect their job performance. They need to know the answers to the following questions:

- Can fire inspectors continue to enforce codes on the property during the appeal process?
- If the property owner wins the appeal, will the decision affect the way fire inspectors enforce that ordinance or section of the code for other properties?
- Is further action necessary? For example, should fire inspectors ask the board to clarify whether it has granted a general or one-time variance?

Figure 1.8 A property owner who disagrees with an inspector's findings may request a ruling from a Board of Appeals. *Courtesy of Lonnie Jackson.*

Note that when the Board grants a general variance by reason of an equivalency, fire inspectors must implement this ruling in the same way with future code enforcement. If, however, the Board of Appeals rules that the code is too vague for enforcement, fire inspectors must take steps to ensure that the vagueness in the code is removed. Appendix A-6 shows an example of a code provision that creates a Board of Appeals; this example is taken from NFPA 1, *Fire Prevention Code.*

RESPONSIBILITY

In researching authority and responsibility for fire inspection code enforcement, fire inspectors may find laws at the federal, state, and local level. If there are different requirements in these laws, generally the most stringent or strict law applies. This may not always be true, however, and fire inspectors should check. Code administration and enforcement should be built on a foundation of current, consistent standards and codes as well as a clear assignment of code enforcement responsibility.

Current Standards and Codes

Fire inspectors should know whether the codes and standards in the local jurisdiction are up to date. Staying aware of changes in occupancy is a challenge in terms of "keeping current" with codes in a community or jurisdiction. Having a good exchange of information with other agencies is one of the positive steps that fire inspectors can take to overcome this problem. Other steps include monitoring business license applications, monitoring building and occupancy permits issued, and conducting an annual occupancy inventory survey. If fire inspection requirements are tied to the issuance of licenses or permits, fire inspectors have a notification mechanism that keeps them aware of the commercial and business changes in the community. The local codes, of course, should then properly reflect requirements for those properties.

Consistent Standards and Codes

Codes and standards enacted at a variety of governmental levels may conflict with each other. Fire codes and standards that include fire-related considerations must be consistent to avoid confusion and duplication of effort. In addition, the interpretation of fire-related codes must be clear and consistent with the purpose and requirements of the fire codes themselves. The following fire-related codes are some that should be assessed:

- Fire Prevention Codes
- Housing Codes
- Zoning Ordinances
- Subdivision Regulations
- Building Codes
- Electrical Codes
- Mechanical Codes
- Gas Codes
- Transportation Regulations
- Health Regulations
- Plumbing Regulations
- Life Safety Codes
- Insurance Codes and Regulations

Clear Assignment of Code Enforcement Responsibility

The enforcement of various codes is often assigned to different agencies. This assignment of enforcement responsibility should be clear and consistent. Lack of communication between agencies can undermine public safety, which is the agency's primary goal.

Most codes allow fire inspectors some degree of latitude in enforcing the code provisions, in terms of interpretation or in terms of judgment, or both.

The willingness to communicate and share information on the part of fire inspectors becomes a vital element of code enforcement. Just as vital is the ability to produce creative, alternate solutions. Fire inspectors should consider this a major priority in maintaining an effective fire inspection program.

ORGANIZATIONAL STRUCTURE

There are two kinds of organizational structures that may affect the inspection activities of fire inspectors:

- The structure of the local fire department or code enforcement division
- The structure of other related, outside agencies, such as departments of planning, zoning, streets, and sanitation

Structure of the Local Department

To function effectively, fire inspectors must understand the structure of the local code enforcement division and the way in which it fits into the structure of the municipal government. Fire inspectors should know who their counterparts are in related agencies (in terms of equivalent authority), to whom the inspectors report, and who is responsible for approving the inspector's judgments.

Figure 1.9 shows a possible organizational structure within which the fire inspection department or division may function. Fire inspectors should be fully aware of this structure to be as ef-

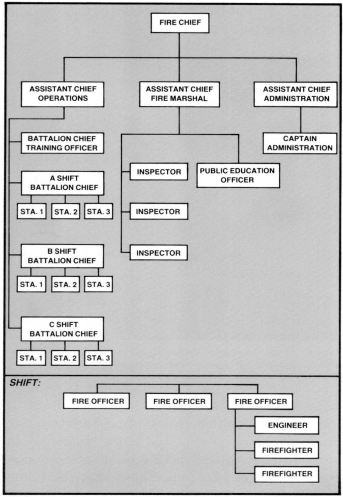

Figure 1.9 A typical organizational chart shows the relationship between the inspection department and other municipal agencies.

fective as possible. It is important to remember that many levels of law, as well as many agencies, must be considered in planning an effective fire inspection program.

Structures of Related Agencies

Fire inspectors must be aware of other agencies whose regulations might affect their actions and activities. These agencies may be mandated by law to prescribe and enforce stricter regulations pertaining to life safety than fire inspectors are allowed to do. These agencies may also be empowered to enforce regulations pertaining to fire safety. It is not uncommon for laws or ordinances to assign responsibilities to an agency, but not to provide the appropriate enforcement powers. It is also not unusual to find laws or ordinances that assign inspection responsibility to one agency and enforcement responsibility to another.

When fire inspectors interact with other agencies involved in life safety, the information that is exchanged can be of benefit to both. Cooperation among agencies enables each to work more efficiently and results in more effective enforcement of codes.

In some areas, the state statutes or court interpretations may not permit more stringent regulation at the local level. In other areas, determining which provision is the more stringent may be a matter of interpretation. The fire inspector may not have the authority to make the final decision in such cases.

Federal Laws

Many federal agencies have set forth regulations designed to ensure the safety of the public. These regulations cover a broad spectrum of activity and include such matters as employee safety, transportation of hazardous materials, patient safety in health care facilities, and minimum housing standards. The federal agency generally assumes the responsibility for enforcement of federal regulations, but in many cases this responsibility can be assumed by the state if it so chooses.

The local fire inspector is not responsible for enforcing federal regulations. However, if the local fire inspector is aware of and understands these regulations, violations can be reported to the proper authority for further action.

For the most part, federally owned buildings located within the local jurisdiction are not required to comply with local codes (Figure 1.10 on pg. 14). The agencies that operate these buildings usually enforce their own fire protection regulations. Fire inspectors should get to know the people who are responsible for these properties and work with them to ensure that fire protection is maintained at a high level.

The federal government's role in the area of fire safety is expanding. Fire inspectors should be aware of the federal acts that may have an effect on local inspection activities — for example, those federal acts regulating flammability standards for materials.

Figure 1.10 Federal agencies enforce their own fire protection regulations and usually are not required to comply with local codes. *Courtesy of Edward Prendergast.*

State Laws

In addition to enforcing federal regulations, states may be empowered to enforce state laws and statutes and to regulate specific fire inspection activities within the state itself. These laws can define and specify the powers and responsibilities of the state fire marshal's office (as well as delegating primary enforcement responsibility to that office), specify building construction and maintenance details in terms of fire protection, and empower agencies to issue regulations. State labor laws, insurance laws, and health laws also have a bearing on fire safety and sometimes encompass fire inspection responsibilities. It is the fire inspector's responsibility to communicate the authority of the local jurisdiction to other agencies with fire-related concerns (Figure 1.11). This kind of communication can form the basis for an effective working relationship between and among agen-

cies, and can also be a means of clarifying lines and levels of authority.

Figure 1.11 Communication between agencies promotes an effective working relationship and more efficient code enforcement. *Courtesy of Mount Prospect, Illinois Fire Department.*

Local Laws and Ordinances

Local laws, although sometimes based on state laws, are more specific and more "personal" in order to address the specific needs of a county, municipality, or a fire protection district. Enabling acts passed by the state allow local government units to adopt state regulations, either by reference (local governmental units follow state laws as drawn) or in the form of enabling acts (these acts allow a local unit of government to adopt state regulations and then to add regulations or ordinances particular to the local unit). Many communities adopt one of the four model fire prevention codes (see Appendix C for a listing) and may modify the code as they feel necessary. Laws should be kept current and should meet the growth and changes in the community. Fire inspectors should be thoroughly familiar with the requirements of both the current and past laws that may apply to their fire inspection duties and also be involved in an updating of those laws. Fire inspectors can use the model codes to help structure more adequate and effective fire prevention and inspection requirements.

Fire safety regulations generally fall into two categories:

- Those that govern the construction and occupancy of a building when it is being planned and constructed (building and safety code regulations)

- Those that regulate activities that are conducted within a building once it has been constructed (fire code regulations)

Fire inspectors should be actively involved in both categories. It is, of course, easier to gain compliance with fire safety requirements during a building's planning or construction phase than it is to require retroactive implementation of fire protection measures once the building has been occupied. Gaining compliance before a structure is occupied requires a close working relationship with the members of the local building, planning, and zoning departments (Figure 1.12). If such a relationship does not exist, the fire inspector should make it a priority to establish a relationship with these departments.

Many jurisdictions require licenses or permits for business operations involving fire protection equipment. An example of this is the licensing of persons or businesses doing fire extinguisher service and/or installing and servicing hood and duct fire extinguishing systems. The fire inspector will need to know which operations require licenses, what the requirements for being licensed are, and how an individual or business goes about obtaining a license.

Efficient and effective fire inspection requires a thorough knowledge of all pertinent fire safety codes and regulations. If requirements in other codes (separate from the fire code) and regulations exist, then often the local fire inspection agency does not have specific enforcement authority for such requirements. Fire inspectors must be aware of these fire safety regulations in order to efficiently and effectively administer the codes they must enforce.

Figure 1.12 Establishing a close working relationship with zoning, planning, and building departments will help ensure code compliance before a structure is occupied. *Courtesy of Springfield, Illinois Fire Department.*

2

Inspection Procedures

This chapter provides information that addresses performance objectives described in NFPA 1031, *Standard for Professional Qualifications for Fire Inspector* (1987), particularly those referenced in the following sections:

Fire Inspector I

3-1 General.

3-1.8

3-19 General Fire Safety.

3-19.1.3

3-19.3 Fire Drills.

3-19.3.1

3-19.3.2

3-20 Inspection and Code Enforcement Procedures.

3-20.1

3-20.2.2

3-20.2.3

3-20.2.4

3-20.2.5

3-20.2.6

3-20.2.7

3-20.2.8

3-20.2.10

10-90

3-20.3 Report Preparation.

3-20.3.1

3-20.3.2

3-20.4 Code Enforcement Equipment.

Fire Inspector II

4-10 Emergency Evacuation Plans.

4-10.1

4-10.2

Chapter 2
Inspection Procedures

Often, the most important decisions and actions performed by fire suppression forces occur before a fire breaks out. These actions — preplanning, developing standard operating procedures, training, maintaining equipment in readiness — are critical if the fire department is to perform efficiently and effectively at the fire scene.

Just as firefighters must plan ahead, so must fire inspectors. Preparing for the inspection, following a standard inspection procedure, and keeping in mind human behavior during a fire are all necessary if the inspector is to perform the job consistently and effectively. This chapter covers all phases of an inspection, from preparation for the inspection to writing final reports after the inspection. Because the fire inspector is often required to witness and critique fire drills in all types of occupancies, the last part of the chapter concentrates on fire drills.

BEHAVIOR DURING A FIRE

Human behavior during a fire, although sometimes erratic and irrational, is predictable. This behavior is a logical attempt to deal with a complex, rapidly changing situation in which minimal information for action is available. Behavior during a fire follows a predictable sequence:

Step 1: Detection of the fire

Step 2: Definition of the situation

Step 3: Coping behavior

Despite the highly stressful situation, most individuals involved in a fire respond in a relatively rational manner. They might notify others, search for the source of the fire, or start to combat the fire. Unfortunately, they can be misled by ambiguous fire clues. In short, their limited knowledge about the rapidly changing environment around them further complicates their chance for survival. It is vitally important that the inspector keep these likely behaviors in mind during an inspection. In this way, the inspector can help to ensure that an occupancy is as safe as possible.

One of the most important aspects of NFPA 101, *Life Safety Code,* is that it outlines standards and special considerations needed to provide a fire- and life-safe environment. Refer to NFPA 101 and to local ordinances for details about inspecting specific occupancies as well as for information about anticipating human behavior in an emergency situation.

PREPARATION FOR AN INSPECTION

Inspectors should be as well prepared as possible for the inspection. It is helpful to review previous inspection reports for the particular occupancy in order to become familiar with the hazards that are likely to be found (Figure 2.1). If inspections

Figure 2.1 Before making an inspection, inspectors should review existing records to become familiar with a particular site. *Courtesy of Mount Prospect, Illinois Fire Department.*

have not been conducted, the building plans and specifications may be available. It is also helpful to review sections of the applicable codes for the particular type of occupancy.

The inspector should always consider personal safety when making an inspection. If the area is under construction, a hard hat and safety shoes should be worn as a preventive measure. No matter what type of occupancy is being inspected, common sense and general safety awareness should be routine procedure to prevent unnecessary injury.

Certain equipment may be needed during the inspection (Figure 2.2). A recommended list of equipment follows.

- Coveralls for crawling into attics and concealed spaces
- Clipboards, inspection forms, and standard plan symbols
- Materials for preparing sketches
- 50-foot (15.24 m) measuring tape
- Flashlight
- Camera, equipped with flash attachment
- Pitot tube and gauges, when waterflow tests are required
- Reference books (Appendix D provides a list of these)

Establishing a favorable atmosphere plays an important role in conducting an inspection. The manner of dress of fire inspectors is a factor that may have a bearing on the amount of cooperation they can expect. Traditionally, inspectors have conducted inspections in a uniform (Figure 2.3), but recently there has been a growing interest in plainclothes dress (Figure 2.4). If the inspector wears a uniform, he or she should follow departmental dress regulations. Whether in a fire department uniform or street clothes, the inspector should be neat and clean.

The attitude that inspection personnel have toward the inspection itself also plays a part in establishing a favorable atmosphere. A positive attitude will usually encourage a positive reaction from the occupant. Fire inspectors should approach the inspection with the attitude of helping to keep the occupancy safe from hazards. Such an attitude will help to promote a high degree of cooperation and compliance from the occupant.

Scheduling the inspection will also help fire inspectors to make a good initial impression. They should make prior contact with the property owner to set an inspection date and time. If it is not possible to keep an appointment, they should notify the owner in advance and arrange another inspection date.

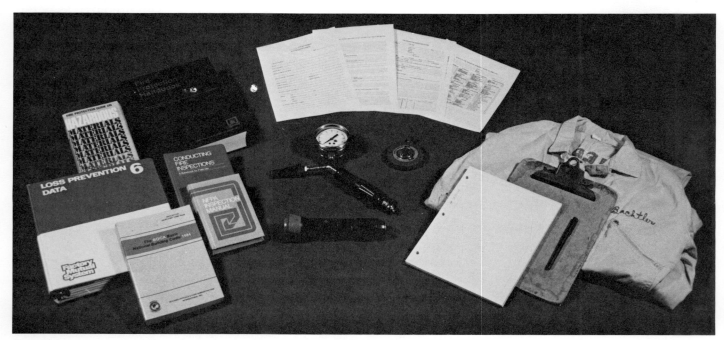

Figure 2.2 Inspectors should have the equipment pictured above during field inspections.

Figure 2.3 Uniforms are the traditional dress of fire inspectors.

Figure 2.4 Nowadays, many inspectors work in street clothes.

Figure 2.5 The inspector should confirm the appointment with the owner/occupant prior to arrival at the inspection site.

CONDUCTING THE INSPECTION

Upon arrival at the inspection site, the fire inspector should contact the owner/occupant and confirm the appointment (Figure 2.5). The inspector should then give a briefing detailing the inspection procedure. Routine inspections should not be attempted without permission from the owner or occupant. It is also advisable to have a representative of the property accompany the inspector during the inspection.

Relating problems and their solutions to the occupant is a vital part of the inspection process. In order to do this efficiently, inspectors must have the ability to express themselves orally. The fundamentals of effective oral communication are enthusiasm, confidence, preparation, and presentation. Through preparation, they gain a knowledge of the subject so they can make intelligent recommendations. To gain the attention of the occupant, inspectors must have confidence in their recommendations and present them with enthusiasm.

In addition to being able to speak effectively, it is very important for fire inspectors to have good listening skills. These skills enable inspectors to determine if the occupant understands what has been discussed. Furthermore, if the occupant participates in developing solutions to the problems, he or she is more likely to comply with the inspector's recommendations.

There may be times when inspectors do not know the answer to a question. Rather than give incorrect information or bluff an answer, it is best to simply say, "I do not know the answer to your question." The inspector should not be ashamed if he or she does not know all the answers. It is helpful to add, "I know where to find the answer and I will let you know." Then make it a point to let the owner/occupant know the information as soon as possible.

There is no set route that fire inspectors must follow when conducting an inspection. However, inspections must be systematic and thorough. No area should be omitted. It is common for the inspection to begin with an examination of the exterior. During the exterior inspection, inspectors collect information about the building dimensions, construction materials, exposures, water sources, hydrants, valves, and the surrounding area. The location of doors, windows, and fire escapes can be noted at this time.

There are several methods used to inspect the interior, such as working from the roof downward, working from the basement upward, or following the manufacturing process from raw materials to finished product (including storage areas). Regardless of the method used, fire inspectors should check each area in order, taking care not to miss any area. They can obtain the key to all locked rooms and closets to facilitate the inspection. Any secret areas or processes that were barred from inspection should be noted and reported to the appropriate supervisor for further action.

In order to make a thorough inspection, fire inspectors must take sufficient time to make notes and a field sketch (Figure 2.6). The field sketch is a rough drawing of the building that is made during the inspection. This drawing should show general information about floor plans, building dimensions, and such related outside information as location of fire hydrants, streets, exposure distances, and permanent objects of importance. It is not important for inspectors to draw the sketch to scale, but it is helpful to draw the sketch in proportion, using a straightedge. This will make it easier to transfer the information onto a final drawing that is part of the inspection report. They should record all hazards and their locations. A complete set of notes and a well-prepared sketch of the building provide information from which a complete report can be prepared.

Photography

Photographs are sometimes necessary when major discrepancies or problems are found during an inspection. Photographs are useful because they provide a record of conditions *at the time of inspection*. Diagrams and sketches show the objects in the area of the relative setting, but nothing has the effect of a photograph — especially if court proceedings follow the inspection.

Many departments tend to shun fire inspection photography. There is a belief that extensive — and expensive — equipment is necessary, and that a person needs to be nearly a professional photographer to obtain useful photographs. This belief is unfounded: although special lenses, filters, and lighting equipment can enhance photographs of the scene, they are not necessary.

Many persons use an "instant-print" camera for initial shots, then follow up with a 35 mm camera. It is best not to place total reliance on "instant-print" photographs, because they lack detail, the prints can fade, and there are no negatives for duplicates. Clear, detailed photographs are important.

Maintaining careful, accurate records is vital to preserve the integrity of the photograph as evidence (Figure 2.7). Records are also helpful for filing photographs. In addition, records of photographs will help photographers in deciding which settings to use — or not to use.

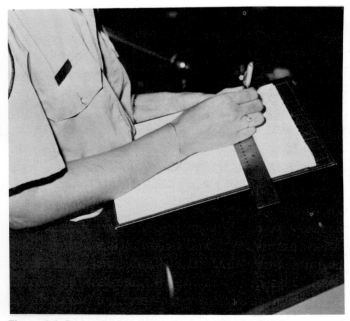

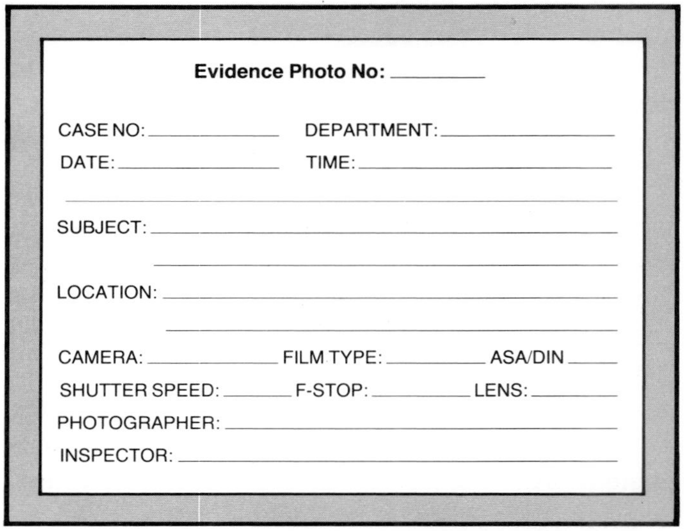

Figure 2.6 During the inspection, the inspector should take notes and make a field sketch.

Evidence Photo No: _____

CASE NO: _____   DEPARTMENT: _____

DATE: _____   TIME: _____

SUBJECT: _____

LOCATION: _____

CAMERA: _____   FILM TYPE: _____   ASA/DIN _____

SHUTTER SPEED: _____   F-STOP: _____   LENS: _____

PHOTOGRAPHER: _____

INSPECTOR: _____

Figure 2.7 All photographs must be carefully tagged and filed.

Each shot made should be recorded, preferably in a small notebook or on a portable tape recorder for later transcription. Note the camera and film used, the aperture and shutter speed settings, and the lighting used. An entry could be similar to that shown in Figure 2.8.

Figure 2.8 Specifics about photographs, such as type of film and aperture setting, must be recorded at the time the photograph is taken.

The shorthand in the last line means: shot number one, aperture setting f8, shutter speed 1/60 second, strobe, looking toward the northwest. The time available and circumstances will dictate to a large degree the amount of information that can be entered when shots are made, but the aperture and shutter settings, the type of lens, and any special lighting or lenses must be recorded at the time.

For a more complete record, draw a sketch of the scene and note the spots from which each photograph is shot. Be sure to indicate north on the sketch.

It is important to protect the integrity of the chain of custody of each roll of film from the time it is put into the camera until the print or color transparency is shown in the courtroom. One must be

able to account for the location of the negatives and prints for their entire existence.

Commercial firms will furnish, if requested, affidavits attesting that the film was processed with other customers' films, was not cut, retouched, or altered in any way, and was returned on a certain date to the same person who authorized the processing. Courts have permitted the introduction of such affidavits to establish custody of the film while it was not under the control of the photographer. The integrity of the U.S. Postal Service, if films were mailed, has never been questioned.

For every film, negative, print, or transparency, a record must be kept of any transfer to anyone other than the authorized storer. Although a person in the fire department would hesitate to question the motives and honesty of the chief or fire marshal, a defense attorney might have no such qualms, so receipts must be demanded of *everyone* who takes the photograph or negative from the custody of the authorized storer.

Closing Interview

When the inspection is complete, the fire inspector should discuss the results with a person in authority for the property (Figure 2.9). The purpose of this closing interview is to note good conditions as well as discuss those conditions that need correcting. The inspector should discuss violations in general terms, indicating that the specific details will be sent in a written report. The reactions

Figure 2.9 The inspector should discuss the findings of the inspection with the owner or an individual in authority.

of the person in authority will vary. Some will welcome the inspection report; these persons usually work immediately to correct the hazards noted. The majority will accept the report as part of the routine of doing business. Most of them will move toward compliance.

Some individuals will display hostility toward the inspector's remarks and recommendations. In these instances, fire inspectors should be polite but firm. They should avoid arguments. In every instance, fire inspectors should use the closing interview to express thanks for any courtesies extended. If a reinspection will be necessary, the inspector should inform the owner/occupant that it will be made. Inspectors must use their judgment, but for those who are reluctant to comply with the codes, it may be necessary to explain the enforcement procedure and the appeal process.

Inspection Drawings

The field sketch is used to make a final inspection drawing. The final drawing(s) should include all the necessary building inspection information. The drawings should be done to scale and should feature standard mapping symbols. (See Appendix

E for standard map symbols. In addition, see Chapter 8, Plans Review for more information about map symbols.) On or attached to the drawing should be a legend explaining the mapping symbols. If an item is included for which there is no standard symbol, a circled numeral should be used. This numeral can be explained in an appendix attached to the drawing.

There are three general types of drawings used to show building information: the plot plan, the floor plan, and the elevation drawing. The plot plan is used to indicate how the building is situated with respect to other buildings and the streets in the area (Figure 2.10). The floor plan shows the layout of individual floors and the roof (Figure 2.11). Most of the building detail can be shown on the floor plan. The elevation drawing is used to show both the number of floors in the building and the grade of the surrounding ground (Figure 2.12).

When the building inspected is small, fire inspectors can often put all the information on one drawing, but as the building size or complexity increases, the number of drawings needed to show the necessary detail will also increase. Every drawing or page of a set of maps should be labeled

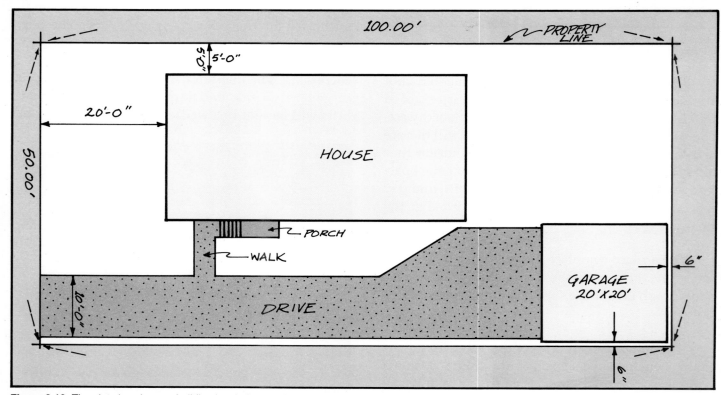

Figure 2.10 The plot plan shows a building in relation to other buildings and streets in the area.

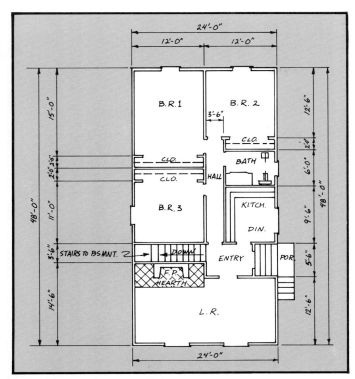

Figure 2.11 The floor plan shows the layout of each floor.

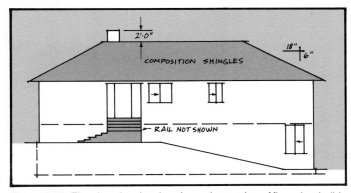

Figure 2.12 The elevation drawing shows the number of floors in a building and the slope of the ground surrounding the building.

with a title that clearly indicates what type of information is included.

After the inspector completes the final drawing, he or she should attach a set of clear, legible notes. In many cases, copies of this material must be given to other departments that need the information.

Written Reports

Written reports not only serve as records of the inspection, but also can be used as a basis for legal action. Without written evidence of an inspection, no proof exists that the fire inspector gave the owner notice of hazardous conditions or the corrective measures to be taken.

Fire inspectors must record every fire inspection, but a formal report is not always necessary. On the majority of inspections, the inspector can easily record data and violations on an inspection form or checklist (Figure 2.13 on pg. 28). Such forms often provide a means by which the inspector can make recommendations to the owner. Situations that require formal reports are those that involve life-threatening hazards, major renovations to comply with the codes, or an extensive list of minor violations.

The inspection report provides information to the owner/occupant. In the contents of the report, fire inspectors should inform, analyze, and recommend. It is sometimes necessary for another person to complete an inspection or to conduct followup inspections; therefore, it is very important that the report be complete.

In an inspection report, fire inspectors are generally concerned with presenting the facts and evidence to prove a point, draw a conclusion, or justify a recommendation. This information must be presented in a businesslike manner. The report should not be opinionated, biased, emotional, or unfair.

The report should be written in letter form, including the following information.

- Name of business
- Type of occupancy
- Date of inspection
- Name of inspector
- Name of business owner/occupant
- Name of property owner
- Name of person accompanying inspector during inspection
- Edition of applicable code, as a reference for future inspections
- List of violations and their locations stated in specific terms (Code section numbers should be referenced)
- Specific recommendations for correcting each violation
- Date of the followup inspection

PLANO FIRE DEPARTMENT
FIRE INSPECTION REPORT

Location _____ Date of Inspection _____

_____ No. Floors _____ Construction _____
Name of Business/Type of Occupancy

Violations requiring correction action are marked below with a cross (X).
Inspection will be made within _____ days.

Inspection Type:
Regular . . . . A ___
Complaint . . . . . B ___
Cert. of Occupancy . . C ___
System . . . . D ___
Home . . . . E ___

1. Aisles
   Blocked . . . . A ___
   Inadequate . . . . B ___
   Not lighted . . . . C ___
2. Alarm System
   None . . . . A ___
   Inadequate . . . . B ___
   Defective . . . . C ___
   Other . . . . D ___
3. Building Repairs
   Dangerous Conditions . . . A ___
   Other . . . . B ___
4. Burning
   No Permit . . . . A ___
   Out of Hours . . . . . . B ___
   Illegal Container . . . . C ___
   Too Close to Building . . . D ___
5. Chemicals
   Improper Storage . . . . . A ___
   No Permit . . . . B ___
6. Combustibles
   Excessive Storage . . . . . A ___
   Improper Storage . . . . B ___
7. Doors
   Blocked . . . . A ___
   Not Self-Closing . . . . . B ___
   Unapproved . . . . C ___
   None as Required . . . . . D ___
8. Ducts
   No Fire Damper . . . . . . A ___
   Other . . . . B ___
9. Electricity
   Overloaded . . . . A ___
   Defective . . . . B ___
   Extension Cords . . . . . C ___
   Other . . . . D ___
10. Elevators
    Emergency Tools/Key . . . A ___
    Improper Enclosure . . . . B ___
    Pit Area . . . . C ___
    Other . . . . D ___

11. Exit Lights/Signs
    Out . . . . A ___
    Needed . . . . B ___
    Wrong Type/Location . . . C ___
12. Extinguishers
    Recharge . . . . A ___
    Hang . . . . B ___
    Repair or Test . . . . . C ___
    Inadequate . . . . D ___
    Wrong Type . . . . E ___
13. Exits
    Insufficient . . . . A ___
    Not Indicated/Lights . . . . B ___
    Blocked or Locked . . . . C ___
    Wrong Type . . . . D ___
    Wrong Hardware . . . . . E ___
14. Explosives/Ammunition
    Improper Storage . . . . . A ___
    No Permit . . . . B ___
    Other . . . . C ___
15. Fire Door
    Blocked/Inoperative . . . . A ___
    No Closing Device . . . . . B ___
    Inadequate . . . . C ___
    None . . . . D ___
16. Fire Escape
    Defective/Repairs . . . . . A ___
    None as Required . . . . . B ___
    Inadequate or Wrong . . . . C ___
17. Flameproofing
    Curtains/Drapes . . . . . A ___
    Decorations . . . . B ___
18. Flammable Liquids
    Improper Storage . . . . . A ___
    No Safety Cans . . . . . . B ___
    Too Close to Heat . . . . C ___
    Improper Dispensing . . . . D ___
    No Permit . . . . E ___
19. Housekeeping
    Area/Yards/Buildings . . . A ___
    Improper Disposal . . . . . B ___
    Excessive Storage of Waste
    . . . . C ___

20. Firewalls/Enclosures
    None . . . . A ___
    Incomplete . . . . B ___
    Door Not to Code . . . . C ___
    Openings Unprotected . . . D ___
21. Open Flame Devices
    Open Top/Sides . . . . . . A ___
    Use Not to Code . . . . . B ___
22. Paint Spraying
    Vented to Outside . . . . A ___
    Booth Not to Code/None . . B ___
    No Sprinkler Heads . . . . C ___
    Other . . . . D ___
23. Fixed Ext. Systems
    Inadequate Coverage . . . A ___
    Needed Service . . . . . . B ___
    Other . . . . C ___
24. Heating
    Repair . . . . A ___
    Combustion Air . . . . . . B ___
    Remove Combustibles . . . C ___
    Room Enclosure . . . . . . D ___
    Other . . . . E ___
25. Sprinklers
    Defective Heads . . . . . . A ___
    Incomplete Coverage . . . B ___
    Storage Too Close to Head . C ___
    Other . . . . D ___
26. Stairs
    No Handrail . . . . A ___
    Blocked . . . . B ___
27. Standpipes-Hose Cabinets
    Threads/Hose . . . . A ___
    Repair/Other . . . . B ___
28. Welding
    No Permit . . . . A ___
    No Fire Curtain . . . . . . B ___
    Too Close to Combustibles . C ___
29. Address
    Not Posted . . . . A ___
30. Other Violations
    See REMARKS below . . . A ___

A COPY OF THIS NOTICE WILL BE ON FILE IN THE OFFICE OF THE BUREAU OF FIRE PREVENTION FOR FURTHER ACTION.

REMARKS

_____ OWNER/AGENT/MANAGER/REPRESENTATIVE _____ INSPECTOR

Emergency Contact #1 _____ Emergency Contact #2 _____

Figure 2.13 The inspector's findings can be recorded on an inspection form.

In the final portion of the report, fire inspectors should offer to answer any questions the owner/occupant may have. If necessary, the inspection report can be sent by registered mail so there is a record showing that the report was received by the owner or occupant.

Proper filing techniques are also an important aspect of record keeping. A report is an accurate account used to describe a specific state or condition. The report becomes a record when it is stored and is capable of being retrieved upon request. All records should be kept in orderly files so they can easily be found when needed.

FOLLOWUP INSPECTIONS

Followup inspections are made to ensure that the recommendations made in the inspection report have been followed. Fire inspectors should confirm the time and date of the reinspection with the occupant on arrival. It is not necessary for the inspector to conduct a complete inspection at this time. He or she needs only to inspect the problem areas included in the inspection report to verify that the hazards have been corrected.

If all hazards have been corrected, the inspector should compliment the owner/occupant for doing such a good job. Then, in order to close out the file, the inspector should send a letter stating that the premises was in full compliance. This letter also gives the inspector a second opportunity to thank the owner/occupant for cooperating.

If some hazards remain to be corrected but the owner/occupant is making a conscientious effort to comply, he or she should be complimented for the progress that has been made. Then the inspector can set a date for another followup inspection. A written update should be added to the inspection files, with the original copy going to the owner/occupant.

If the hazards have not been corrected and it is apparent that the owner/occupant is making no effort to correct them, the inspector should issue a final notice with a date for another inspection. The final notice should inform the owner/occupant exactly what legal action will be taken if full compliance is not attained by the date specified.

MODIFICATION OF REQUIREMENTS, APPEALS PROCEDURE, AND JUDICIAL REVIEW

Many municipalities allow the chief of the fire prevention bureau to modify various provisions of the respective code. To receive consideration for a code modification, the applicant must generally make a formal written request to the authority having jurisdiction. The appropriate office then analyzes the request to ensure that the general intent of the code is being observed and that public safety is maintained. Following the decision of the official, the applicant receives a signed copy of the decision. Usually, detailed records of the decision will be available at the fire prevention bureau office. An example of a request for the modification of a fire code is provided in Figure 2.14 on pg. 30.

If the applicant feels that the decision reached was unfair, or that a fire prevention code was unfairly enforced or was misinterpreted by the enforcement officer, the individual may file a request for a review by the Board of Appeals. Applications for review must normally be submitted within seven days of the notice or order of the inspecting official. The board will then accept or reject the appeal and file a notice of denial or acceptance, as shown in Figure 2.15 on pg. 31. The Board of Appeals usually consists of five to seven members who have previous experience in the field of fire prevention or building construction. Rules and regulations used by the board during its hearing generally are made public. A schematic of a typical appeals procedure is provided in Figure 2.16 on pg. 32.

Enforcement Procedures

The actions taken to ensure code compliance vary from jurisdiction to jurisdiction and are dictated by the adopted codes. It is important for inspectors to fully understand the enforcement procedures to be followed and to keep complete, accurate records of all action taken. Notifying the responsible party of the violations found is usually the first step taken in the enforcement procedure. This step is followed by a reinspection after a predetermined period of time. If voluntary compliance is not gained, most jurisdictions issue some type of sanction such as a citation, fine, court summons, or stop work order. The procedures for issuing sanc-

CITY OF _____

FIRE DEPARTMENT

Appeal for Modification of Fire Code

Address all communications to Department of Building

Meeting Date: _____

Petition must be on file at the office of the Board of Appeals _____,
one week before meeting date.

Petitioner(s) must be present at meeting.

Approvals on proposed construction are null and void unless permit is obtained within 6 months.

Address of Job _____ Legal Description: Lot _____ Block _____ Tract _____
　　　　　　　No.　　　Street

Between Cross Streets _____ and _____

Owner's Name _____

Petitioner's Name _____

Petitioner is: _____ Owner _____ Contractor _____ Architect _____ Engineer

Address of Petitioner _____
　　　　　　　　　No.　　　　　Street

Phone No. _____ City _____ P.O. Zone

Status of Job: _____ Not Started _____ Under Construction _____ Finished

Permit No. _____ Plan Check No. _____
　　　　　　　　　　　　　　(If permit has not been issued)

Specific ordinance modification desired: _____

_____

_____

(Additional sheets or data may be attached.)
Date: _____ Owner: _____
　　　　　　　　　　　　　Signature
(It is understood that only those points specifically mentioned are affected by action taken on this appeal.)

Plot on Reverse Side Required for Yard Modification

Figure 2.14 A sample request for modification of fire code form is shown above.

CITY OF _____

FIRE DEPARTMENT
BOARD OF APPEALS

NOTICE OF DENIAL

Date: _____

Petition No. _____

Petition of: _____

Premises affected: _____

_____

Referring to the above petition for a variation from the requirements of the _____

_____ ordinance so

as to permit _____

_____

_____

After a public hearing, the Board of Appeals voted that the petition be and it hereby is denied, and cannot again be considered on its merits within two years from this date.

Signed: _____

_____

_____

_____

Board of Appeals

Figure 2.15 An owner's appeal will be accepted or rejected by the Board of Appeals.

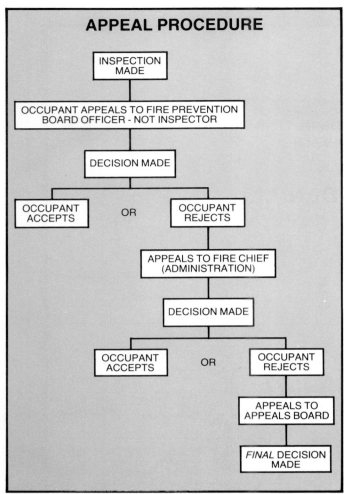

APPEAL PROCEDURE

Figure 2.16 The steps in the appeals process are shown.

tions differ; however, each jurisdiction should have a written program that details the process. Appendix F is an example of a citation program.

Taking Cases to Court

In preparation for court, fire inspectors should gather all pertinent information such as photographs, notes taken during the inspection, and exhibits to be used for demonstration purposes. The inspector should be prepared to state the date, time, and circumstances under which photographs were taken; identify the photographs; state that they accurately depict what was observed; and be able to describe the pictures factually. Questions regarding the admissibility of photographs in evidence should be cleared in advance with legal counsel. The day before the trial, the inspector should reinspect the facility, taking new notes and photographs to reinforce the case. This new infor-

mation can also be used to help determine the appropriate fine if compliance has been obtained before the trial.

Fire inspectors may serve as witnesses in the courtroom or as advisors to the prosecutor. As witnesses, they should confine their testimony to the facts, avoiding, "hearsay" and irrelevant statements. They should remain impartial and calm, never entering into arguments with the attorney for the defendant. As advisors, they can assist the prosecuting attorney with information about the fire ordinances and technical terms as well as the facts of the case.

Because of the complex nature of code enforcement, inspectors often need help from outside sources. Appendix C contains a list of groups that are involved in code formulation or administration. With each organization is a list of the codes they produce or their major publications. Inspectors can locate additional sources of information in the *Code Administrations Resource Directory,* prepared for the Federal Emergency Management Agency in 1981 by Lee M. Feldstein, International Association of Fire Chiefs.

Complaint Handling

Public involvement is an essential factor in a successful fire prevention program. When citizens call to report a hazard, fire inspectors must treat them with courtesy and express genuine interest in their complaint. This procedure will encourage their continued participation in fire prevention. The person taking the call should record all pertinent information. Departmental complaint forms are useful in obtaining complete and accurate data. Complaints that do not require immediate corrective action can be forwarded to the fire prevention bureau. If the complaint requires immediate action, it should be forwarded to the responsible party for further action.

When investigating a complaint, fire inspectors do not need to give the owner advance notice. The inspector should show appropriate identification and explain the purpose of the inspection. If the inspector does find code violations, he or she should take the appropriate departmental action to correct the violation(s). Inspectors should be prepared to deal with negative attitudes, as they are frequently encountered during complaint inspections.

FIRE DRILLS

Fire exit drills should be conducted in all types of occupancies. Occupants must be instructed in the correct techniques of exiting the building by more than one route. For business, educational, and assembly occupancies, the instruction and drill may be very basic, involving nothing more than an efficient evacuation of the structure. Drills for other occupancies, such as health care, industrial, and correctional facilities will be more complex. These drills involve the movement of critical or bedridden patients, perimeter security of inmates, or process shutdown and industrial fire brigade response. All of these factors will have to be evaluated prior to and during the drill.

Fire inspectors should assign personnel familiar with the premises to check exits for usability, search for any stragglers, count the evacuated occupants, and prevent reentry by any unauthorized personnel until the building has been declared safe. One of the biggest decisions to be made, and perhaps the most difficult for some, is when to evacuate the building. Any area that is affected by smoke, heat, or flames, should be evacuated, followed by evacuation of the entire building. Upon hearing an alarm, personnel should shut off equipment as required and evacuate to an assembly location. Alternate routes of travel should be established, and all employees should be trained to use both routes.

The responsibility for planning and executing fire drills lies with the occupancy's fire loss prevention and control management staff. Plans for drills need to be discussed with both middle and line management so that they understand and cooperate in the effort. Following an exit drill, line and staff managers should meet to critique the effectiveness of the drill and the evacuation plan. Exit drills should be conducted at least monthly and at different times of the day. In facilities operating more than one shift, drills should be held on all shifts. In occupancies such as nursing homes, hospitals, hotels, and stores, drills usually involve staff personnel only to avoid alarming patients, guests, and customers.

Fire exit drills are particularly important in educational occupancies. Many states have laws that give certain individuals responsibility for conducting fire drills. These laws further dictate the frequency of drills and other details. Drills in educational occupancies should ensure that all persons in the building actually participate. Emphasis should be placed upon orderly evacuation under proper discipline rather than upon speed. If weather conditions may endanger the health of children during winter months, weekly drills may be held at the beginning of the school term to complete the required number of drills before the onset of cold weather.

Drills should be executed at different times of the day: during changes of classes, when the school is at assembly, during recess or gymnastic periods, or during other special events. If a drill is called during the time classes are changing, the pupils should be instructed to form a line and immediately proceed to the nearest available exit in an orderly manner. Complete control of the class is necessary so teachers can quickly and calmly control the students, form them into lines, and direct them as required. If there are pupils who are incapable of holding their places in a line moving at a reasonable speed, they should be moved independently of the regular line of march.

Monitors should be appointed from the more mature pupils to assist in the proper execution of all drills. The monitors should be instructed to hold doors open in the line of the march or to close doors when necessary to prevent the spread of smoke and fire. Teachers and other members of the staff should have responsibility for searching restrooms or other rooms. Each class or group should proceed to a predetermined point outside the building and remain there while a check is made to see that all are accounted for and that the building is safe to reenter. Assembly points should be far enough from the building to avoid danger from fire in the building, interference with fire department operations, or confusion among classes and groups. No one should be permitted to reenter until the drill is complete. Fire inspectors should check on the frequency of exit drills and the time required to vacate the building.

Hospitals and nursing homes require special evacuation procedures. A total evacuation is not warranted every time; the extent of the evacuation is determined by the severity of the emergency. Evacuation plans should have a series of steps or phases:

Phase I — Evacuating a Single Room
Personnel follow a procedure explained by the acronym "REACT"
"R" — Remove those in immediate danger.
"E" — Ensure that the room door is closed.
"A" — Activate the fire alarm if it has not already been activated.
"C" — Call the fire department.
"T" — Try to extinguish or control the fire.

Phase II — Evacuating an Entire Zone
Personnel close all doors and smoke barrier doors, evacuate all rooms adjacent to the fire area, and then all rooms remaining in the fire zone. Occupants are moved into safe refuge areas on the fire floor.

Phase III — Evacuating a Floor and Zones Above
The entire fire floor is evacuated using stairwells or elevators when it is safe to do so. Then zones above the fire floor are evacuated.

Phase IV — Evacuating the Building
The fire floor is evacuated, then all floors above the fire floor, followed by floors below the fire floor. Patients are then transported to other buildings or facilities.

Fire drills and evacuation procedures for correctional facilities provide a unique problem due to the inherent security features of the building. During an emergency, the lives of the inmates, visitors, and security personnel depend upon the quick and effective actions of facility employees. Fire inspectors must make sure personnel are properly trained in appropriate evacuation routes and fire reporting procedures. Fire department access and coordination of fire department actions with those of security and police agencies must be an important part of emergency planning.

The vast majority of fires in correctional facilities are deliberately set by inmates. The most common reasons for setting fires are to cause malicious damage, increase chances of escaping, commit suicide, or to show force during a riot. In a correctional facility, maintaining security is of primary importance. Thus, occupants are either protected where they are or are evacuated to a secure area or refuge.

Correctional facilities should maintain an effective fire detection system, key control system, and a written emergency plan, copies of which should be posted for inmates to read. There should be at least two means of access to each main cellblock. Inmates should be released by a reliable means, whether electrical, mechanical, or by master keys. Personnel should ensure that prisoners remain segregated according to their detention classifications (general population protective custody, and so on).

Hotels and motels provide different problems for fire inspectors with regard to fire exit drills and evacuation procedures because of the temporary nature of the occupancy. Often, the occupants of these structures have never been in the building before and will remain there only overnight or for a few days. Therefore, fire inspectors cannot rely on occupants being familiar with the building if they should have to evacuate. For this reason, a reliable protective signaling system must be installed in accordance with applicable standards. The fire alarm system should include an occupant notification system so that personnel can quickly relay evacuation instructions to the occupants. Management should post evacuation procedures in all rooms, at all fire alarm stations, and by exits (Figure 2.17). Fire inspectors must train employees in evacuation procedures and may use them as "fire wardens" to direct the flow of evacuees to ensure a complete and efficient evacuation. Fire drills must be held frequently enough to keep all personnel familiar with the emergency evacuation plan and their assigned duties.

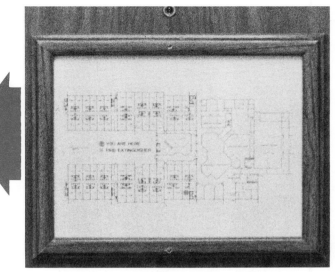

Figure 2.17 Evacuation procedures for hotels and motels must be posted prominently.

HOME FIRE SAFETY PROGRAMS

Inspectors should identify and implement home fire safety programs. Programs such as EDITH (Exit Drills In The Home) or "Learn Not to Burn" have gained nationwide publicity. Both of these programs are designed to instruct the public in emergency procedures to follow if they are involved in a fire. Additional information about the basic rules of home fire safety and formulating escape plans can be found in IFSTA **Essentials of Fire Fighting.**

Homeowners often do not practice residential fire drills, with tragic results. Fire inspectors can easily document the outcome of such apathy by reviewing current fire fatality statistics for residential occupancies. The NFPA has developed a program known as "Operation EDITH." (Exit Drills In The Home). This program emphasizes the need for practicing escape drills from the home, installing and maintaining smoke detectors, and developing a written evacuation procedure for the home (Figure 2.18). This procedure should include two means of escape from every room. For two-story residences, owners may have to use fire escape ladders, which may be purchased for a nominal fee. Residents should establish a predetermined meeting place for family members and procedure for notifying the fire department. Fire inspection offi-

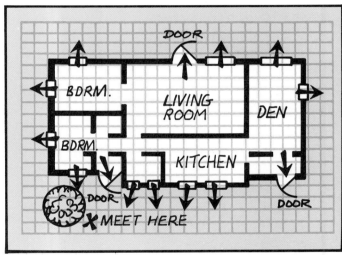

Figure 2.18 Operation E.D.I.T.H. emphasizes the need for written evacuation procedures for the home.

cials should be familiar with the recommended practices for residential escape plans, as they may be called upon to assist in developing such a plan or to witness or critique a drill.

NOTE: This text contains no recommendations for half-way houses. The Committee for Life Safety, NFPA, is in the process of adopting such a standard which will be in Chapter 21 of NFPA 101 *Life Safety Code.* Persons interested should watch for the availability of this information.

3

Principles of Combustion and Fire Cause Determination

This chapter provides information that addresses performance objectives described in NFPA 1031, *Standard for Professional Qualifications for Fire Inspector* (1987), particularly those referenced in the following sections:

Fire Inspector I

3-9 Fire Behavior

3-9.9

3-12 Flammable and Combustible Liquids.
(see NFPA 30, *Flammable and Combustible Liquids Code***.)**

3-12.1.5

3-20.6 Fire Cause Determination.

Fire Inspector II

4-11 Fire Cause Determination.

Chapter 3
Principles of Combustion and Fire Cause Determination

To better understand and recognize unsafe conditions and practices, the fire inspector must have a basic understanding of the combustion process and fire hazards. This chapter discusses the combustion process, the phases of fire, heat transfer, and dangerous conditions brought on by fire. Common fire hazards and their control are also discussed.

In some areas, fire inspectors are responsible for conducting fire cause investigations. The last section of this chapter, Fire Cause Determination, discusses the principles of fire cause determination and methods for preserving physical evidence.

THE COMBUSTION PROCESS

Fire is a chemical process known as combustion. It is frequently defined as the rapid oxidation of combustible material accompanied by a release of energy in the form of heat and light. The same elements that are necessary for fires to exist — oxygen, fuel, and heat — are also necessary for animal life to exist. However, when these elements are brought together in certain proportions, there is a fire. Most fires involve a fuel that is chemically combined with the oxygen normally found in atmospheric air. Atmospheric air contains 21 percent oxygen, 78 percent nitrogen, and 1 percent other gases.

Sources of Heat Energy

Heat is a form of energy caused by the movement of molecules. All matter, regardless of temperature, has constantly moving molecules. When a body of matter is heated, the speed of the molecules increases, and thus the temperature increases. Anything that sets the molecules of a material in motion produces heat in that material.

There are four general categories of heat energy:

Chemical Heat Energy
- Heat of Combustion
- Spontaneous Heating
- Heat of Decomposition
- Heat of Solution

Electrical Heat Energy
- Resistance Heating
- Dielectric Heating
- Induction Heating
- Leakage Current Heating
- Heat from Arcing
- Static Electricity Heating
- Heat Generated by Lighting

Mechanical Heat Energy
- Frictional Heat
- Frictional Sparks
- Heat of Compression

Nuclear Heat Energy
- Nuclear Fission and Fusion

Refer to IFSTA **Essentials of Fire Fighting** for a detailed explanation of each of these categories.

Fuel may be found in any of three states of matter: solid, liquid, or gas. All fuels must be converted to a gaseous state before combustion occurs. *This conversion is the first phase of the combustion process.* Fuel gases are evolved from solid fuels by *pyrolysis.* Pyrolysis is defined as the chemical decomposition of a substance through the action of heat. Fuel gases are evolved from liquids by *vaporization*.

A solid is a substance that has definite shape and size. The larger the surface area of the solid, the faster the material will heat. The physical position of a solid fuel is also important in terms of fire spread: a vertically shaped solid fuel will spread the fire more rapidly than a horizontally shaped solid fuel due to increased heat transfer through convection.

A liquid is a substance that assumes the shape of its container and is not compressible. Liquids burn when heat causes them to vaporize. The temperature at which a liquid vaporizes is known as its *flash point*. Obviously, liquids with low flash points present great fire hazards. Liquids with low flash points, such as gasoline and acetone, are also known as volatile liquids.

The density of liquids in relation to water is known as specific gravity (water has value of one). Liquids with a specific gravity less than one are lighter than water, while those with a specific gravity greater than one are heavier than water. Most flammable liquids have a specific gravity of less than one.

A gas is a substance that tends to assume the shape of its container but has no specific volume. Unlike liquids, however, gases are compressible. Gases burn more readily than liquids or solids because they are already in the state needed for combustion. If a gas is lighter than air, it will rise and dissipate. If a gas is heavier than air, it will seek and collect in the lowest available level. The movement of gases is also affected by terrain and wind.

The second phase in initiating combustion is the mixture of the fuel vapor with an oxidizer (air) within the flammable range. Every flammable or combustible substance has a flammable range. This range is the minimum and maximum percentage of air/vapor mixtures that will support combustion. The flammable range varies with the fuel and with the ambient temperature.

The *flash point* refers to the minimum temperature at which a *liquid* fuel gives off sufficient vapors to form an ignitable mixture with the air near the surface. At this temperature, the ignited vap-

ors will flash but will not continue to burn. For liquids, the *fire point* refers to the temperature at which a liquid fuel will continue to burn. For all other fuels, the *ignition temperature* is the temperature at which the fuel will continue to burn.

The Fire Triangle

The fire triangle, as shown in (Figure 3.1), has been used previously to explain why fires are created and why they go out. The triangle illustrates that oxygen, fuel, and heat in certain amounts create a fire and that if any one of the three elements is removed, a fire cannot exist.

The Fire Tetrahedron

In recent years, the fire triangle has become inadequate to explain the burning process. Many of the new chemicals and materials burn and react differently than materials produced earlier. Of the several theories advanced about the burning process, the one most generally accepted is the "fire tetrahedron," as illustrated in Figure 3.2. This theory has not done away with the fire triangle; it has simply added a fourth component. The tetrahedron resembles a pyramid: The three standing sides of the pyramid represent oxygen, fuel, and heat. The fourth side, or base, represents the chain reaction that occurs when materials burn. For a complete study of these two theories of the burning process, see IFSTA **Fire Ventilation Practices**.

NOTE: The terms *heat* and *temperature* should not be confused. Heat is a form of energy and temperature is the measurement of the intensity of the heating of a material.

The most common measurement of heat is the British Thermal Unit (Btu) (joules: 1 Btu = 1.55 kj). One Btu is the amount of heat required to raise one pound of water one degree Fahrenheit. Another familiar unit of heat is the calorie (Cal). (One calorie = 4.19 j.) One calorie is the amount of heat required to raise one gram (1/254 of a lb.) of water one degree Celsius (°C). One Btu is equal to 252 calories.

Scales of temperature are based on the freezing and boiling points of pure water. On the Fahrenheit (°F) scale, the melting point of ice is 32° and its boiling point is 212°. On the Celsius (°C)

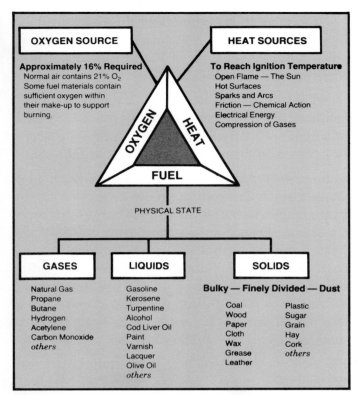

Figure 3.1 The fire triangle shows the three components necessary for burning to occur.

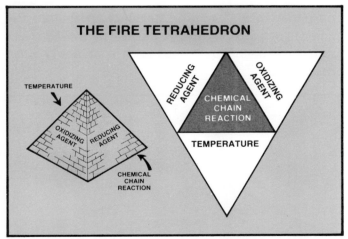

Figure 3.2 The fire tetrahedron includes the chemical reaction as a component of burning.

scale, 0° is the melting point of ice and 100° is the boiling point of water.

Phases of Fire

When matter (fuel) burns, it undergoes chemical change, and all the fuel elements are transformed into another form or state. There are four products of combustion: heat, gases, flame, and smoke. Fire itself has three progressive phases: in-cipient, free burning, and smoldering. These are further described as follows:

Incipient Phase — The oxygen content in the air has not been significantly reduced, and the fire is producing water vapor H_2O, carbon dioxide (CO_2), perhaps a small quantity of sulfur dioxide (SO_2), and other gases. During this phase, some heat is being generated and will increase as the fire progresses.

Free-Burning Phase — All free-burning activities of the fire are included during this phase. Oxygen-rich air is drawn into the flame as convection (the rise of heated gases) carries the heat to the uppermost regions of the confined area. The heated gases spread out laterally from the top downward, forcing the cooler air to seek lower levels, and eventually igniting all the combustible material in the upper levels of the area. The temperature in the upper region can exceed 1300°F (700°C). As the fire progresses, it consumes the free oxygen until there is insufficient oxygen to react with the fuel. Two specific conditions can also exist during this phase:

- *Steady State* — A fully involved fire condition with a supply of air to feed the fire and carry the product of combustion away.

- *Clear Burning* — That phase of burning accompanied by high temperature and complete combustion. Thermal columns will normally occur with high air speeds at the base of the fire.

Hot Smoldering Phase — The flame may cease to exist if the area of confinement is sufficiently airtight. Burning is reduced to glowing embers, and the room is completely filled with dense smoke and gases. The intense heat (over 1000° [537°C]) will have vaporized the lighter fuel fractions, such as hydrogen and methane, from the combustible material in the room. These fuel gases increase the hazards to the firefighter and create the possibility of a backdraft explosion. The fire reduced to the smoldering phase needs only a supply of oxygen to burn rapidly or explode.

The missing element of combustion in the smoldering phase of burning is oxygen. In this phase, burning is incomplete because not enough oxygen is available to sustain combustion. However, the heat from the free-burning phase remains, and the unburned carbon particles and other flammable products of combustion will burst into rapid, almost instantaneous, combustion when more oxygen is supplied. Oxygen is introduced into the atmosphere by improper ventilation. As air rushes in, the stalled combustion resumes at such devastating speed that it truly qualifies as an explosion (Figure 3.3). This condition is known as *backdraft*.

One warning sign of a possible backdraft hazard is dense, black (carbon-filled) smoke puffing from openings in a structure. The possibility of a backdraft can be reduced through proper ventilation. Ventilation is the systematic removal of heated air, smoke, and gases from a structure, followed by the replacement of a supply of cooler air. This cooler air facilitates other fire fighting operations. Proper ventilation performs the following:

- Aids rescue operations
- Speeds attack and extinguishment
- Reduces property damage
- Reduces mushrooming (the lateral spread of fire)
- Reduces the dangers of backdraft

For more information regarding backdraft explosions and correct ventilation techniques, consult IFSTA **Essentials of Fire Fighting** or **Fire Ventilation Practices.**

The terms *flameover* and *flashover* are often confused but they refer to different conditions, both of which are likely to occur during the course of a fire. These differences are explained as follows:

Flameover— As a fire continues to burn, heated gases rise to the ceiling. When a portion of the gases trapped around the ceiling burst into flame, the flame spreads across the ceiling until all the heated gas has ignited. The term "rollover" is also used because the flame "rolls" over the entire length of the ceiling like a wave. This flame spread is similar to the lighting of a book of matches: as one match ignites, it continues to light the next match until all the matches are lit.

Flashover — Flashover occurs when the entire contents of the fire area become heated to their ignition temperature. When they reach this point, simultaneous ignition occurs and the area becomes engulfed in flames. The reaction is similar to placing paper into a burning fireplace. Although the paper is not directly in the flame, the intense heat from the fire will cause the paper to suddenly burst into flame.

BACKDRAFT
- High Oxygen
- Moderate Heat
- Smoldering Fire
- High Fuel Vapor Concentrations

Introduction of Oxygen Causes Fire of Explosive Force

A I R

Figure 3.3 A backdraft explosion is caused by air rushing into stalled combustion.

Heat Transfer

Heat flows from a hot substance to a cold substance. If two materials are in contact, the colder material will absorb heat until both objects are the same temperature. Heat can travel by one or more of three common methods: conduction, convection, and radiation.

Conduction — The transfer of heat from one body to another body by direct contact of the two bodies or by an intervening heat-conducting medium (Figure 3.4).

Convection — The transfer of heat by the movement of air or liquid. As air or water is heated, it expands, becoming lighter and moving upward (Figure 3.5 on pg. 44). As the heated air moves upward, cooler air takes its place at lower levels. (**NOTE:** Although it was once thought to be a separate form of heat transfer, direct flame contact is now considered to be a type of convection.)

Radiation — The transmission of heat waves through air. Heat waves travel much like light waves. Radiated heat travels through air until it reaches an opaque object (Figure 3.6 on pg. 44). As the object is heated, eventually it reaches its ignition temperature and catches fire.

FIRE HAZARDS

A "fire hazard" may be defined as any material condition or act that contributes to the start of a fire or that increases the extent and severity of a fire. Since fire requires heat, fuel, and oxygen, the job of the fire inspector lies in identifying and controlling situations where these elements could unite, thus starting a fire. Fuel hazards exist when fuels are present in a form that can be easily ignited. Any heat source is also potentially dangerous. Oxygen, of course, cannot be removed because it is necessary for life.

NOTE: Certain reactive materials will burn in an atmosphere of nitrogen or will produce their own oxygen.

Fire inspectors often measure fire hazards in terms of the relative ease with which the fuel may be ignited. These measurements are helpful in determining what types of safeguards and practices to apply to the hazard involved. For example, a pile

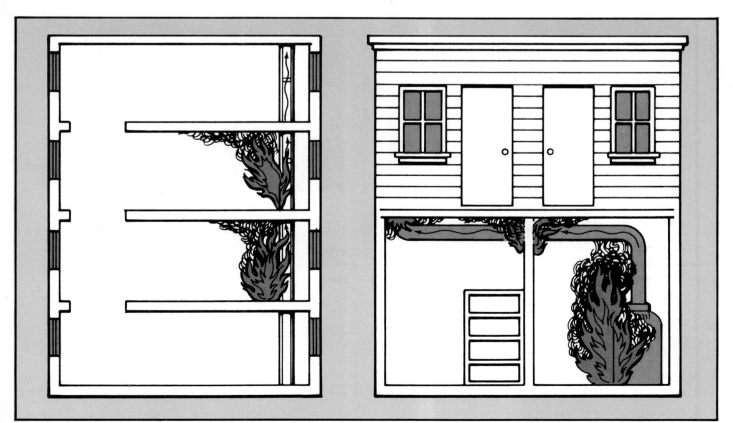

Figure 3.4 Heat is transferred by conduction when an intervening medium carries the heat from floor to floor or room to room.

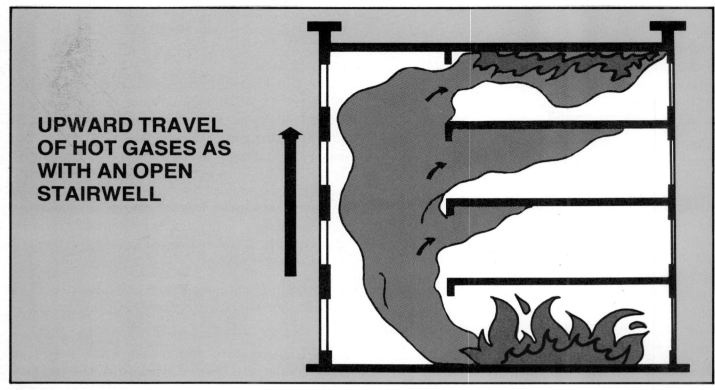

UPWARD TRAVEL OF HOT GASES AS WITH AN OPEN STAIRWELL

Figure 3.5 Heat is transferred by convection when air or water rises as they become heated.

Figure 3.6 Heat is transferred by radiation when heat waves hit an object and heat it.

of shavings or an open pan of gasoline — fuel supplies — can be hazardous because they are present in an easily ignitable form (Figure 3.7). A burning cigarette or a flame from a welding torch — heat sources — are hazardous because their heat is sufficient to ignite many combustible materials (Figure 3.8).

Fuel hazards are measured as follows:
- Flammable gases are measured by their flammable or explosive limits.
- Flammable liquids are measured in terms of flash points.
- Solids are measured relative to their ease of ignition and rate of flame spread.

Figure 3.7 These fuel supplies can be easily ignited.

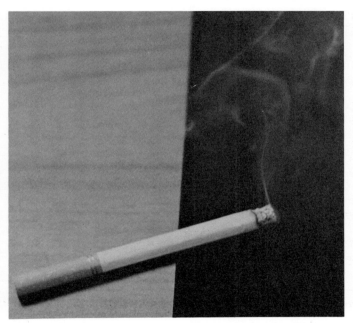

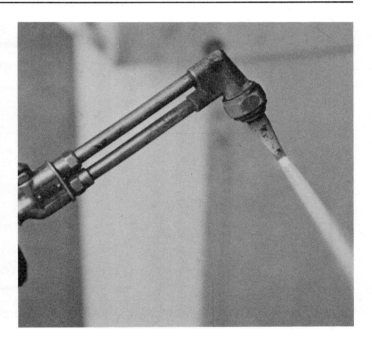

Figure 3.8 These sources of heat can ignite nearby combustibles.

Fire hazards can be considered to be cause-effect phenomena brought on as a direct result of violators, failure to follow sound fire prevention practices, and inherent problems of the occupancy. The effects of a fire resulting from one of these hazards can be measured as incident/loss consequences, particularly production loss and life and property loss (Figure 3.9 on pg. 46). With this information in mind, the fire inspector must identify the hazards and then act to eliminate or control them.

Types of Fire Hazards

Fire hazards have been classified into two groups: common hazards and special hazards. A common fire hazard is a condition that is likely to be found in almost all occupancies and is not generally associated with any specific occupancy, process, or activity.

A special fire hazard is a fire hazard arising from the processes or operations that are peculiar to the individual occupancy. Precautions against special hazards are covered by published standards and well-formulated rules. Fire inspectors can resolve unique problems not covered by those standards or rules in several ways. They must analyze the basic components and elements of the special hazards, compare them to similar hazards, and evaluate their loss potential based on personal experience and a general knowledge of the way in

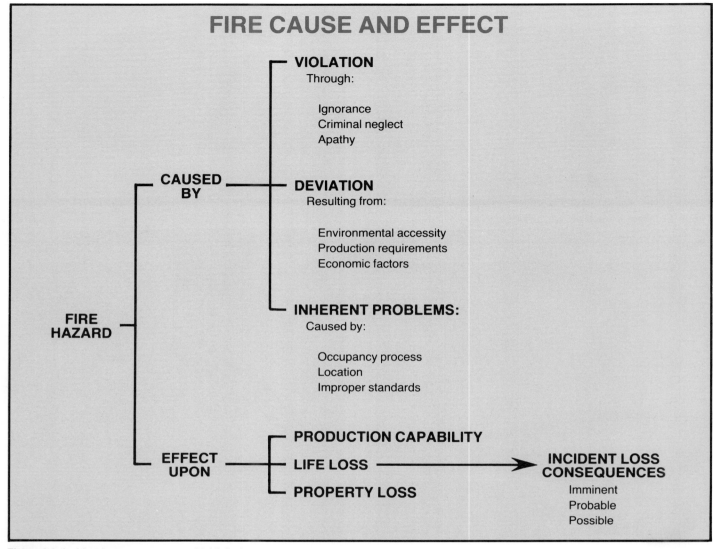

Figure 3.9 Incident/loss consequences highlight the potential losses from fire.

which fires start. Some materials and processes that constitute a special hazard include:

- Painting
- Welding
- Chemicals
- Acids
- Flammable liquids
- Explosives
- Flammable gases
- Combustible dusts

Flammable liquids are used widely in dip tanks, ovens, and dryers as well as in mixing, coating, spraying, and degreasing processes. These processes involve a variety of special hazards (Figure 3.10). Flammable gases are commonly used as fuel and for special purposes such as welding and cutting (Figure 3.11). Processing certain materials, such as metal, rubber, cork, fertilizers, and drugs, produces flammable dusts. These dusts present problems for fire inspectors in terms of preventing explosions and the spontaneous ignition of the finely divided materials. Combustible fibers and chemical processes also present problems.

Occasionally, materials and processes with hazardous features are found in occupancies that are normally relatively free of special hazards. A hotel, for example, may contain a private laundry, a woodworking and upholstering shop, or an imple-

Figure 3.10 Flammable liquids are used in industrial paint spray booths. *Courtesy of Des Plaines, Illinois Fire Department.*

ment warehouse equipped for repairing and painting.

Levels of Hazards

The level of hazard present in an occupancy or structure is determined by the authority having jurisdiction such as the fire chief, fire marshal, or fire inspector. Hazard levels are based on the characteristics of the contents of a structure and the processes or operations conducted in the structure. The classification of hazards is as follows:

- Low Hazard — Contents of low combustibility in which no self-propagating fire is likely to occur. The only probable danger requiring the use of emergency exits will be from panic, fumes, smoke, or fire from some external source.

- Ordinary Hazard — Contents that would burn moderately fast or that produce a considerable amount of smoke. Poisonous fumes or explosions are not a concern during a fire.

- High Hazard — Contents that would burn extremely fast or that would result in poisonous fumes or explosions during a fire.

FIRE CAUSE DETERMINATION

Since flames consume most of the evidence during a fire, the cause of most fires cannot be determined without a careful investigation. In some

Figure 3.11 Welding tanks are filled with compressed flammable gas. *Courtesy of Illinois Fire Inspectors Association.*

jurisdictions, fire inspectors are also responsible for conducting fire investigations.

All fires are caused by a combination of a fuel supply and a heat source in the presence of oxygen. These elements can be brought together in a number of ways:

- A hazardous act, such as welding in the presence of an open container of gasoline

- An accident, such as when a piece of equipment fails or the rays of the sun heat a combustible material to its ignition source

- A deliberate act (arson)

Regardless of the cause, it is the *combination* of heat, fuel, and oxygen that results in a fire. For example, when a welding operation ignites a nearby open container of gasoline, the cause is neither welding nor the presence of an open container of gasoline. The cause is welding *in the proximity of* an open container of gasoline. The hazardous act resulted in a fuel supply (the gasoline) coming into contact with a heat source (the welders torch). Figure 3.12 shows some fuel supplies, heat sources, and hazardous acts that frequently interact to cause a fire.

NOTE: The terms "fire cause" and "fire hazard" do not mean the same thing; nonetheless, fire service personnel often use them interchangeably. A fire *hazard* is a condition that can contribute to the start of a fire or that will increase the extent of a fire. A fire *cause* is any action, whether accidental or deliberate, that joins a fuel supply with a heat source at sufficient temperature to sustain combustion.

The needs and purposes for fire cause determination and fire investigation are numerous. A professionally conducted fire investigation will determine the following:

- Point of origin
- Cause of fire: fuel ignited, form of heat of ignition, source of heat of ignition, and human involvement

- How the fire spread
- Flame spread characteristics
- Flame pattern characteristics
- Who or what was responsible for the fire
- Need for further investigation
- Corrective action needed

Proper fire cause determination is the shared responsibility of the firefighters, the company officer, and when necessary, the fire investigator. The primary objective is discovering the cause of the fire. There are many clues a trained investigator will recognize if the evidence is left undisturbed.

One of the most important requirements for proper fire cause determination is the firefighter. Although the chief of the fire department has the legal responsibility for fire cause determination, the firefighters at the scene can draw valuable conclusions from the physical characteristics of the fire and relay these observations to the investigator. Firefighters can also assist during the sal-

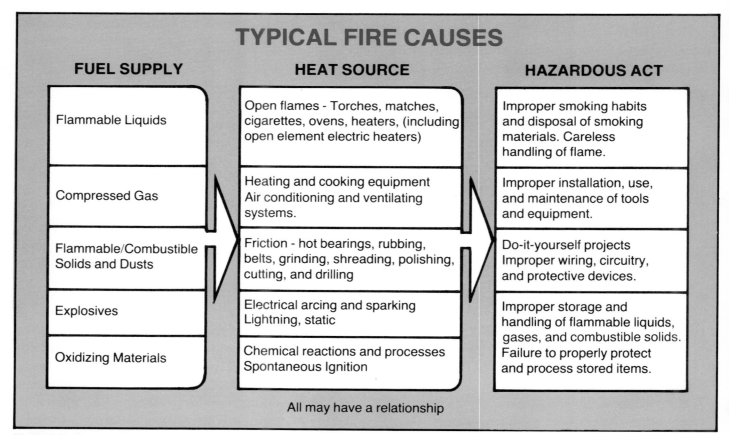

Figure 3.12 Carelessness and improper handling of potential dangerous materials can contribute to the start of fire.

vage and overhaul period by not disturbing the debris any more than is necessary.

The first-arriving company officer will file the report and interview occupants, owners, and witnesses. The officer and the firefighters will work together to determine the cause of the fire. If their investigation cannot discover the cause, they can call in a trained investigator.

NOTE: Calling for an investigator does not mean that the fire was deliberately set, only that the fire department could not determine the cause of the fire.

Preserving Physical Evidence

It is extremely important that firefighters preserve physical evidence as much as is possible. The firefighter's role in protecting evidence is relatively simple compared to its importance: report everything to a superior officer, and do not disturb anything needlessly.

The fire officer, in the absence of the investigator, should take the following actions:

- Suspend salvage and overhaul, and secure the scene. Keep unneeded firefighters out of the area.
- If the evidence is in danger of destruction, preserve it in the best way for that type of evidence.
- Sketch, mark, and label the location of the evidence at the fire scene. In addition, have photographs taken.
- Record the time the evidence was found, where it was found, and the name of the person who found it.
- Be sure a constant watch is kept on the evidence if it is left in place. If the evidence must be moved, put it in a secure place accessible only to the officer or where a responsible firefighter can stand watch.

A small kit that can be used to preserve evidence for the investigator should be kept in a locked tool box and used only for fire investigations. The kit should include the following items:

- Several 1-gallon (4 L) paint cans, uncoated, obtained directly from the manufacturer (some laboratories require glass containers)
- Small evidence boxes
- Plastic protectors for sheets of paper; these can be obtained from a stationery store
- Small glass vials with a cotton swab (such as a Q-tip) in each
- Special plastic evidence bag that is impervious to hydrocarbons
- Paper envelopes of various sizes
- Carpet knife
- Penlight
- Putty knife and tablespoon for collecting samples
- Carbide-tipped metal scriber for marking evidence containers
- Set of felt-tipped marking pens with permanent ink
- Evidence tags
- Pressure-sensitive evidence labels
- 50- or 100-foot (15 m or 30 m) steel measuring tape

Other items that are useful include:
- Heavy-duty coveralls
- Rubber gloves
- Flat-nose spade
- Camera with flash and color film
- 12-inch (304.8 mm) ruler
- Hand cleaner and paper towels
- Lock and hasp with screws
- Common hand tools (screwdrivers, pliers, etc.)

Evidence that is moved must be handled carefully. The following are some good methods for handling and moving evidence:

Glass or bottles for fingerprints — Pick up broken glass carefully, with two fingers at the edges. Pick up bottles by inserting a finger into the neck or by gripping the upper extremity of the neck. Stand glass exhibits upright and store them in a warm, dry place. Do not wrap glass objects in a handkerchief because any fingerprints will smudge.

Ash and most fire debris—Push ash and most other debris into a can, a glass container, or a special evidence bag.

Flammable liquid containers—The heat of a fire does not necessarily destroy fingerprints, so handle flammable liquid containers carefully and seal them in metal or plastic containers.

Flammable liquids—Liquid samples should be put into a can, glass vials, or special plastic evidence bags. Hydrocarbons will usually destroy or vaporize through ordinary plastic. Great quantities of liquids are not needed by testing labs.

Charred documents—Charred documents found in a container that can be moved easily should be left in the container (examples are wastebaskets, small file cabinets, and bindings). Loose charred documents must be handled very carefully and kept away from drafts. One method of preservation is to slip a sheet of paper, cardboard, or thin metal underneath the document and put it between layers of soft tissue in a box. Support the sides with cotton batting. The best protection, when available, is two panes of glass sandwiching the document.

Tire tracks or footprints—If there is an inescapable likelihood that tire tracks or footprints are going to be destroyed, photograph them from several angles, using side lighting with a scale beside them. Be sure to get a general photograph of the print showing its relative location in respect to the rest of the area. In other cases, cover the tracks with boxes to prevent dust accumulation until molds can be made.

4

General Fire Safety

This chapter provides information that addresses performance objectives described in NFPA 1031, *Standard for Professional Qualifications for Fire Inspector* (1987), particularly those referenced in the following sections:

Fire Inspector I

3-18 Principles of Electricity.

3-18.1

3-18.2

3-18.3

3-18.4

3-19 General Fire Safety.

3-19.1.2

3-19.1.4

10-90

Chapter 4
General Fire Safety

Fire inspectors will encounter some general fire safety matters that apply to all occupancies: ensuring good housekeeping practices, regulating smoking, controlling the use of flammable decorations, controlling static electricity and other electrical hazards, regulating open burning, maintaining access for fire equipment and personnel, ensuring that all fire incidents are reported, and performing fire safety standby at public events.

HOUSEKEEPING PRACTICES

Housekeeping practices, particularly the storage and disposal of trash and debris, often present serious problems in terms of fire hazards. Such problems are a result of ignorance, negligence, and apathy.

It is of primary importance that trash and debris be stored safely and disposed of properly. Trash containers should be constructed of a non-combustible material and should not be easy to tip over (Figure 4.1). Fire inspectors must consider how many waste containers are needed and where they should be placed. There must be a sufficient number of containers available in convenient locations; in addition, trash must be removed frequently enough to prevent interior waste containers from being overfilled.

Containers should be available for specific types of waste, such as smoking materials, oily rags, and highly combustible materials, so that these items are disposed of safely. Furthermore, these containers should not be located near heating equipment or flammable liquids, or directly under combustible objects such as tables or work benches. Flammable industrial waste should be

Figure 4.1 Waste containers should be distributed throughout the premises to collect trash and other debris.

discarded in containers that have self-closing, tightly fitting lids. The name of the particular waste should be clearly marked on the outside of the container (Figure 4.2 on pg. 54). Containers for oil or greasy materials should have airtight lids and should be supported by legs so that the bottom is several inches (millimeters) off the floor (Figure 4.3 on pg. 54).

Exterior trash disposal sites also require the fire inspector's attention. These sites must be located far enough away from buildings and exterior operations that they will not pose an exposure problem if a fire occurs. They must also be conveniently located to allow for easy disposal of interior waste. The method of permanent disposal, such as

Figure 4.2 Special containers for flammable materials should have the name of the material clearly marked on the outside.

Figure 4.3 Containers for oily or greasy materials should have airtight lids. *Courtesy of Justrite Manufacturing, Inc.*

controlled burning or landfill disposal, will have an effect on these items.

Fire inspectors must evaluate other housekeeping practices for potential fire hazards. Cer-

tain items, such as wastepaper, partly filled paint cans, and discarded furniture, should be disposed of promptly to help reduce the potential for fire in the building. Roofs, yards, courts, vacant lots, and open spaces should all be kept free of wastepaper, weeds, litter, or combustible waste and rubbish of any kind. All exterior vegetation, such as grass and vines, should be cut down or removed when it presents an exposure hazard.

Fire inspectors should check the basic structure of the occupancy, including walls, floors, ceilings, and drains, to be sure that hazardous chemicals, flammable liquids, fuel by-products, or dusts have not accumulated or impregnated these components. Dust that accumulates on beams, ledges, or other surfaces is extremely hazardous because it can lead to a dust explosion. (NOTE: Dust should be vacuumed rather than wiped or blown off. For more information about combustible dusts, see Chapters 9 and 10.)

Fire inspectors should check individual work areas and operations to be sure that chips, filings, droppings, dusts, and by-products do not accumulate in the immediate areas. Workbenches and tables must be kept clean of shavings. Operations using flammable or combustible liquids should have approved drip pans to catch any drippings (Figure 4.4). These pans should be emptied and maintained on a regular basis in a manner similar to that used for common waste containers. Mops used to clean up hazardous drippings should have

Figure 4.4 Drip pans are used to catch drippings from flammable or combustible liquids. *Courtesy of Mount Prospect, Illinois Fire Department.*

detachable handles so the mop heads can be stored in tightly covered metal cans. These cans should be constructed of sturdy metal with 3- or 4-inch (76 mm or 102 mm) legs and have self-closing lids and carrying handles (Figure 4.5). It is also important that any flammable liquids be stored in approved containers and in a safe and orderly manner.

Figure 4.5 Mop heads used to clean up flammable liquids should be stored in airtight containers. *Courtesy of Justrite Manufacturing, Inc.*

SMOKING

Smoking materials present a major problem for fire inspectors because of the difficulty in enforcing smoking regulations. This is a particular problem in such occupancies as department stores, warehouses, lumberyards, and museums where there is a great deal of valuable property, but there are relatively few people. Regulated smoking areas must be clearly marked with an ample number of signs. The more common of these signs include the following:

> **NO SMOKING**
> **SMOKING PROHIBITED EXCEPT IN THIS AREA**
> **DESIGNATED SMOKING AREA**

The symbol in Figure 4.6 is also becoming an increasingly popular means of signifying a no smoking area.

Figure 4.6 Areas where smoking is not permitted should be clearly marked.

Where smoking is permitted, fire inspectors must make sure that those who smoke do so in a safe manner. There must be an adequate number of ashtrays that are conveniently located. Ashtrays should *not be* placed near curtains, drapes, or ventilation equipment. Ashtrays should be made of noncombustible materials and contain grooves or snuffers that hold the cigarette securely. The sides should be steep enough that users must place smoking materials entirely within the ashtray. They should also be small enough that people cannot use them for trash containers. Cigarette lighters are recommended to reduce the hazards from disposing of matches improperly.

OPEN BURNING

All types of open burning should be controlled by local ordinances that require individuals or businesses to obtain a permit or some type of authorization before they can burn trash or rubbish (Figure 4.7 on pg. 56). After receiving a permit, they then have several responsibilities. They must burn the trash or rubbish only at designated places and times. Preferably, they should use only ap-

CITY OF PLANO

Burn Permit
CONSTRUCTION SITE WARMING FIRE

Date: _____

Warming fires shall be permitted on construction sites provided they meet the following conditions:

1. Maintain at least 20 feet clearance from structures or flammable materials. Warming fires shall not be permitted inside structures.

2. Warming fires shall be attended by responsible personnel at all times.

3. Warming fires shall be built in metal containers (5-gallons maximum size) and hardware cloth spark screens provided.

4. Warming fires shall use wood fuel *only*.

5. Fire extinguishment methods shall be provided for warming fires.
   (water buckets, metal covers)

6. Warming fires shall be extinguished by responsible personnel before leaving site(s).

7. Non-compliance to the above conditions shall result in abolishment of warming fires at the discretion of the Plano Fire Department.

_____
Signature of Applicant   Applicant's Address   Phone No.

_____
Company's Name   Company's Address   Phone No.

                                       _____
                                       Fire Department Inspector

Figure 4.7 Open burning must be strictly regulated and controlled. *Courtesy of Plano, Texas Fire Department.*

proved incinerators. Fire inspectors should make sure that the incinerators are in good shape and that spark arresters are in place.

Fire inspectors should evaluate the location selected for open burning to make sure that it is sufficiently remote from buildings and outside storage so there is little or no danger from flames, sparks, or flying brands. If the person responsible for burning the trash uses approved incinerators, these containers should be at least 15 feet (5 m) from any structure. Open fires should be at least 50 feet (15 m) from all structures, and the responsible person should make provisions to prevent the fire from spreading any closer. In addition, those responsible for burning trash should stay with the fire constantly until it is extinguished. At all times, someone should have a garden hose or other equipment readily available for extinguishing the fire.

FLAMMABLE DECORATIONS

When considering decorations with regard to fire safety, fire inspectors must remember that the degree of hazard involved depends upon the ease with which the decoration can be ignited by cigarettes, sparks, electrical defects, and similar heat sources. Fire inspectors can often provide valuable information concerning reputable firms that have both the experience and the knowledge to effectively treat flammable decorations with fire retardants. It may also be beneficial to perform a flame test on decorative material. WARNING: The test material may be highly flammable. Perform flame tests in such a way that they cannot result in a fire that can get out of control. The test material or a sample of the material should be removed for testing at a safe location.

To perform a flame test, apply a small flame from a ¾-inch (19 mm) paraffin candle to the material for one minute. The material should not flash, support combustion, or continue to flame for more than two seconds. The material should not glow for more than 30 seconds following removal of the test flame.

In occupancies where flammable decorations are used, fire inspectors must make sure that several precautions are observed. Highly flammable materials, such as straw, leaves, cotton batting, dried vines, trees, shrubbery, artificial flowers, and foam or plastic materials, must be treated with a fire-retardant substance before they can be used in mercantile and health care occupancies. Paper or other combustible materials that are used near electric light bulbs must be treated with a fire retardant. In many cases, noncombustible materials or treated fabrics can be used instead of more combustible decorations to ensure a safe decor in almost any occupancy.

ELECTRICAL SAFETY
Principles of Electricity

Electrical energy can produce unwanted and unexpected fires. These fires are caused by one of the by-products of electrical energy: heat. Even in an ordinary household lamp, less than 10 percent of the electricity (current) is converted into light; over 90 percent is wasted as heat. Unless electrical energy is used in a very efficient manner, it will almost always produce heat as an unwanted by-product.

Inspectors should have a basic knowledge of electrical theory so they can understand what types of electrical energy cause fires and why. In order to understand how electricity can cause a fire, it is important to know what electricity is and why it can be so dangerous if abused.

Basic Electrical Theory

The concept of electricity is best understood by making an analogy to water. Electricity and water have many of the same basic characteristics. Both are measured by quantity, flow, and resistance.

Basically, electricity involves the transfer or movement of electrons between atoms. Materials that allow a free movement of a large number of electrons are referred to as conductors. Copper, gold, silver, and aluminum are excellent conductors and are among the more common materials used in wiring. Electrical energy moves through these conductors as a result of electrons moving from atom to atom within the conductor. Each of the migrating electrons moves to a neighboring atom and pushes away one or more electrons that in turn move to the next atom. This same process occurs again and again. This phenomenon results in the flow of electrical energy. Poor conductors, such as glass, dry wood, or rubber, possess very few free electrons and thus impede the flow of electrical energy through the material. Semiconductors are materials that are neither good conductors nor good insulators, and therefore may be used as either in some applications.

Static electricity refers to the presence of a nonflowing electrical charge that may develop on almost any surface. Normally, each atom in a given material has a balanced number of electrons surrounding it. If any of these electrons is removed or lost due to friction or some other means, there will be more positively charged protons than negatively charged electrons. If this material is then brought against or close to a normally charged material, there will be a momentary electric current, known as a static discharge. (For more information about static electricity, see Static Electricity on the following pages.)

The flow of electricity through a wiring system is like the flow of water through a water distribution system. Water flow has three major components: quantity (gpm or L/min), pressure (psi or kPa), and resistance (friction loss). An electrical flow is also measured by quantity, pressure, and resistance. The quantity of electricity is expressed in terms of amperes (amps). Electrical pressure is termed voltage, and it is measured in terms of volts. Resistance in an electrical circuit is referred to as ohms. Electrical resistance will occur if a poor conductor is used or if a circuit or switch is opened. If a circuit or switch is opened, the resistance is infinite, and no current can flow, just as if a valve in a water system were closed. The relationship between water and electrons is shown in Table 4.1 below.

TABLE 4.1
RELATIONSHIP BETWEEN
WATER AND ELECTRONS

Measurement	Water	Electrons
Quantity	gpm (L/min)	amperage
Pressure	psi (kPa)	voltage
Resistance	friction loss	ohms

The relationship between these three variables may be expressed mathematically by Ohm's Law, which states:

$$E = IR \text{ or } I = E/R \text{ or } R = E/I$$
Where E = pressure in volts
I = current in amperes
R = resistance in ohms

By manipulating the variables in the equation, any value for a variable may be determined if the other two variables are known.

Electrical Hazards

In order to safely work around and inspect electrical installations and appliances, inspectors must first be familiar with electrical hazards. Perhaps the greatest threat posed involves the possibility of electrical shock. The seriousness of the shock depends largely on the amount of current (voltage) that passes through the body, the type of current (alternating or direct), the path the electricity takes through the body, and the length of time of contact. The damage produced by an electrical shock depends upon the number of vital organs affected by the electricity. Body resistance to the passage of electricity varies from approximately 1,000 to 500,000 ohms, if the person's skin is dry and unbroken. Currents of 100 to 200 milliamperes may be lethal, and currents of 100 milliamperes or less may cause ventricular fibrillation of the heart (rapid, ineffective contractions of the heart). After being exposed to 200 milliamperes or more, people may lose consciousness and suffer severe burns.

The majority of accidents that result in injury, fires, or damaged machinery occur during maintenance and repair operations. Thus, all such operations should be performed by qualified and authorized personnel. Before overhauling or repairing any electrical equipment, repair personnel must open and tag the main supply or cutoff switches to indicate that they have been placed out of service (Figure 4.8). They should keep fuse boxes and junction box covers securely in place except when they are actually working on the equipment. Interlocks, overload relays, and fuses should remain in service at all times and not be altered in

Figure 4.8 Repair personnel must tag electrical equipment before beginning repairs or maintenance.

any way. Fuses should be removed and replaced only after the circuit has been deenergized. Replacement fuses should have the same voltage and current rating as the original fuses. Whenever possible, work on energized circuits should be avoided.

The proper grounding of electrical equipment is extremely important because a poor or improperly connected ground is more dangerous than no ground. Improper grounding is extremely dangerous because the appliance user develops a false sense of security. Incorrectly wired grounds literally energize the shell of the appliance, thus providing a means for the user to be shocked. Three-wire cords with polarized plugs and ground pins are highly recommended.

The use of batteries for electrical power also produces various hazards. The major hazard associated with using batteries is the possibility of a person receiving acid burns during connection and refill operations. This hazard can be greatly reduced by utilizing proper protective clothing. Another serious hazard is present during battery recharging. Hydrogen gas is produced during this operation, which can ignite explosively. Fire inspectors must be sure that smoking and open

flames are prohibited near these operations and that personnel understand proper procedures, particularly for shutting down chargers and unhooking cables.

Fires involving or caused by electrical systems and appliances should be a major concern of fire inspectors. To prevent such fires, proper housekeeping and general cleanliness in and around electrical equipment is essential. An electrical arc can easily ignite excess grease, oil, or carbon dust. Personnel must properly dispose of the rags used for wiping off electrical equipment in tightly closed metal containers. If a fire should occur, occupants should be instructed to deenergize the circuit, notify the fire department, control or extinguish the fire if it can be done safely, and report the incident to the proper personnel. If the interior layers of the insulation of electrical cables are burning, the cables must be cut and separated to ensure complete extinguishment. Electrical components, such as rectifiers that are constructed of selenium, produce poisonous fumes of selenium dioxide, thereby introducing an additional problem into the fire situation.

The use of manual or electrical tools around electrical equipment requires additional precautions. Hand tools should be equipped with nonconducting handles. Fuse pullers should be sized appropriately for the job by matching the puller rating with the amperage and voltage of the fuse to be pulled. Portable power tools must be inspected carefully before use to ensure they are clean, properly lubricated, and safe to use. All switches should operate properly, and the flexible cords should be free of defects. All portable tools must be properly grounded.

Before performing work on any electrical equipment or installation, personnel should first study the applicable schematics and wiring diagrams to determine which circuits they must deenergize to eliminate current in the equipment. Inspectors should realize that deenergizing main supply circuits by opening switches does not necessarily deenergize all circuits supplying energy and that some devices can store electrical energy. When an inspector must evaluate an installation in a damp or wet location, he or she should have a dry platform and/or a rubber mat to walk on. Before inspecting electrical equipment, inspectors should remove all rings, wrist watches, badges, metal insignia, or other jewelry. Inspection personnel **MUST NOT** activate switches, or alter or switch electrical components.

Static Electricity

Sometimes materials take on an electrical charge when they undergo a process of physical contact and separation. This electrical charge is known as static electricity. Static electricity is hazardous if a charge is present on the surface of a nonconductive material and does not immediately dissipate. In the presence of a mechanism for generating a static charge, a spark can occur when there is no good electrical path between the two materials.

Many common processes generate static charges. Some examples include: 1) nonconductive fluids flowing through pipes; 2) liquids breaking into drops and the drops then hitting liquid or solid surfaces; 3) air, gas, or steam flowing from an opening in a hose or pipe; 4) pulverized materials traveling through chutes or pneumatic transfer devices; 5) belts in motion; 6) moving vehicles.

Fire inspectors must answer four questions concerning static electricity as an ignition source:

- Is there a source that generates a dangerous amount of static electricity?

- Is there a conductor that will accumulate the charges generated and maintain a suitable difference of electrical potential between the materials involved?

- Will there be a spark discharge of sufficient energy to create a hazardous situation?

- Is there a mixture present that may ignite?

Fire inspectors cannot prevent static electricity from being generated, but they can implement control measures that will dissipate the generated charges. One method involves controlling the humidity of an area. Maintaining a relative humidity of 60 percent to 70 percent greatly reduces the static electricity problems associated with the manufacturing of paper, cloth, and fiber. Unfortunately, this method is not practical for all occupancies; furthermore, it is somewhat ineffective in controlling static electricity on heated materials and oils.

Bonding and grounding are two other means of dissipating static charges. Bonding involves connecting two objects that conduct electricity with something that is also a conductor (Figure 4.9).

Figure 4.9 Bonding involves connecting two objects that conduct electricity with something that is also a conductor.

This procedure reduces the difference in electrical potential between the two objects that conduct electricity. Grounding involves connecting an object that conducts electricity to the ground with something that is a conductor (Figure 4.10). In this case, the procedure reduces the difference in electrical potential between the object and the ground.

Figure 4.10 Grounding involves connecting an object to the ground with another object that is a conductor.

In either procedure, the bonding or grounding wires (the conductor) must be large enough to carry the greatest amount of current that may be produced in a given situation. Solid conductors may be used for connections that are permanent; however, if the connection will be made and broken frequently, flexible conductors should be used. They may be insulated or noninsulated. Temporary connections should use battery, magnetic, or other clamps that provide metal-to-metal contact, while permanent connections should be pressure-type brazed or welded clamps.

Another acceptable means of dissipating static charges is to ionize the air in the immediate vicinity where the charges accumulate. Several methods can be used for ionization. In the first method, a static comb (a metal bar equipped with a series of fine needle points) that is grounded is brought close to the charged body. The air ionizes at the needle points and the accumulated charge "leaks" away, or dissipates. Static combs are suitable for operations involving paper, fabrics, or power belts. In the second method, the electrical charges are neutralized with a high voltage device. This device produces a conducting, ionized atmosphere in the area of the charged surface and thereby provides a means by which the charges may dissipate. This method is applicable for processes involving cotton, wool, silk, paper, or printing. In a third method, an open flame is used to dissipate the electrical charges. This method is frequently used in the printing industry to remove static charges from sheets of paper as they come off the press. However, the method is somewhat more hazardous than the others because the flame has the potential for igniting materials it comes in contact with.

MAINTAINING ACCESS FOR FIRE APPARATUS AND PERSONNEL

During the course of fire inspections, the inspector should check for any conditions that will affect fire department accessibility during emergency operations, either outside or inside the premises.

Access to the Area Outside a Structure

The fire inspector must be alert for conditions that will seriously hinder or completely block fire department accessibility. These include parking problems, locations of dumpsters, installation of fences and gate landscaping, construction work, and the like (Figure 4.11).

Figure 4.11 Fire department efforts can be seriously hampered when a vehicle, such as a truck, blocks access to emergency equipment. *Courtesy of Edward Prendergast.*

Access to the Inside of a Structure

Ornamental walls, sun screens, and new or false fronts are examples of items that can permanently block access to the inside of buildings (Figure 4.12). In some instances, fire inspectors are powerless to prevent the use of these items. On the other hand, some codes require that occupancies be accessible under certain conditions (Figure 4.13), and fire inspectors should be familiar with these provisions. For example, NFPA 101, *Life Safety Code*, requires that certain occupancies have windows if they do not have sprinklers. Furthermore, the Code specifies the minimum size of the window opening and the maximum distance that the bottom of the window can be from the floor. If all the windows of the structure are sealed, then it qualifies as a windowless structure and a fire inspector can order automatic sprinklers to be installed.

Figure 4.12 A sunscreen can permanently block access to the interior of a building. *Courtesy of Edward Prendergast.*

Figure 4.13 This false wall shows that accessibility can be maintained with little alteration to the building's appearance. *Courtesy of Springfield, Illinois Fire Department.*

Whenever fire inspectors find permanent changes in access to the inside of a structure, they should notify the fire department. If fire inspectors find that required exits, exitways, or stairwells are blocked, they should immediately take steps to correct the situation. Frequently, if the inspector meets with the owners or management of the structure and explains the hazards of the situation, the problem will be corrected voluntarily.

REPORTING FIRE INCIDENTS

While conducting a routine fire inspection, the inspector may learn that minor fires have occurred that were unreported. The fire inspector should encourage managers to report all fires promptly so they can be checked by fire officials. It should be stressed that calling the fire department immediately is important to avoid a potentially serious incident. Reporting all incidents also enables fire officials to maintain an accurate data base of fire occurrences. Reporting all fire incidents may also detect an arson problem before it becomes more serious and can establish reports of incidents for prosecution purposes.

Every inspector should be able to define the procedure that citizens should take to report fires and other emergencies correctly. The information should include the following:

Telephone
- Dial Numbers
  - 911
  - Fire Department Direct
  - "0"perator
- Report Type of Incident
- Address
- Cross Street
- Your Name and Location

Fire Alarm Box
- Send Signal
- Stay at Location

Fire Alarm Station
- Send Signal
- Notify Fire Department

For additional information about fire communications, signaling methods, and alarm-initiating devices, consult IFSTA **Essentials of Fire Fighting** in the Communications sections.

PERFORMING FIRE SAFETY STANDBY IN PUBLIC ASSEMBLY OCCUPANCIES

Fire inspectors are sometimes called upon to stand by at major public events where large crowds will be present. The duties of the fire inspector at these functions are as follows:

- Check the files to see when the last regular fire safety inspection of the premises was made and review the recommendations that were made. If the premises is past due for inspection or there were major problems, a complete fire safety inspection should be made.

- Check with the person in charge of the event and determine if any hazardous event or display (in terms of fire safety) will take place. If so, the inspector should receive a detailed briefing describing exactly what is proposed. The inspector can then determine if the display will be allowed or what special precautions, if any, are necessary.

- Check equipment brought in to be sure that it meets code requirements and, where appropriate, that the equipment is of an approved type.

- Check the "temporary" wiring used for the event.

- Be sure that all exits are unlocked and the exit lights are on.

- Check emergency lights to be sure they are working.

- Work with the building management to be sure that the occupancy limit is not exceeded.

- Just before the event begins, have an announcement made calling attention to the location of exits.

- Keep aisles open. Prevent people from standing or sitting in the aisles.

- If smoking is not allowed, enforce this rule.
- In case of any significant fire, transmit an immediate alarm, then direct an orderly evacuation. Make special efforts to control panic.
- Fight the fire if possible.

5

Building Construction for
Fire and Life Safety

This chapter provides information that addresses performance objectives described in NFPA 1031, *Standard for Professional Qualifications for Fire Inspector* (1987), particularly those referenced in the following sections:

Fire Inspector I

3-17 Heating and Cooking Equipment.

3-17.1

3-17.2

3-17.3

3-17.4

3-19.1.6 Interior Finishes.
(See NFPA 101, *Life Safety Code®* **and NFPA 220,** *Standard on Types of Building Construction.***)**

3-19.1.6.1

3-19.1.6.2

3-19.1.6.3

3-19.1.7 Building Construction.

3-19.1.7.1

3-19.1.7.2

3-19.1.7.3

3-19.1.7.4

3-19.1.7.6

3-19.1.7.7

3-19.1.7.8

3-19.1.7.9

3-19.1.7.10

3-19.1.8 Building Equipment.

3-19.1.8.1

3-19.1.8.2

3-19.1.8.3

3-19.2 Decorations, Decorative Materials, and Furnishings.

3-19.2.1

3-19.2.2

Fire Inspector II

4-7 Heating and Cooking Equipment.

4-7.1

4-7.2

4-7.3

4-7.4 Industrial Ovens and Furnaces.

4-7.4.1

4-7.4.2

4-8 Safety to Life.
(See NFPA 101, *Life Safety Code.***)**

4-8.2.1

4-8.2 Interior Finishes.

4-8.2.2

Chapter 5
Building Construction For Fire And Life Safety

Fire can alter the structural systems of a building in many ways. For example, if fire heats through structural members, it will weaken their load-bearing ability. The load may then be transferred to adjacent members. Since these members may not be designed to support additional loads, structural failure may result.

The effect of fire on steel beams also illustrates how structural members can be affected. Steel structural components may begin losing their integrity at temperatures as low as 600°F (316°C). Further, a steel beam will initially tend to expand upon exposure to a fire. This expansion imposes horizontal forces upon adjacent columns or walls and can lead to collapse.

NOTE: See IFSTA **Building Construction** for further information about construction principles, assemblies and their fire resistance, building services, door and window assemblies, and special types of structures as they relate to fire protection.

This chapter discusses a number of features related to construction which the fire inspector must be aware of. First discussed are construction classifications. These classifications are based upon NFPA 220, *Types of Building Construction* and are commonly used in other NFPA Codes and Standards, such as NFPA 101, *Life Safety Code.* Of course, the inspector must be familiar with other construction classifications if they are applicable. The next section, Testing, describes various fire-resistance tests applied to roof coverings, construction assemblies, and materials used for interior finishes. The third section, Fire Safety Components, describes such features of fire protection as fire doors, walls and partitions, draft curtains,

smoke and heat vents, and fire stops. The section on Heating, Ventilating, and Air Conditioning Systems covers these building features as they relate to fire safety. The section on Cooking Equipment discusses restaurant-type cooking equipment as well as industrial ovens and furnaces. Flammable and Combustible Liquids Storage Rooms deserve special mention because of their inherent fire danger. The last section, Evaluating Buildings for Structural Safety, discusses life safety hazards that result when structures are misused or are allowed to deteriorate.

CONSTRUCTION CLASSIFICATIONS

Building construction is classified in different ways in different codes. In general, construction classifications are based upon materials used in construction and upon hourly fire-resistance ratings of structural components. NFPA 220, *Types of Building Construction*, divides construction types into five basic classifications: Type I through Type V. Each Classification is further broken into subtypes by use of a three-digit Arabic number code or several letters (for example, Type I-443):

- The first digit refers to the fire-resistive rating (in hours) of the exterior bearing walls.

- The second digit refers to the fire-resistive rating (in hours) of structural frames or columns and girders that support loads of more than one floor.

- The third digit indicates the fire-resistive rating (in hours) of the floor construction.

A listing of construction types and the degree of fire resistance of each is shown in Table 5.1. Construction classifications are further explained as follows:

Type I Construction (443 or 332)
Type I construction has structural members, including walls, columns, beams, floors, and roofs, of noncombustible or limited combustible materials having the degree of fire resistance shown in Table 5.1.

Type II Construction (222, 111, or 000)
Type II construction is similar to Type I construction except the degree of fire resistance is lower. Note that in subtype 000, noncombustible materials with no fire-resistance rating are used.

Type III Construction (211 or 200)
Type III construction has exterior walls and structural members that are of approved noncombusti-

TABLE 5.1
FIRE RESISTANCE REQUIREMENTS FOR TYPE I THROUGH TYPE V CONSTRUCTION
(NFPA 220, STANDARD ON TYPES OF BUILDING CONSTRUCTION)

	Type I		Type II			Type III		Type IV	Type V	
	443	332	222	111	000	211	200	2HH	111	000
EXTERIOR BEARING WALLS —										
Supporting more than one floor, columns or other bearing walls	4	3	2	1	0[1]	2	2	2	1	0[1]
Supporting one floor only	4	3	2	1	0[1]	2	2	2	1	0[1]
Supporting a roof only	4	3	1	1	0[1]	2	2	2	1	0[1]
INTERIOR BEARING WALLS —										
Supporting more than one floor, columns or other bearing walls	4	3	2	1	0	1	0	2	1	0
Supporting one floor only	3	2	2	1	0	1	0	1	1	0
Supporting a roof only	3	2	1	1	0	1	0	1	1	0
COLUMNS —										
Supporting more than one floor, bearing walls or other columns	4	3	2	1	0	1	0	H[2]	1	0
Supporting one floor only	3	2	2	1	0	1	0	H[2]	1	0
Supporting a roof only	3	2	1	1	0	1	0	H[2]	1	0
BEAMS, GIRDERS, TRUSSES & ARCHES —										
Supporting more than one floor, bearing walls or columns	4	3	2	1	0	1	0	H[2]	1	0
Supporting one floor only	3	2	2	1	0	1	0	H[2]	1	0
Supporting a roof only	3	2	1	1	0	1	0	H[2]	1	0
FLOOR CONSTRUCTION	3	2	2	1	0	1	0	H[2]	1	0
ROOF CONSTRUCTION	2	1½	1	1	0	1	0	H[2]	1	0
EXTERIOR NONBEARING WALLS	0[1]	0[1]	0[1]	0[1]	0[1]	0[1]	0[1]	0[1]	0[1]	0[1]

Those members listed that are permitted to be of approved combustible materials.
[1]Requirements for fire resistance of exterior walls, the provision of spandrel wall sections, and the limitation or protection of wall openings are not related to construction type. They need to be specified in other standards and codes, where appropriate, and may be required in addition to the requirements of this Standard for the construction type.
[2]"H" indicates heavy timber members; see text for requirements.

ble or limited-combustible materials. Interior structural members, including walls, columns, beams, floors, and roofs, are wholly or partly constructed of wood. The wood used in these members is of smaller dimensions than are required for heavy timber construction. In addition, structural members must have fire-resistance ratings that are not less than those set forth in Table 5.1.

Type IV Construction (2HH)
In Type IV construction, exterior walls, interior walls, and structural members that are portions of these walls are of approved noncombustible or limited-combustible materials. Other interior structural members, including columns, beams, arches, floors, and roofs are of solid or laminated wood without concealed spaces. The wood used in these interior structural members must be of the dimensions required for heavy timber construction. The specific requirements for these members can be found in NFPA 220, Chapter 3.

Type V Construction (111 or 000)
Type V construction has exterior walls, bearing walls, floors, roofs, and supports made wholly or partly of wood or other approved materials of smaller dimensions than those required for Type IV construction. In addition, structural members have fire-resistance ratings that are not less than those set forth in Table 5.1. **NOTE:** For an explanation of the way fire-resistive ratings are determined, fire inspectors should refer to NFPA 251, *Standard Methods of Fire Tests of Building Construction and Materials.*

The three model building codes used in the United States and the major Canadian building code use other classifications of a similar nature. Because these may differ somewhat from the classifications described above, the inspector should be familiar with the applicable codes.

TESTING OF BUILDING COMPONENTS

Since no material is immune to damage from severe and prolonged exposure to fire temperatures, it is necessary to specify the desired fire resistance of materials to be used in a building. This is done on the basis of the expected fire loading or occupancy factors. (**NOTE:** fire loading and occu-

pancy classifications are discussed in the next chapter.)

Roof Coverings

The combustibility of the surface of a roof is a basic concern to the fire safety of an entire community. Roofs that can be easily ignited by flaming brands have been a frequent cause of major fires. The danger of easily ignited roof coverings was recognized hundreds of years ago and some of the first fire regulations in the United States imposed restrictions on combustible roof materials.

Test methods have been developed to evaluate the fire hazards of roof coverings. NFPA 256, *Methods of Fire Tests of Roof Coverings* describes the appropriate procedures. The test evaluates the flammability of the roof covering, the protection it provides to a combustible roof deck, and the potential for producing flaming brands. Roof materials are classified as Class A, Class B, and Class C. To receive one of the classifications, the roof covering is given a series of fire tests of varying degree of severity.

In order to test roof coverings, several test decks of each type of roof are constructed. The intermittent flame exposure test, the spread of flame test, the burning brand test, the flying brand test, and the rain test are conducted on each type of test roof. The tests involving flame require a gas burner for the flame source, a wind tunnel, and an air blower.

INTERMITTENT FLAME EXPOSURE TEST

In this test, a gas flame of specified temperature is applied to the test deck in an on-and-off fashion. The length of application varies for each class of roof.

The roof covering is then observed to determine

- If sustained flaming occurs on the underside of the roof deck
- Whether flaming or glowing brands are produced
- If portions of the test samples are displaced
- Whether portions of the roof deck are exposed and fall away

SPREAD OF FLAME TEST

In this test, the gas flame is applied continuously for 10 minutes to Class A and Class B test decks, and for 4 minutes to Class C test decks. During and after the test, observations are made to determine the distance to which flame spreads, whether flaming or glowing brands are produced, and if portions of the test sample are displaced.

BURNING BRAND TEST

The burning brand test is used to determine whether burning brands are likely to ignite roofing materials. Wood brands are constructed of varying sizes, depending upon the class of roof. The appropriate brands are then ignited and secured to the test decks. (For Class C test decks, the burning brands are applied at 1- to 2-minute intervals.) For all types of roofs, the tests are continued until the brands are totally consumed. Observations are made to determine if

- Sustained flaming occurs on the underside of the test deck

- Flaming or glowing brands are produced

- Any part of the test sample is displaced

- Any part of the roof deck is exposed or falls away

FLYING BRAND TEST

In this test, a flame is applied to the roof covering for a specified time with a 12-mph (19.3 km/h) air current. Observations are made to see if flying brands will develop.

RAIN TEST

The rain test is used to determine if the fire-retardant abilities of the roof are adversely affected by rain. In this test, spray nozzles are mounted above the test decks. These nozzles deliver an average of 0.7 inches (17.8 mm) of water per hour. All test decks are exposed to 12 one-week cycles; each cycle consists of 96 hours of water application followed by 72 hours of drying time. When the rain test is complete, the intermittent flame test, burning brand test, and flying brand test are repeated.

After all roof covering tests have been conducted, roof coverings are classified based upon test results:

- A Class A covering is one that is effective against a severe fire exposure, affords a high degree of fire protection to the roof deck, does not slip from position, and does not present a flying brand hazard.

- A Class B roof covering is one that is effective against a moderate fire exposure, affords a moderate degree of fire protection to the roof deck, does not slip from position, and does not present a flying brand hazard.

- A Class C covering is effective against a light test exposure, provides a light degree of fire protection to the roof deck, does not slip from position, and does not present a flying brand hazard.

Class A materials possess the best fire retardant properties, Class C the least. For example, asphalt-asbestos felt-assembled sheets of four-ply thickness have a Class A rating, but the same material of a single thickness would have a Class C rating.

Construction Assemblies

Fire resistance is defined as the ability of a structural assembly to maintain its load-bearing ability under fire conditions. In the case of walls, partitions, and ceilings, it also means the ability of the assembly to act as a barrier to the fire. The fire resistance of a structural component is a function of various properties of the materials used. This includes their combustibility, thermal conductivity, chemical composition, and dimensions.

The fire-resistance ratings of materials are determined by fire test procedures simulating fire conditions using the standard time-temperature curve (Figure 5.1). Fire-resistance ratings are given for assemblies of structural elements such as floors, floor-ceiling assemblies, columns, walls, and partitions. It is necessary to test such assemblies as erected in the field so values can be assigned that are meaningful to the fire protection engineer or building designer. (Unfortunately, not all possible assemblies have been tested and listed.)

In the standard fire test, the furnace temperatures are regulated to conform to the time-temperature curve. The temperature rises to 1,000°F (538°C) in five minutes, to 1,700°F (927°C) at one

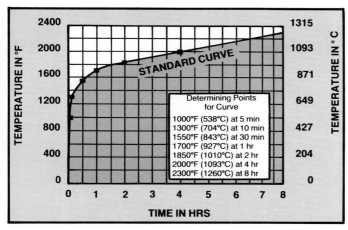

Figure 5.1 The time-temperature curve indicates how rapidly a fire builds up initial heat and then levels as the temperature continues to rise over an extended time.

Determining Points for Curve
1000°F (538°C) at 5 min
1300°F (704°C) at 10 min
1550°F (843°C) at 30 min
1700°F (927°C) at 1 hr
1850°F (1010°C) at 2 hr
2000°F (1093°C) at 4 hr
2300°F (1260°C) at 8 hr

hour. The fire-resistive rating is the period of time that the assembly will perform satisfactorily when exposed to this standard test fire. Failure of an assembly is determined by one of several criteria. These include failure to support the load, passage of flame through the assembly, and excessive increase in temperature on the unexposed side of an assembly. NFPA 251, *Fire Tests of Building Construction and Materials* contains specifications of the test procedures.

Although an assembly may fail at any time, fire resistance is expressed in certain standard intervals such as 15 minutes, 30 minutes, 45 minutes, 1 hour, 1½ hours, 2 hours, 3 hours, and 4 hours. The fire-resistance ratings form the basis upon which types of building construction are recognized in building codes. However, the standard test is a laboratory index of performance for various materials. Not all field conditions can be duplicated in the laboratory. Obviously, size restrictions do not permit testing entire buildings. The field performance may deviate from the laboratory results. This is possible when building components have not been installed with the same craftsmanship used in the laboratory. For example, joints in wall or ceiling assemblies may not have been carefully fitted. Furthermore, some of the lighter weight building materials may not be properly maintained, such as when holes are made through plaster assemblies.

Some laboratories that perform fire-resistive tests are

Underwriters Laboratories, Inc.
National Bureau of Standards
Factory Mutual System
Forest Products Laboratory
Portland Cement Association
Southwest Research Institute

Probably the best known of these is Underwriters Laboratories, Inc., which publishes test results annually in the *Fire Resistance Directory*.

Construction assemblies, such as roof assemblies, fire doors, and fire windows, are tested under a variety of fire conditions to determine the structural integrity of the assemblies during and after fire exposure. Building construction techniques and materials are tested under two standards: NFPA 251, *Fire Tests of Building Construction and Materials,* and NFPA 255, *Test of Surface Burning Characteristics of Building Materials.* Fire doors are tested under NFPA 252, *Fire Tests of Door Assemblies,* to determine relative fire integrity. Tests for doors consist of a fire endurance test and a hose stream test. Acceptable criteria require that the door remain in the opening throughout both the fire endurance test and the hose stream test. Windows are tested under NFPA 257, *Fire Tests of Window Assemblies.* These tests are made to determine the ability of a window or light-transmitting assembly to remain in an opening during a specified level of fire exposure for 45 minutes. Like the test for door assemblies, tests for window assemblies involve both fire endurance and hose stream tests.

Flame Spread Ratings

An important consideration in overall building fire safety is the combustibility of the materials used for interior finish, that is, the speed and extent to which flame can travel over interior surfaces. Several disastrous fires over the years have tragically illustrated that a combustible interior finish can contribute greatly to loss of life. Examples include the following:

● LaSalle Hotel, Chicago (1946, 61 dead)

● Winecoff Hotel, Atlanta (1946, 199 dead)

● Rhythm Night Club, Natchez, Mississippi (1940, 240 dead)

- Cocoanut Grove Night Club, Boston (1942, 492 dead)
- Providence College Dormitory, Providence, RI (1977, 10 dead)
- Beverly Hills Supper Club, Southgate, KY (1977, 165 dead)

An important consideration during inspections is overall building fire safety. The speed and extent to which flame travels over interior surfaces are primary concerns in evaluating occupancies for life safety. Therefore, it is extremely important that the fire inspector be knowledgeable in testing procedures and building materials. Inspectors familiar with fire test methods can help ensure that a building is reasonably safe. Interior finish contributes to fire impact in four ways:

- It affects the rate of fire buildup to a flashover condition.
- It may contribute to fire extension through flame spread over its surface.
- It may add to the intensity of a fire by contributing additional fuel.
- It may produce smoke and toxic gases that can contribute to life hazard and property damage.

Materials that exhibit high rates of flame spread, contribute substantial quantities of fuel to a fire, or produce hazardous quantities of smoke or toxic gases would be undesirable.

Naturally, the surfaces of some materials, such as wood veneers, will pose more of a fire hazard than other materials such as plaster. Once materials have been evaluated, restrictions can be placed on the use of those materials that are the most hazardous.

The most widely used test for determining the surface burning characteristics of interior finishes is the Steiner Tunnel Test. This test was developed by A.J. Steiner, an engineer at Underwriters Laboratories, Inc. This test also has the designations ASTM E84, NFPA 255, and UL 723. When materials are subjected to this test, their surface combustibility can be expressed in a number known as a "flame spread rating." The flame spread rating provides a means of determining the relative hazard presented by interior surface finishes as compared to standard materials.

The "Tunnel" in the Steiner Tunnel Test is a horizontal furnace 25 feet (7.6 m) long (Figure 5.2). The interior of the furnace is 17½ inches (445 mm) wide and 12 inches (305 mm) high. The top of the tunnel furnace is removable. The specimen to be

Figure 5.2 The horizontal tunnel furnace is used to determine flame spread. *Courtesy of Underwriters Laboratories, Inc.*

tested is attached to the underside of the furnace top and the assembly is lowered into place. A gas burner located at one end of the tunnel produces a gas flame that is projected against the test materials. The flame is adjusted to produce approximately 5,000 Btus (5 270 kJ) per minute. The extent of flame travel along the material is observed through view ports along the side of the tunnel.

To derive the numerical flame spread rating, the flame travel along the test material is compared to two standard materials: asbestos cement board and red oak. Asbestos cement board has a flame spread rating of 0 and red oak has a flame spread rating of 100. Obviously, the higher the flame spread rating, the more hazardous the material.

Building codes usually classify interior finish materials according to their flame spread rating (Figure 5.3). The *Life Safety Code* uses the following classifications, as shown in Table 5.2.

Figure 5.3 The flame spread rating is stamped on the back of each sheet of paneling. *Courtesy of Des Plaines, Illinois Fire Department.*

TABLE 5.2
FLAME SPREAD RATINGS

Class of Material	Flame Spread
A	0-25
B	26-75
C	76-200

A building code will specify what class of material can be used for interior finish in various occupancies or portions of a building. For example, corridors in a hospital might be limited to Class A (25) flame spread. In an office building, where the occupants are usually ambulatory, Class B (26-75) flame spread might be permitted.

In addition to the flame spread rating, the tunnel test provides two other measures of combustibility. These are the fuel contributed and the smoke developed. The measure of fuel contributed by the specimen in the test is rarely used in building codes and is not listed for many modern materials. The smoke-developed rating is sometimes used in codes.

The smoke-developed rating of the tunnel test is not a measure of the toxicity of the products of combustion of a particular material. It is a measure of the visual obscurity created by the smoke generated by a given material. This is determined by passing a beam of light through the exhaust end of the tunnel furnace which impinges on a photo-electric cell.

Thus, an interior finish material might be classified in the tunnel as follows:

Acoustical Ceiling Tile
Flame Spread 25
Smoke developed 10
Fuel contributed 5

Tunnel furnaces exist in several laboratories across the nation. In addition, several manufacturers of building materials have test furnaces for product development purposes. Because the flame spread rating cannot be accurately determined in the field, the inspector must research the ratings of materials noted during the inspection. These ratings can be obtained from the *Building Materials Directory* published by Underwriters Laboratories, Inc.

FIRE SAFETY COMPONENTS IN BUILDING CONSTRUCTION
Fire Walls and Partitions

A fire wall is a wall with a specified degree of fire resistance that is designed to prevent the spread of fire within a structure or between two structures.

The fire wall should extend from the foundation of a structure through and above combustible roofs so that it will stop flame spread on the roof covering and to adjacent surfaces. This protection is accomplished by topping the fire wall with a parapet (Figure 5.4).

Figure 5.4 A parapet is an extension of a fire wall above the roof that is designed to prevent radiant heat from fire from igniting adjacent surfaces.

Because fire walls are designed to protect against the spread of fire, no combustible construction is permitted to penetrate these walls. Properly designed fire walls will also be self-supporting so they can maintain their stability in the event of structural collapse on either side of the wall.

Self-supporting fire walls are most often found in one- and two-story industrial occupancies and storage facilities that have unprotected steel frame construction. The minimum height will be specified by the respective building code, and the maximum height will be limited by economics and practicality. Free-standing fire walls are designed to be totally self-supportive under vertical loads; however, strong horizontal loads, such as seismic (earthquake) forces, wind forces, or failure of attached portions of the structure, may cause them to collapse.

There are several important items that fire inspectors must check when performing an inspection in a structure with fire walls. The walls must be of the proper fire resistivity. The code being enforced should specify the required rating of fire walls; typical requirements will range from two to four hours. Walls should be checked for cracks, crumbling bricks, rotten mortar joints, and, in older structures, wooden lintels that carry masonry walls. Penetrations of the wall for the purposes of utility extension should be limited where possible, and all holes around these services must be completely resealed. If ducts penetrate fire walls, proper fire dampers must be in place within the ducts.

Doors in the fire walls must be examined to determine if they are defective, inoperative, or blocked open. Any of these conditions may render the wall ineffective. The inspector should also evaluate manufacturing or operating processes in the structure to determine if there is a possibility of a flash fire or explosion that could destroy or damage the wall.

A fire partition usually has a smaller degree of fire resistivity than a fire wall, and extends from one floor to the underside of the floor above or to the underside of a fire-rated ceiling assembly. Fire partitions are built with noncombustible or protected combustible materials, and are supported by structural members having a fire-resistance rating equal to or greater than that of the partition. The type of construction, the size of the area protected, and the severity of the fire hazard will govern the degree of fire resistance required, which generally ranges from ¾-hour to 2 hours.

Fire Doors and Door Assemblies

The number and size of openings in fire walls and fire partitions should be limited to the minimum number possible without interfering with building operations. Doorways and other openings in fire walls reduce the walls' effectiveness as fire barriers; therefore, fire doors are installed to retain as much fire integrity as possible.

A rated fire door is a door that has complied with the requirements of the American Society for Testing and Materials (ASTM) Test E152, Fire Test of Door Assemblies. These tests are usually conducted by nationally recognized laboratories such as Underwriters Laboratories. NFPA 252,

Fire Tests of Door Assemblies, also describes the methods of fire tests applicable to various types of doors and door assemblies. Fire doors must be labeled as such. Labels or classification marks may be of metal, paper, or plastic, or may be stamped or diecast into the door. Labels are generally located on the edge of the door, most commonly on the hinged edge of the door. Some doors have the label attached to the top edge of the door. Inspecting the label in this case requires the use of a chair, step ladder, or hand-held mirror.

A fire door, or more correctly, a fire door assembly, includes a fire door, the frame, the hardware, and other accessories. When these elements are assembled, they are capable of providing a specified degree of fire resistivity. Standard fire doors are either horizontal or vertical sliding, single or double swinging, or overhead rolling. They may or may not be counterbalanced. Fire doors designed to serve as exit enclosures, such as doors into stairwells, must be side hinged.

To provide an effective barrier against fire, fire doors must be closed during the fire. Self-closing doors automatically return to the closed position after being opened. Automatic-closing fire doors include those that normally remain open but that close when a fire situation occurs. Closing mechanisms include fusible links that, when they melt, allow the door to close. Electronic closing devices cause the door to close after heat detectors, smoke detectors, or sprinkler systems are activated. Fire doors that protect people may be held open only by closing mechanisms that are smoke detector activated.

When fire doors that serve as part of an exit are kept closed, they may be latched but not locked against egress; otherwise, people will not be able to use them. Fire inspectors must also make sure that fire doors are not wedged open. The door openings and the immediate area must be kept clear of all obstructions that might interfere with closing the door.

Horizontal sliding fire doors are mounted close to the wall and require little floor space and room for operation (Figure 5.5). Materials stored against these doors may, however, render them inopera-

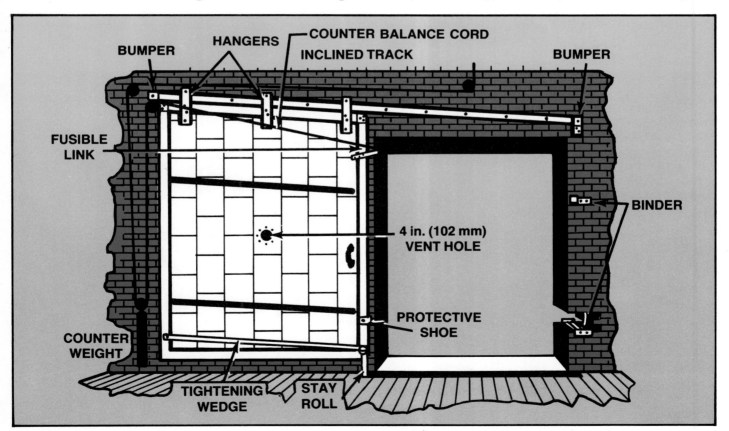

Figure 5.5 Horizontal sliding fire doors are often used to close large openings in fire walls between sections of a large building.

tive. Horizontal sliding doors are generally used on stair enclosures, in hallways and corridors where they will be opened and closed frequently, and to close large openings in fire walls between sections of a large building. These doors are normally kept open. Swinging doors that fit into a rabbeted jamb generally fit tighter than sliding fire doors; however, an automatic closing device is easier to employ in a sliding door. Vertical sliding doors are also normally kept open and arranged to close automatically. Overhead fire doors may be installed where space limitations prevent the installation of other types of fire doors (Figure 5.6). Fire inspectors must make sure that rolling doors are checked regularly to ensure that objects do not prevent them from closing completely. Counterbalanced fire doors are generally used on openings to elevators. They are mounted on the face of the wall inside the elevator shaft.

Fire doors used in stairwells and corridors sometimes have windows (Figure 5.7). These windows are acceptable if the door has been tested and rated with the window in place. The windows in fire doors to stairwells must be wired glass and no longer than 100 square inches (65 416 mm^2). The wire in the glass distributes the heat, thereby lowering thermal stress and increasing the overall strength of the assembly. The glass will undoubtedly crack; however, the wire is designed to prevent the glass from falling out.

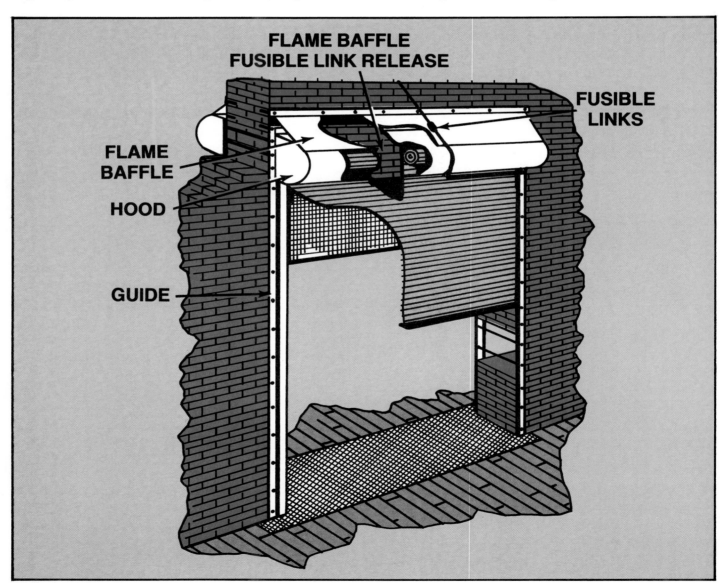

Figure 5.6 Rolling steel doors are often used in fire walls to provide automatic protection.

doors rated for 1½ hours are used to protect openings for 2-hour rated fire enclosures, vertical openings in buildings, and openings in walls separating buildings or fire areas requiring a 2-hour fire-resistance rating. Openings in 1-hour rated fire enclosures will be protected by 1-hour fire doors. Fire doors rated at ¾-hour are used to protect corridor openings and openings in room partitions. Doors rated at ½-hour and ⅓-hour are also available; they are used mainly where the primary concern is smoke control.

Maintenance of fire doors is relatively simple, but necessary. The doors must have a clear path for closing and should move easily. Fusible links and/or stay rollers must not be painted. All components of the assembly must be listed by Underwriters Laboratories (UL) or approved by Factory Mutual Laboratories (FM) for a specific use. Installations should be installed and maintained in accordance with local building codes and with NFPA 80, *Fire Doors and Windows*.

Draft Curtains (Curtain Boards)

Draft curtains, or curtain boards, are designed to limit the mushrooming effect of heat and smoke as it rises in large areas of buildings that are not otherwise subdivided by partitions or walls (Figure 5.8). By allowing the heat and smoke to concentrate in the area that is curtained off, venting systems will activate more quickly to remove the heat and smoke from the building. This feature makes draft curtains extremely beneficial during fire

Figure 5.7 Certain fire doors may have windows, if the door has been tested and rated with the window in place.

A major problem with a window in a fire door is that the window easily permits radiant heat to pass through it. For this reason, windows are prohibited in doors with a rating of 1½ hours or higher.

Fire doors are given hourly fire-resistance ratings in much the same manner that walls and partitions are rated. Currently, 3-hour fire doors are used to protect openings in walls separating buildings or fire areas that require a 3-hour or more fire-resistance rating. When a fire wall requires a 4-hour rating, authorities may require the use of two 3-hour doors, one on each side of the opening. Fire

Figure 5.8 Draft curtains are used in large areas to prevent the spread of heat and smoke.

situations. The quicker response time of detection and suppression systems, coupled with effective ventilation of the area, greatly aids in rapid fire extinguishment.

Draft curtains can be constructed of any sturdy noncombustible material that will resist the passage of smoke through it. Draft curtains should be fastened at the ceiling; their depth should be at least 20 percent of the ceiling height in the area being protected. Generally, curtain boards will not extend below 10 feet (3 m) from the floor. To ensure maximum effectiveness, the distance between curtain boards should not exceed eight times the ceiling height.

Smoke and Heat Vents

Removal of the toxic gases produced in a fire situation has often been the determining factor in successful rescue and effective fire control. Studies have proven that venting a building involved in fire maintains visibility longer; however, such venting also results in more rapid fuel consumption by the fire. Smoke and heat vents are designed to evacuate heat and toxic gases and are not meant to be used as a substitute for properly designed sprinkler systems. Smoke and heat vents are applicable for large areas, single-story structures, windowless and underground buildings, and high-rise buildings where life safety is critical or where firefighters cannot ventilate by the usual means. When installed properly and in the appropriate number, these vents will direct the flow of smoke and heated gases away from access points and escape routes, release them from the building, and restrict the spread of fire and smoke. All installations should be in accordance with NFPA 204M, *Smoke and Heat Venting*.

Venting devices should be designed and installed to operate automatically during a fire situation, in order to eliminate the possibility of failure due to human error. Stationary shutters, slots, breakable glass, hinged dampers, fusible links, counterweights, and smoke- and heat-actuated solenoids are all considered automatic features of venting systems. Venting systems may include the use of monitors, continuous gravity vents, unit-type vents, sawtooth skylights, exterior wall windows, or power roof ventilators (Figure 5.9).

Monitors are vents that resemble small buildings on the top of structures. They have sloped sides of glass or metal that open outwards following the melting of a fusible link and are especially effective when used over areas divided by draft curtains. Continuous gravity vents consist of a continuous narrow slot opening protected by a weather hood. If these vents have shutters, the shutters should open automatically in a fire. These vents are most commonly found in older industrial occupancies. Energy conservation measures limit their use in new construction.

Unit-type vents are usually small, with an effective venting area of only 16 to 100 square feet (1.48 m^2 to 9.29 m^2). These devices consist of a metal frame and housing with hinged dampers, or plastic domes that open when spring-loaded levers release. Sawtooth roof skylights may be used as vents if the glass is of ordinary strength or if the skylight has a movable sash that can be automatically operated in a fire situation. Exterior wall windows may be considered as part of a venting system if the openings are high on the wall near the ceiling and eaves. Venting from such windows is dependent on glass breaking due to heat buildup. To be effective, this type of ventilation must also include openings near the ground to provide make-up air. Power roof ventilators are activated by fire conditions and eliminate the need for large roof vents. This style of ventilation system is especially effective for fires that produce a lot of smoke but little heat. Underground areas can also be vented effectively with these devices. Wherever possible, power roof ventilators should operate automatically. The opening mechanisms should be controlled by a number of means, including fusible links, sprinkler systems interface, or smoke detection interface. When possible, the operation should be independent of electrical power.

Fire and Smoke Dampers

A fire damper is a device that automatically interrupts air flow through all or part of an air handling system, thereby restricting the passage of heat. Fire dampers have a fire-resistive rating of 1½ hours and include single blade, multiple blade, and interlocking blade types. Fire dampers are usually activated by fusible links, and most are de-

Figure 5.9 Smoke and heat vents are designed to evacuate heat and toxic gases. *Courtesy of The Bilco Company, New Haven, Connecticut.*

signed for vertical installation. Fire dampers are listed by UL, and their use is governed by NFPA 90A, *Installation of Air Conditioning and Ventilation Systems*, NFPA 90B, *Installation of Warm Air Heating and Air Conditioning Systems,* and by locally adopted codes.

A smoke damper is a device that restricts the passage of smoke in an air handling system. It operates automatically upon activation of a smoke detector. Smoke dampers are required in air handling ducts that pass through smoke barrier partitions. They are supposed to stop the flow of smoke within the duct. Fire dampers may be used as smoke dampers; however, they must be modified so that a smoke detector will cause them to operate. Smoke dampers typically have no fire-resistance ratings.

Fire Stops

Fire stopping is a means of preventing or limiting the spread of fire in hollow walls or floors, above suspended ceilings, in penetrations for plumbing or electrical installations, or in cocklofts and crawl spaces. The most common form of fire stopping for wood-frame construction involves using pieces of wood, 2 inches (51 mm) thick, placed in walls, partitions, and ceilings between the studs at each floor level and at the upper end of the stud channel in the attic. These wooden fire stops cut off the draft within walls and prevent the spread of fire and smoke within the concealed space. Other fire stopping applications should be of a noncombustible material that will not readily break down due to the heat of fire and smoke. Acceptable materials include sheet metal, gypsum board, brick, plaster, mineral fiber insulation, cement grout, ceramic fiber boards, and, in some cases, sand. Special devices and systems should be employed to prevent fire travel around cable, conduit, pipe, and cable tray penetrations. One accepted fire stopping method involves using fire resistant, silicone elastomer insulation. The insulation is applied as a foam at the point of penetration and hardens to seal the opening.

Heating, Ventilating, and Air Conditioning Systems

Heating, ventilating, and air conditioning systems (HVAC) present unique concerns for fire inspectors, as these systems are present in one form or another in all occupancies. To effectively inspect these systems from a fire safety standpoint, inspectors must become familiar with the various types and operating conditions of a variety of systems. These systems are covered by a number of standards, including NFPA 90A, *Installation of Air Conditioning and Ventilating Systems*, for systems in buildings greater than 25,000 cubic feet (707 m³). These systems may provide a variety of services including the filtering or washing of air, cooling and dehumidifying, or heating and humidifying. Equipment for these systems includes fans, heaters, filters, and heat exchangers, all of which should be located in a room separated from the rest of the building by 1-hour rated construction. Make-up intakes should be selectively located to prevent fire, fire gases, or smoke from being drawn into the system. Intakes should also be provided with ½-inch (13 mm) wire mesh screens to prevent combustible materials from being drawn into the system.

Fire inspectors will generally encounter three types of heating appliances: central heating appliances, unit heaters, and room heaters. There are four basic types of central heating appliances: boilers, warm air furnaces, floor furnaces, and wall furnaces.

CENTRAL HEATING APPLIANCES

Boilers may be of the steam or hot water type. They are classified as low pressure if they operate at 15 psi (102 kPa) or less, or high pressure (hot water) if they operate at less than 160 psi (1 088 kPa) or less than 250°F (121°C) (Figure 5.10). Boilers are regulated under the American Society of Mechanical Engineers (ASME) *Boiler and Pressure Vessel Code,* and should be equipped with various safety devices to prevent overpressure conditions from developing. Fire boilers should be equipped with controls and interlocks that will

Figure 5.10 Boilers should be equipped with safety devices that are designed to prevent overpressure conditions. *Courtesy of Edward Prendergast.*

shut off the burner or electric heating elements in the event of low water conditions, and when predetermined pressures and temperatures have been reached.

The fire problems associated with boilers include mounting and clearances from combustible materials as well as explosion potential. These explosions occur when fuel is not ignited by combustion controls and therefore accumulates in the firebox, or when the pressure in the system becomes too high.

Warm air furnaces operate on a variety of principles. Gravity furnaces operate primarily by circulation of air due to gravity. Another type of gravity system has a fan incorporated into the construction of the furnace to overcome any internal resistance to air flow and circulation. An additional type of warm air furnace utilizes gravity with a booster fan that helps circulate the air. The final type of forced air units is equipped with a fan that is the primary means by which air is circulated (Figure 5.11). Because these units contain plenums and ductwork that may become hot

Figure 5.11 Warm air furnaces use fans to help circulate air. *Courtesy of Des Plaines, Illinois Fire Department.*

Figure 5.12 Combustibles must not be placed on or close to floor furnaces.

enough to ignite adjacent unprotected woodwork, they require the use of appropriate clearances and insulation. Automatic controls should also be provided that will shut off the heating elements when the warm air plenum reaches 250°F (121°C). Fire problems with warm air furnaces are principally due to inadequate clearances, lack of proper limit controls, heat exchanger burnout, improper installation, or improper maintenance procedures.

Most floor furnaces are designed and approved for installation underneath combustible floors (Figure 5.12). However, they require proper clearances from miscellaneous combustibles and temperature limit controls. Fire inspectors should warn individuals using these devices not to cover registers or place clothing over them for purposes of drying.

Wall furnaces are self-contained, indirect fired gas or oil heaters installed in or on a wall. These units may incorporate a direct vent or may be vented via another vent or chimney.

UNIT HEATERS

The second major type of heating appliance is the unit heater. Unit heaters are self-contained units that are automatically controlled. They may be mounted on the floor or suspended from a ceiling or wall (Figure 5.13). The heating element and fan are enclosed in a common operating unit.

Figure 5.13 Unit heaters may be placed on the floor or suspended from a ceiling. *Courtesy of Des Plaines, Illinois Fire Department.*

ROOM HEATERS

The third major type of heating appliance is the room heater. Room heaters use circulation or radiant heat as the heating medium and are self-contained units designed to directly heat the surrounding area. Wood and coal stoves, electric and gas logs, and open front heaters are examples of room heaters (Figure 5.14). They usually incorporate manual or thermostatically controlled drafts and require that nearby floors and walls be properly protected. Solid fuel room heaters, such as wood and coal stoves, have poor fire records, largely due to inadequate clearances from combustibles and to inadequate maintenance. The hazards involved include overfiring, careless handling of ashes and live coals, inadequate container clearances, and accumulations of creosote in flue piping. Kerosene heaters and salamanders present problems because of their portability; as a result, people misuse and improperly place the units (Figure 5.15). In tightly enclosed rooms, kerosene heaters may generate enough carbon monoxide, due to incomplete combustion, to cause asphyxiation. Salamanders, which are often found on construction sites and in unheated buildings, may be as crude as a barrel in which scrap materials are burned. The hazards are very similar to those described for kerosene heaters.

Figure 5.14 Wood-burning stoves can be used as room heaters. *Courtesy of Alan Richard.*

Figure 5.15 Kerosene heaters and salamanders can be dangerous if they are improperly placed or inadequately vented. *Courtesy of Edward Prendergast and Springfield, Illinois Fire Department.*

Primary safety controls are a basic requirement on all fuel-burning heating units. These controls stop fuel from flowing to the unit in the event of an ignition or flame failure. Other common controls that fire inspectors may encounter include air-fuel interlocks, atomizer-fuel interlocks, pressure regulation interlocks, oil temperature interlocks, manual restart mechanisms, remote shutoffs, and safety shutoff valves.

Air conditioning and ventilation systems provide their own unique fire hazards. Air conditioning systems are a form of mechanical refrigeration. A gas, such as freon, is compressed in coils into a liquid, then allowed to expand back into a gas. During the expansion process, the freon cools. Air is blown across the coils to cool the surrounding air.

Most commercial refrigerants have a classified level of toxicity and flammability, as determined by the American National Standards Institute (ANSI) Standard, B79.1. Table 5.3 lists these refrigerants in groups that define a refrigerant's flammability. A Group 1 refrigerant, for example, is not very flammable, while a Group 3 refrigerant is highly flammable. This grouping is based only on the refrigerant's flammability; the toxicity of the refrigerant is an individual characteristic and is not classified in the grouping. Thus, the fire inspector may find an air conditioning system that uses a nonflammable but toxic refrigerant.

To clean the air and remove particulate dust and pollens, filtering devices are used. Filtering devices include fibrous media unit filters, renewable media filters that are cleaned and reused, and electronic air cleaners that consist of static precipitators. All of these devices may create hazardous fire conditions. Underwriters Laboratories classifies filters into two categories based upon flame propagation and smoke development. Class I filters, when clean, will not contribute fuel to a fire and emit very small quantities of smoke when attacked by flame. Class 2 filters, when clean, will burn moderately and emit moderate quantities of smoke.

Duct work for HVAC systems can provide a means for smoke, gases, heat, and flame to spread throughout the entire area served by the duct system. Problem areas include exit corridors and crawl spaces that are used as plenums. In addition, HVAC systems may lack smoke detection activated controls and/or smoke dampers in walls, ceilings, or partitions. Crawl spaces used as plenums are especially hazardous because they are usually constructed of combustible materials, are often used for storage, and usually include no provisions for the control of fire spread. The duct work itself may be combustible, and is classified according to combustibility and smoke development. Class O ducts are manufactured of materials having flame spread and smoke development ratings of zero. Class 1 ducts are constructed of materials with a flame spread rating of no more than 25, with no evidence of continued or progressive combustion, and a smoke development rating of 50 or less. Class 2 ducts are built from materials having a flame spread rating of 50 or less, with no evidence of continued or progressive combustion, and a smoke development rating of 50 or less on the interior surface and 100 or less on the outside surface.

Provisions must be made to control the flow of smoke within the building from HVAC systems. In fact, HVAC systems are sometimes used to

TABLE 5.3
HAZARDS OF COMMON REFRIGERANTS

Refrigerant	Toxicity
Group 1 (Nonflammable to weakly flammable)	
Carbon Dioxide (R-744)	Slightly Toxic
Monochlorodifluoromethane (R-22)	Slightly Toxic
Dichlorodifluoromethane (R-12)	Nontoxic
Dichlorofluoromethane (R-21)	Slightly Toxic
Dichlorotetrafluoromethane (R-114)	Nontoxic
Trichlorofluoromethane (R-11)	Slightly Toxic
Methylene Chloride (R-30)	Slightly Toxic
Monochlorotrifluoromethane (R-13)	Nontoxic
Trichlortrifluoromethane (R-113)	Slightly Toxic
Group 2 (Flammable, except sulfur dioxide)	
Ammonia (R-717)	Toxic
Methyl Chloride	Moderately Toxic
Methyl Formate (R-611)	Toxic
Sulfur Dioxide (R-764)	Very Toxic
Group 3 (Highly Flammable)	
Butane (R-746)	Slightly Toxic
Ethane (R-170)	Slightly Toxic
Propane (R-290)	Slightly Toxic
Ethylene (R-1150)	Slightly Toxic
Isobutane (R-601)	Slightly Toxic

evacuate smoke and gases from a fire-involved structure. Most buildings utilize a system of passive smoke control: the compartmentation concept is used, fans shut down, and fire and smoke dampers close to prevent the further spread of fire and smoke. Smoke dampers must be provided in ducts that pass through any required smoke barriers, and must operate automatically upon the detection of smoke. Smoke detectors should be provided in all systems rated over 2,000 cubic feet per minute (CFM) (944 L/sec). (**NOTE:** Some building codes specify 15,000 CFM (7 080 L/sec).

Upon activation, these detectors may shut down the system, sound alarms, operate smoke control dampers, activate fire suppression equipment, or activate various smoke control functions. Smoke detectors should be located in the main supply ducts, downstream of air filters and cleaners, and in the main return ducts prior to the point where the duct discharges exhaust from the building or joins the fresh air intake ducts.

An active smoke control system may be separate from a building's HVAC system. Alternately, it may use the building's HVAC system to create and maintain differential pressure that prevents smoke from moving from one area to another and exhausts combustion products from the structure.

When the HVAC system is used, the system operates in a special mode that no longer delivers supply air to the fire area. Air drawn from the fire area is discharged to the outside without recirculating or contaminating the air intake. At the same time, the return air supply to all or part of the building shuts down while the system continues to supply air from the outside to this space. In this type of smoke control system, fire dampers are usually omitted. Active smoke control systems are especially applicable in high-rise structures because a lengthy time is needed for evacuation, fire apparatus cannot reach upper floors, and there are strong air flow patterns in vertical shafts (stack effect phenomenon) (Figure 5.16). The system must be capable of maintaining smoke- and heat-free exit routes for occupants. It must also allow for sufficient evacuating time to either leave the structure or to move to various refuge areas.

Figure 5.16 In high-rise buildings, open vertical shafts promote the spread of smoke to upper levels.

Cooking Equipment

Cooking equipment, inherent to restaurants, motels, convention centers, and various other occupancies, presents several fire hazards that fire inspectors must be aware of. Cooking equipment includes ranges, deep fat fryers, steamers, broilers, hot plates, griddles, and portable ovens. For all types of equipment, the most common fire hazard is the potential for ignition of adjacent combustible materials (Figure 5.17). NFPA 54, *National Fuel Gas Code,* provides recommendations for the safe installation of gas-fired cooking devices. NFPA 96, *Installation of Equipment for the Removal of Smoke and Grease-Laden Vapors from Commercial Cooking Equipment,* addresses ventilation practices for fixed cooking equipment.

Exhaust systems for cooking equipment are a necessity as well as a major fire problem. Grease carried into ducts by hot gases condenses on the interior of the ducts and often is ignited. Because oils and fats used in frying may provide an ignition source for the grease, grease removal systems are often used. These systems consist of extractors, noncombustible grease filters, or specialized fans designed to remove grease vapors and provide a fire barrier. The ducts should maintain a minimum air velocity of 1,500 feet per minute (547.2 m/min) and must be installed with appropriate clearances

Figure 5.17 The most common hazard associated with cooking equipment is the potential for igniting nearby objects.

from combustibles. They should be constructed of relatively strong materials with liquidtight seals and discharge directly to the outside of the building. Separate exhaust systems should utilize separate ducts, and there should be a manual control for the fan motor, located near the hood. In addition, they should have an automatic heat-actuated shutdown device. All electrical equipment should be installed in accordance with NFPA 70, *National Electrical Code®*.

Fixed extinguishing systems should be installed where required by NFPA 96 (Figure 5.18). All installations should be in accordance with the applicable standards:

- NFPA 12, *Carbon Dioxide Extinguishing Systems*
- NFPA 13, *Installation of Sprinkler Systems*
- NFPA 16, *Installation of Deluge Foam-Water Sprinkler Systems and Foam-Water Spray Systems*

Figure 5.18 Here, an extinguishing system is installed above a deep fryer and grill area.

- NFPA 17, *Dry Chemical Extinguishing Systems*

National Electrical Code® and NEC® are Registered Trademarks of the National Fire Protection Association, Inc., Quincy, MA.

- NFPA 17A, *Wet Chemical Extinguishing Systems*

A manual means of activating a fire extinguishing system is required in addition to automatic activation devices. The manual actuator should be located along the escape path. When it is activated, it should simultaneously activate all systems protecting a single hazard and shut off all fuel and heat to the protected area. The provision and installation of portable fire extinguishers should be in accordance with NFPA 10, *Standard for Portable Fire Extinguishers*.

Industrial ovens and furnaces vary in size, complexity, location, operating systems, and purpose. The methods by which they operate — electrically, fuel fired, and heat transfer — all have their own hazards. The process hazards in the occupancy may involve various combustible dusts, flammable solvents and liquids, and/or combustible solids. These process hazards exist as a result of roasting, coating, or heat treating various materials. Overheating resulting from human error, faulty equipment, or improper safeguards may result in fire and/or explosions. Overheating can be prevented by temperature controls and by careful monitoring; fire can be prevented by insulation or separation from the source of heat.

When industrial ovens or furnaces are being installed, there are a number of safety factors that are of concern to the inspector. These are

- Location of hazardous contents in the building
- Proximity to combustible structural members
- Capacity and location of HVAC systems
- Maximum temperature required for the particular process
- The way in which materials are handled, moved, or stored after heating.

The most important fire safety practices for industrial ovens and furnaces are sound construction and installation, good housekeeping practices, automatic sprinkler protection, automatic humidity control, good air circulation, and proper ventilation. Hydrants or hose connections should be located close by as well as portable fire extinguishers recommended for that heating system.

Flammable and Combustible Liquids Storage Rooms

Flammable and combustible liquids storage rooms should have interior walls, ceilings, and floors that are constructed of materials with a minimum fire-resistive rating of 2 hours. The doorway openings in the walls should have noncombustible, liquidtight, raised sills or ramps at least 4 inches (101 mm) in height to prevent spilled liquids from flowing into adjoining areas. With the exception of drains, all floors should be liquidtight. Further, the room should be liquidtight at the places where the walls join the floor. The storage room should have mechanical or gravity ventilation that removes vapors from the atmosphere. All electrical wiring and equipment used in rooms storing Class I flammable liquids should be appropriate for Class I, Division 2 locations, as specified by NFPA 70, *National Electrical Code*. Aisles at least 3 feet (0.9 m) wide should be maintained in the facility, and in places where Class I or II liquids are stored in 30-gallon (132 L) or larger containers, they should be stored no more than one container high. Further information about the storage, handling, and use of flammable and combustible liquids can be found in Chapters 9 and 10, and in NFPA 30, *Flammable and Combustible Liquids Code*.

Evaluating Buildings for Structural Safety

Structures can be designed specifically for the loads that will be placed upon them. To do this, structural engineers must possess a scientific knowledge of many elements of construction: internal and external forces that act on a structure, properties of construction materials, and the structural members themselves. This knowledge, plus generations of experience, has given structural engineers a very sophisticated basis for design practices. The ability to design very specific structures enables construction materials to be used more economically. It also results in faster construction, thus increasing productivity. There are potential disadvantages, however, in building a structure for a very specific use. The major disadvantage is that it can make a building more susceptible to structural abuse. When the forces on a structure are changed from those for which it was designed, or if structural members are modified or deteriorate, failure may occur.

The fire inspector must be aware of the following types of structural abuse:

- Subjecting the structure to loads for which it was not designed
- Structural modifications by unqualified workers or contractors
- Deterioration of structural members due to age, weather, or lack of maintenance
- The forces associated with the violence of a fire

One of the items specified by a building code is the building's live load as a function of the intended use of a building. Changing the occupancy of a building, such as converting a school building into a warehouse, can subject the structure to loads greater than it was designed for. Other examples include mounting a heavy air conditioner cooling tower on a roof or using a roof truss to support a machinery hoist.

Structural modifications can take the form of removal of portions of bearing walls, cutting openings in bearing walls, and cutting away a portion of a structural member, such as a beam or column, to accommodate a run of pipe.

The life of a building may be 75 to 100 years or more. Over that time, the forces of nature (wind, rains, temperature changes, ground settling) can alter the structure. These forces include the erosion of mortar in brick walls, rust and corrosion of exposed metals, and rotting of wooden structural members. Most damage of this type, however, occurs slowly and can be corrected by ordinary building maintenance and repair.

A fire inspector encountering instances of structural abuse during fire inspection should refer the matter to the local building department for structural inspection. If there is evidence that conditions are serious enough to pose an imminent danger, such as sagging or cracked floors, immediate steps may have to be taken to protect the public.

6

Occupancy Classifications
and Means of Egress

This chapter provides information that addresses performance objectives described in NFPA 1031, *Standard for Professional Qualifications for Fire Inspector* (1987), particularly those referenced in the following sections:

Fire Inspector I

3-19.1 General.

3-19.1.1

3-19.1.5 Means of Egress.
(See NFPA 101, *Life Safety Code*®**)**

3-19.1.5.1

3-19.1.5.2

3-19.1.5.3

3-19.1.5.4

3-19.1.5.5

3-19.1.5.6

3-19.1.5.7

3-19.1.5.8

Fire Inspector II

4-8.1 Means of Egress.

4-8.1.1

4-8.1.2

4-8.1.3

4-8.7 Occupant Loads.

4-8.7.1

4-8.7.2

Chapter 6
Occupancy Classifications And Means of Egress

The previous chapter discussed building construction and features of fire safety as they relate to life safety. An equally important aspect of life safety is the ability to evacuate a structure rapidly and safely in the event of a fire or other emergency. Many lives have been lost during emergencies because exits were blocked, locked, improperly marked, or otherwise inaccessible. Even properly designed and maintained means of egress cannot function efficiently if the total occupant load has been exceeded and too many people are trying to exit at the same time. The fire inspector's role in the area of life safety is a vital one. He or she must ensure that each occupancy is correctly designated and that the means of egress meet the specifications set forth by NFPA 101, *Life Safety Code,* Chapters 5 - 30.

This chapter begins with a discussion of occupancy classifications as set forth in the *Life Safety Code.* The second section, Means of Egress, includes means of egress components, occupant load, and exit capacity. The third section, Fire Load, describes the concept of fire loading and its relationship to life safety.

OCCUPANCY CLASSIFICATIONS

For the purposes of inspection and code enforcement, NFPA 101, *Life Safety Code,* classifies buildings according to the type of occupancy or intended end use. There are 10 occupancy classifications:

- Places of Assembly
- Educational Occupancies
- Health Care Occupancies
- Detention and Correctional Occupancies
- Residential Occupancies
- Mercantile Occupancies
- Business Occupancies
- Industrial Occupancies
- Storage Occupancies
- Occupancies in Unusual Structures

Places of Assembly include, but are not limited to, all buildings or portions of buildings used for gathering together 50 or more persons for such purposes as deliberation, worship, entertainment, dining, amusement, or awaiting transportation (Figure 6.1). Assembly occupancies are further classified according to the number of persons that the structure can accommodate:

- Class A — Facilities capable of handling 1,000 persons or more

Figure 6.1 Churches are examples of assembly occupancies.

- Class B — Facilities capable of handling 300 to 1,000 persons

- Class C — Facilities capable of handling 50 to 300 persons

Educational Occupancies include all buildings used for educational purposes up through the twelfth grade by 6 or more persons for 4 hours per day or more than 12 hours per week (Figure 6.2). Some examples are part-day nursery preschools, kindergartens, and other schools whose purpose is primarily educational even though the children are of preschool age. Other occupancies within educational occupancies are governed by regulations for the specific occupancy. For example, auditoriums in schools are governed by the regulations for assembly occupancies.

Health Care Occupancies are used for medical treatment or care of persons suffering from physical or mental illness, disease, or infirmity; or for the care of infants, convalescents or infirm aged persons (Figure 6.3). Included are hospitals, nursing homes, residential-custodial care facilities, supervisory care facilities, and ambulatory health care centers.

Detention and Correctional Occupancies provide sleeping facilities for four or more residents and are occupied by persons who are generally prevented from protecting themselves because of security measures not under their control. Included are jails, detention centers, correctional institutions, reformatories, houses of correction, prerelease centers, and other residential restrained-care facilities where occupants are confined or housed under some degree of restraint or security (Figure 6.4).

Residential Occupancies are those in which sleeping accommodations are provided for normal residential purposes and include all buildings designed to provide sleeping accommodations. In-

Figure 6.2 Some educational facilities, such as this vocational-technical school, may be occupied during the evening hours in addition to normal school hours.

Figure 6.3 Health care facilities present severe life safety concerns due to the immobility of many of the occupants within them.

cluded in this classification are hotels, apartments, dormitories, lodging and rooming facilities, and one- and two-family dwellings (Figure 6.5).

Mercantile Occupancies include stores, markets, and other rooms, buildings, or structures used to display and sell merchandise. Some examples are supermarkets, department stores, drugstores, auction rooms, and shopping centers (Figure 6.6). Minor merchandising operations in buildings predominantly of another occupancy are subject to the requirements of the predominant occupancy.

Figure 6.4 Correctional facilities are designed, in theory, to limit mobility of the occupants within them. This is a serious life safety concern. *Courtesy of Edward Prendergast.*

Figure 6.5 Large apartment complexes feature multiple living units within each building and multiple buildings within each complex.

Figure 6.6 Strip shopping centers, such as these, often present access problems for arriving fire personnel.

Business Occupancies are those used for transacting business (other than covered under mercantile), for keeping accounts and records, and for similar purposes. Some examples are doctors' and dentists' offices, outpatient clinics for the ambulatory, general offices, city halls, town halls, courthouses, college and university instructional buildings housing classrooms for less than 50 persons, and instructional laboratories (Figure 6.7). A minor office occupancy incidental to the operations of another occupancy is considered to be a part of the predominant occupancy.

Industrial Occupancies include factories making products of all kinds and properties devoted to such operations as processing, assembly, mixing, packaging, finishing, decorating, and repairing. These occupancies include laboratories, dry cleaning plants, power plants, pumping operations, smokehouses, laundries, creameries, gas plants, refineries, sawmills, and college and university noninstructional laboratories (Figure 6.8).

Storage Occupancies include all buildings or structures used primarily for storing or sheltering goods, merchandise, products, vehicles, or animals. Examples are warehouses, cold storage facilities, freight terminals, truck and marine terminals, bulk oil storage facilities, parking garages, aircraft hangars, grain elevators, barns, and stables (Figure 6.9).

Unusual structures include any building or structure that cannot be properly classified in any of the other classifications by reason of some function not encompassed or some unusual combination of functions necessary to the purpose of the building or structure.

MEANS OF EGRESS

The "way out" or the means of egress is one of the most important factors to be considered in determining whether the design and construction of an occupancy is safe. NFPA 101, *Life Safety Code*, defines a means of egress as "a continuous and unobstructed way of exit travel from any point in a building or structure to a public way." A public way is defined as a street, alley, or similar parcel of land essentially open to the outside and which is used by the public. The *Life Safety Code* sets forth rigid requirements for means of egress from each type of occupancy.

A means of egress consists of three distinct components: the exit access, the exit, and the exit discharge. The *exit access* is that portion of a means of egress that leads to an exit. Hallways, corridors, and aisles commonly serve as exit access. The *exit* is that portion of a means of egress that is separated from all other spaces of the building structure by construction or equipment and that provides a protected way of travel to the exit dis-

Figure 6.7 Municipal buildings are an example of business occupancies.

Figure 6.8 Large industrial buildings, such as this one, may actually have many different operations going on within the same building. *Courtesy of Edward Prendergast.*

Figure 6.9 The large amount of jet fuel stored in the planes that occupy aircraft hangars poses a major fire hazard to the building.

charge. A door leading directly outside or through a protected passageway to the outside, a horizontal exit, smokeproof tower, and interior and exterior stairs and ramps are examples of exits. The *exit discharge* is that portion of a means of egress that is between the end of an exit and a public way. An alley that joins a street or sidewalk is a typical exit discharge.

In order to fulfill its intended function during an emergency, each component of a means of egress must be kept free from obstructions at all times. No furnishings or decorations may be al-

lowed to obstruct an exit or exit access. Hangings and decorations, too, cannot be placed so that they conceal any exit. Further specifications for means of egress in different occupancies can be found in NFPA 101, Chapters 8-30. Of particular importance is Chapter 5, Means of Egress.

There are many questions that must be answered when determining whether the means of egress meets *Life Safety Code* requirements:

- Are the components of a means of egress of a type allowed for the specific occupancy classification?

- What is the total exit capacity?
- Is the travel distance to the nearest exit within the specified maximums for the particular occupancy classification?
- Are all the exits accessible and recognizable?
- Are the means of egress properly illuminated and marked?
- Do exit doors open easily and are they equipped with panic hardware if required?
- Are the components of the means of egress free of obstructions?
- What is the maximum number of occupants allowed for the particular occupancy and building?
- Are the interior finishes and decorations used within specified flame spread and smoke development limits for the particular occupancy?

Each of these elements of means of egress will be discussed in the sections that follow.

Means of Egress Components

DOORS

Two of the most important functions of doors, in terms of life safety, are to act as a barrier to fire and smoke, and to serve as components in a means of egress. When doors serve as components in a means of egress, they must be constructed so the way of exit travel is obvious. In general, doors should open in the direction of travel. Each door opening must be wide enough to accommodate the number of people expected to travel through the door in an emergency. In new buildings, each door opening serving as a component in a means of egress must be at least 32 inches (810 mm) wide. In existing buildings, each door opening must be at least 28 inches (710 mm) wide.

The floor on both sides of a door must be substantially level and must have the same elevation on both sides of the door. When a building is occupied, the exit doors must open easily in the direction of travel. Panic hardware is not required for all exit doors. Situations for which panic hardware is required are detailed under specific occupancy classifications. When panic hardware is required, occupants should be able to open the door by applying a force of not more than 15 pounds (66.72 N).

(NOTE: For more information about doors, consult NFPA 101, *Life Safety Code*, Chapter 5, and Chapter 5 of this manual under Features of Fire Protection.)

STAIRS

Exit stairs are a critical component of the means of egress in multistoried buildings. Stairways must be at least 44 inches (1 120 mm) wide, unless the total occupant load of all the floors served by stairways is less than 50. In this case, stairways may be 36 inches (910 mm) wide. Stair treads must give good footing, and landings should be provided to break up any excessively long individual flights (Figure 6.10). Railings are recommended for both sides of the stairs, and stairs that are exceptionally wide should have one or more intermediate rails (Figure 6.11).

Figure 6.10 The narrow width of these stairs requires that railings be installed only on each side of the stairs.

In order to provide a protected path of travel and to qualify as an exit, all interior stairs must be separated from other parts of the building by proper construction. The construction that encloses the exit must have at least a 1-hour minimum fire-re-

Figure 6.11 Wider staircases require that an additional railing be installed in the center of the stairs.

sistance rating when the exit connects three stories or less. This minimum applies whether the stories connected are above or below the story at which exit discharge begins. If the exit connects four or more stories, the separating construction must have at least a 2-hour fire-resistance rating. Any opening to the exit must be protected by a self-closing fire door. The fire door must have a 1-hour fire rating when used in a 1-hour rated enclosure. In a 2-hour rated enclosure, the fire door must be rated at 1½ hours. The only permissible openings in an exit are those that allow people to enter the exit from the building and those that empty to the exit discharge.

Exit stairs may be inside or outside a building. NFPA 101 generally requires that exit stairs inside a building be enclosed. Outside stairs are permitted to serve as exit stairs if they comply with the Code for outside stairs. Other requirements for exit stairs can be found in NFPA 101, *Life Safety Code,* Chapter 5, and in Chapters 8-30 under specific types of occupancies.

NOTE: Exterior exit stairs should not be confused with fire escape stairs. Fire escape stairs are discussed later in this section.

SMOKEPROOF ENCLOSURES

Smokeproof enclosures are stairways that are

designed to limit the penetration of smoke, heat, and toxic gases. Smokeproof enclosures provide the highest degree of fire protection of stair enclosures recommended by NFPA 101. All smokeproof enclosures must discharge into a public way, into a yard or court having direct access to a public way, or into an exit passageway. The stairway must be enclosed from its highest point to its lowest point by fire barriers. These barriers must have a 1-hour rating in buildings that are three stories or less and 2 hours in buildings that are four stories or higher. Access to a smokeproof enclosure is made through a vestibule or outside balcony. This arrangement prevents smoke from entering the stairwell when corridor doors are opened. The stair enclosure may be made smokeproof by the use of natural ventilation, mechanical ventilation, or by pressurizing the stair enclosure (Figure 6.12).

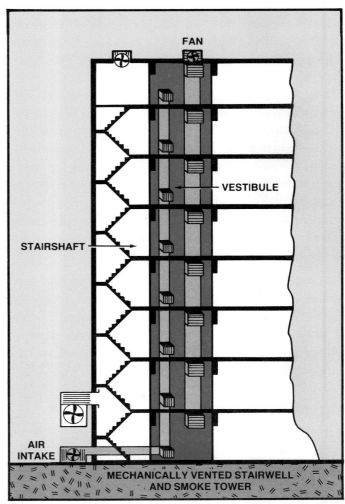

Figure 6.12 Mechanically ventilated smokeproof towers are an efficient way to protect stairwells.

HORIZONTAL EXITS

Horizontal exits are becoming popular in high-rise and institutional-type buildings. These types of exits require a separating wall or partition having at least a 2-hour fire-resistance rating. Horizontal exits may be substituted for other exits if they do not make up more than 50 percent of the total exit capacity of the building. Certain exceptions are permitted for health care facilities and detention facilities. A space used as an area of refuge making use of a horizontal exit must also have a standard exit, such as a stairway or door, that leads outside. The area of refuge must be large enough to shelter occupants from the fire area as well as those in the area of refuge, allowing at least 3 square feet (0.28 m^2) per person.

RAMPS

A ramp can be used as a component in a means of egress; in fact, ramps are preferable to stairs in certain occupancies, particularly schools and institutions. Ramps are easier to traverse for persons who are elderly, handicapped, or infirm. There are two main classes of ramps: Class A and Class B. Class A ramps must be at least 44 inches (1 120 mm) wide, and with a maximum slope of 1 to 10. Class A ramps are not limited to a certain height between landings. This type of ramp is especially useful in certain health care occupancies for moving patients in wheelchairs or cots. Class B ramps must be at least 30 inches (760 mm) wide, with a maximum slope of 1 to 8. Several exceptions to these requirements are detailed in Chapters 8-30 of the *Life Safety Code*.

Interior exit ramps must be enclosed and protected by construction in a manner similar to stairs. Exterior ramps must offer the same degree of protection as exterior exit stairs, with several exceptions. These exceptions are detailed in the *Life Safety Code*.

EXIT PASSAGEWAYS

An exit passageway serves as an access to or from an exit. It is not to be confused with an exit access corridor, which is a component of an exit. Exit passageways have stricter construction protection requirements than do exit access corridors. Probably the most important use of an exit passageway is to satisfy the requirement that exit stairs shall discharge directly outside multistory buildings. Thus, if it is impractical to locate the stair on an exterior wall, an exit passageway can be connected to the bottom of the stair to convey the occupants safely to an outside exit door. In buildings of extremely large area, such as shopping malls and some factories, the exit passageway can be used to advantage where the distance of travel to reach an exit would otherwise be excessive. An exit passageway must be wide enough to accommodate the total capacity of all exits that discharge through it.

ESCALATORS

Escalators and moving walks are not allowed as part of a means of egress. The only exceptions are for those escalators already approved in existing buildings. Even then, they must be capable of moving only in the direction of travel and must be enclosed if they are to be considered an exit.

FIRE ESCAPE STAIRS

Fire escape stairs cannot be used as any part of a means of egress in new construction, and may constitute only one-half of the required means of egress in existing buildings. For a number of reasons, fire escape stairs are not considered to be reliable exits. Often, they are poorly maintained and may be wet or icy, creating an additional hazard for those who must use them. Fire escape stairs are an unusual means of exiting; as such, individuals often hesitate to use them. Further, persons with a fear of heights may find fire escape stairs extremely difficult to use, slowing the progress of others who are trying to use the stairs.

To avoid trapping occupants, fire escape stairs must be exposed to the fewest number of door or window openings as possible. Windows may be used as access to fire escapes in existing buildings if they meet certain criteria concerning the size of the window. These windows must open with a minimum of effort. Access to fire escape stairs must be directly to a balcony, landing, or platform (Figure 6.13).

FIRE ESCAPE LADDERS

As with fire escape stairs, fire escape ladders cannot ordinarily constitute any portion of a means of egress.

Figure 6.13 Access to fire escape stairs must be directly to a balcony, landing, or platform.

FIRE ESCAPE SLIDES

Escape slides may be used as a means of egress where they are specifically authorized. They must be of an approved type and must be rated at one exit unit per slide with a rated capacity of 60 persons per unit.

Capacity of Means of Egress

OCCUPANT LOAD

Occupant load is defined by NFPA 101, *Life Safety Code* as "the total number of persons that may occupy a building or portion thereof at any one time." For purposes of life safety, the inspector must assess occupant load in terms of the number of people who can safely occupy and exit a building at one time.

The occupant load is determined by dividing the gross area of the building or the net area of a specific portion of the building by the area in square feet (or square meters) projected for each person. The amount of floor area projected for each person varies with the type of occupancy. Chapters 8-30 of NFPA 101, *Life Safety Code*, provide information about allowable occupant loads for each type of occupancy. Some examples from New Assembly Occupancies will illustrate the different amounts of space designated for each person:

- An assembly area of concentrated use without fixed seating, such as an auditorium, church, chapel, dance floor, or lodge room — 7 square feet (.65 m^2) per person

- An assembly area of less concentrated use, such as a conference room, dining room, drinking establishment, exhibit room, gymnasium, or lounge — 15 square feet (1.4 m^2) per person

- Libraries: in stack areas, 100 square feet (9.3 m^2) per person; in reading rooms, 50 square feet (4.6 m^2) per person

- Assembly areas with fixed seating — the occupant load shall be determined by the number of fixed seats installed

Under certain circumstances, the occupant load in assembly areas may be increased above the list specified, provided necessary aisles and exits are available (see NFPA 101, *Life Safety Code*, Chapter 8). Even with additional means of egress, the maximum density allowed is 5 square feet (0.46 m^2) per person and is subject to approval by the authority having jurisdiction.

In assembly areas where people are admitted into the building for the purpose of waiting for seating, admission is permitted provided maximum square footage per person is not exceeded, and the required means of egress are not occupied for the purpose of waiting. A lobby or similar space is acceptable for such purposes. Exits should be provided for each person waiting based on 3 square feet (.027 m^2) per person.

Determining Occupant Load

The *Life Safety Code* recognizes two methods of determining the occupant load of a structure or part of a structure. In general, the most common method used is the floor area method. The floor area method consists of dividing the net floor area of the occupancy by the amount of area allowed per occupant based on the occupancy classification. Both figures will be in units of square feet or square meters. The equation is:

$$\text{Occupant Load} = \frac{\text{Net Floor Area (ft}^2 \text{ or m}^2)}{\text{Area per Occupant (ft}^2 \text{ or m}^2)}$$

The second method of determining occupant load is to calculate the maximum number of people that the existing building exits are designed to accommodate. This number will be based upon figures from the *Life Safety Code*.

Means of egress is measured by units of exit width. A single unit of exit width is 22 inches (560 mm). This measurement was based on studies showing that the average shoulder width of a World War I soldier was 22 inches (560 mm). The *Life Safety Code* also designated 12 inches (305 mm) as one-half unit of exit width. Any width less than 34 inches (860 mm) is considered to be only one unit of exit width. In new buildings, each door opening must be at least 32 inches (810 mm) wide. In existing buildings, each door opening must be at least 28 inches (710 mm) wide.

The usefulness of this second method comes into play when it is determined that the existing units of exit width will permit a higher occupant load than the floor area method. In this case, the occupant load originally determined by the floor area method may be exceeded up to the amount calculated by the second method *if* the authority having jurisdiction approves this deviation and *if* the area per person is not reduced below 5 square feet (0.46 m^2). Local policy will dictate whether or not this is an acceptable practice. The following examples will demonstrate both methods of calculating occupant load.

Upon surveying an existing business occupancy, the following information has been gathered regarding the building.

Size: 100 x 150 feet (30 x 45 meters), one story
Number of exits: two, each 36 inches (916 mm) wide
Code in effect: NFPA 101, *Life Safety Code*

Floor Area Method

Step 1: Multiply the building's length and width to determine the net floor area.
100 x 150 = 15,000 ft^2 net floor area
30 x 45 = 1 350 m^2 net floor area

Step 2: Consult Chapter 27, Existing Business Occupancies, to determine that the area allowed for each person is 100 ft^2 (9 m^2).

Step 3: Use the following formula to determine the occupant load:
$$\text{Occupant Load} = \frac{\text{Net Floor Area}}{\text{Area Per Person}}$$

$$\text{Occupant Load} = \frac{15,000 \text{ ft}^2}{100 \text{ ft}^2} = 150 \text{ persons}$$

$$\text{Occupant Load} = \frac{1\,350 \text{ m}^2}{9 \text{ m}^2} = 150 \text{ persons}$$

Unit of Exit Width Method

Step 1: Determine from Chapter 27, Existing Business Occupancies, that a level unit of exit width (22 inches [560 mm]) enables up to 100 people to exit.

Step 2: Note that each door is 1½ units wide (22 inches + 12 inches = 34 inches [560 mm + 305 mm = 865 mm]). Since 100 people can exit through a single unit of exit width, 150 people can exit through each door (100 x 1.5 units of exit width = 150 persons).

Step 3: Multiply the number of persons per exit by the number of exits:
150 people per exit x 2 exits = 300 persons
Occupant Load

After computing the occupant load using both methods, notice that the floor area method yields an occupant load of 150 people; according to the units of exit width method, however, there are enough exits for a load of 300 people. If the authority having jurisdiction approves, the final occupancy load for this building may be raised to 300 people. Since the number of people allowed in the occupancy may be doubled, the area per person will be cut in half, from 100 square feet (9 m^2) to 50 square feet (4.5 m^2). This figure is still well above the minimum of 5 square feet (0.46 m^2) per person required by the *Life Safety Code.*

When computing occupant load, always be aware that the occupant load determined by the units of exit width method should always be equal to, or greater than, the occupant load determined by the floor area method. If this is not the case, the occupancy load must be reduced to the level that can be handled by the existing exits or additional exits must be installed.

DETERMINING THE CAPACITY OF AN EXIT

To determine the capacity of an exit, it is necessary to compute the exit capacity for each of the three components of a means of egress: the exit access, the exit, and the exit discharge. (Note that each exit component must be measured in clear width at its narrowest part.) The smallest of the three components is the total capacity of the exit. For example, a means of egress consisting of a corridor (exit access), exit stairway (exit), and an alley (exit discharge) would be evaluated as follows:

> Corridor: 42 inches (1 044 mm) clear: 1½ units x 100 persons/unit = 150 persons
> Stairs: 44 inches (1 116 mm) clear: 2 units x 75 persons/unit = 150 persons
> Alley: 12 feet wide (3.66 m): 6.5 units x 100 persons/unit = 650 persons

The smallest number from the three computations is 150 persons, so the capacity of the exit is 150 persons.

DETERMINING TOTAL EXIT CAPACITY

Exit capacity is based on units of exit width. (Remember that a unit of exit width is 22 inches [560 mm]). The number of required units of exit width varies with each occupancy classification.

In order to compute the total exit capacity of an occupancy, it is necessary to first determine the capacity of each exit and then total figures from each exit. This is the procedure for level egress. For

nonlevel egress (egress down stairs or Class B ramps), the exiting capacity of the lower floors must be sufficient to accommodate the persons who may be arriving from upper floors. The examples that follow will further illustrate the methods of determining total exit capacity.

Level Egress

The example used will be a new business occupancy with three 36-inch (914 mm) wide exit doors, all of the side-hinge swinging type.

Step 1: Consult NFPA 101, Chapter 26. This chapter states that each unit of level exit width may only provide for 100 persons.

Step 2: Note that each door is 1½ units wide (1½ units of exit width is 34 inches [864 mm], and each door in this occupancy is 36 inches [762 mm] wide). Each exit, therefore, has a rated exit capacity of 150 persons:
100 x 1½ units of exit width = 150 persons

Step 3: Since there are three doors, the total exit capacity is determined by multiplying the number of doors by the maximum number of persons who can exit through each door:
150 persons x 3 doors = 450 persons total exit capacity

Nonlevel Egress

The example used will be a new apartment building, four stories high, with a basement.

Step 1: Consult NFPA 101, Chapter 18. Note that two remote means of egress are required on each level. Each stairwell must be 44 inches (1 118 mm) wide, which is 2 units of exit width, and 75 persons can travel down one nonlevel unit of exit width.
75 persons x 2 units of exit width = 150 persons down each stairwell

Step 2: Since there are two exits on each floor, 150 persons x 2 exits = 300 persons total exit capacity per floor.

Step 3: Stairs must be able to accommodate the largest number of people from any single floor above. Of course, everyone from all floors will not be in the same part of the stairwell at the same time. For example, by the time the people from the third floor reach the second floor, the people from the second floor will already have exited the building.

NUMBER OF EXITS REQUIRED

The number of exits required for an occupancy is specified by each occupancy chapter of NFPA 101. Usually, at least two exits are required, with each exit as remote from the other as possible.

There are, however, exceptions for which one exit is allowed. A small business with less than 30 employees might be permitted to have only one exit.

MAXIMUM TRAVEL DISTANCE TO AN EXIT

NFPA 101 establishes the maximum allowable travel distance to the nearest exit and limits the length of dead-end corridors. Travel distance refers to the total length of travel necessary to reach the protection of an exit. Maximum allowable travel distances vary with each occupancy clas-

TABLE 6.1
MAXIMUM TRAVEL DISTANCE TO EXITS

Occupancy	Dead-End Limit, Ft (m)	Travel Limit to an Exit, Ft (m) Unsprinklered	Sprinklered
Places of Assembly	20** (6.1)**	150 (45.7)	200 (70)
Educational	20 (6.1)	150 (45.7)	200 (70)
Open plan	N.R.*	150 (45.7)	200 (70)
Flexible plan	N.R.*	150 (45.7)	200 (70)
Health Care			
New	30	100 (30.5)	150 (45.7)
Existing	N.R.*	100 (30.5)	150 (45.7)
Residential			
Hotels	35 (10.7)	100 (30.5)	150 (45.7)
Apartments	35 (10.7)	100 (30.5)	150 (45.7)
Dormitories	0 0	100 (30.5)	150 (45.7)
Lodging or rooming houses, 1- and 2-family dwellings	N.R.*	N.R.*	N.R.*
Mercantile			
Class A, B, and C	50 (15.2)	100 (30.5)	150 (45.7)
Open Air	0 0	N.R.*	N.R.*
Covered Mall	50 (15.2)	200 (70)	300 (91.4)
Business	50 (15.2)	200 (70)	300 (91.4)
Industrial			
General, and special purpose	50 (15.2)	100 (30.5)	150† (30.5)†
High hazard	0 0	75 (22.9)	75 (22.9)
Open structures	N.R.*	N.R.*	N.R.*
Storage			
Low	N.R.*	N.R.*	N.R.*
Ordinary hazard	N.R.*	200 (70)	400 (121.9)
High hazard	0 0	75 (22.9)	100 (30.5)
Open parking garages	50 (15.2)	200 (70)	300 (91.4)
Enclosed parking garages	50 (15.2)	150 (45.7)	200 (70)
Aircraft hangars, ground floor	20 (6.1)	Varies‡	Varies‡
Aircraft hangars, mezzanine floor	N.R.*	75 (22.9)	75 (22.9)
Grain elevators	N.R.*	N.R.*	N.R.*
Miscellaneous occupancies	N.R.*	100 (30.5)	150 (45.7)

*No requirement or not applicable.
†A special exception is made for one-story, sprinklered, industrial occupancies.
‡See Paragraph 15-4.2 of *Life Safety Code* for special requirements.
**In Aisles.

sification. The maximum travel distances specified in Table 6.1 are distances for reaching any *one* of the exits, not all of the exits.

RECOGNIZABLE AND ACCESSIBLE EXITS

NFPA 101 requires that exits be easily recognizable. Hangings or draperies cannot be placed over exit doors, and mirrors cannot be placed on exit doors. Where there are similar adjacent doors that are not exits, they must be marked "No Exit." Brightly illuminated signs, such as advertising signs, placed near an exit sign can obscure the exit sign and therefore are not permitted.

All exits are required to be accessible at all times. When two or more exits are required, at least two exits shall be remote from each other. These exits must also be arranged to minimize any possibility that both may be blocked by any one fire.

PROPER EXIT ILLUMINATION AND MARKING

Illumination requirements vary with each occupancy classification. When illumination is required, it must be continuous during occupancy. Floors have to be illuminated at not less than 1 footcandle (10.76 lumens [lx]) measured at the floor. A reduction to one-fifth footcandle (2.15 lx) is permitted in auditoriums, theaters, concert or opera halls, and other places of assembly during performances.

Any required illumination must be arranged so that the failure of any single lighting unit, such as a defective light bulb, will not leave any area in darkness. Battery operated units may not be used for primary illumination.

Emergency lighting may also be required in certain occupancies (Figure 6.14). The emergency lighting system has to provide the proper amount of illumination (1 footcandle [10.76 lumens] [lx]) if normal lighting is interrupted.

The requirements for marking exits may also vary according to the occupancy classification. Generally, illuminated exit signs are required (Figure 6.15). Exit signs must be placed so that no point in the exit access is more than 100 feet (30.5 m) from the nearest visible sign. The letters on exit signs must be at least 6 inches (152 mm) high and

Figure 6.14 Emergency lighting may be required in certain types of occupancies.

Figure 6.15 Exit signs should include directional arrows that indicate the direction of the exit.

the principal strokes of the letter at least ¾-inch (19 mm) wide.

Low-level exit signs may be used in certain occupancies, although they are not required by the *Life Safety Code* at present. The use of these signs is based on the observation that the area near the ceiling is the first to be obscured by smoke. If occupants are crawling toward an exit, exit signs placed near the floor will be more visible. These signs are not meant to replace standard exit signs, but rather to serve as a supplement.

FIRE LOAD

The arrangement of the materials in a building directly affects fire development and severity and must be considered when determining the possible duration and intensity of a fire. Therefore, occupancies are classified not only by the type of activities conducted within them, but also by fire load. The concept of fire loading is *not* a part of the *Life Safety Code*.

Fire load is defined as the *maximum heat* that can be produced if all the combustible materials in a given area burn. Maximum heat release is the product of the weight of each combustible multiplied by its heat of combustion. In a normal building, fire load is calculated for the structural components, interior finish, floor finish, and combustible contents of the building.

Certain occupancy groups have similar fire loading characteristics. The National Bureau of Standards (NBS) conducted studies in 1942 and in 1957 to determine the average fire load for various occupancies. Based on their evaluations, basic fire load classifications were derived for "typical" occupancies. They are as follows:

SLIGHT: TYPICAL FIRE LOAD 5 PSF (Pounds per square foot [24 kg/m^2])

- Well-arranged office, metal furniture, noncombustible building
- Welding areas containing few combustibles
- Noncombustible power house
- Noncombustible buildings, small amount of combustibles

MODERATE: TYPICAL FIRE LOAD 10 PSF (49 kg/m^2)

- Cotton and waste paper storage (baled) and well-arranged noncombustible building
- Paper manufacturing, noncombustible building
- Noncombustible institutional building with combustible occupancy

MODERATELY SEVERE: TYPICAL FIRE LOAD 10 to 15 PSF (49 to 73 kg/m^2)

- Well-arranged combustible storage, noncombustible building

- Machine shops with combustible floors

SEVERE: TYPICAL FIRE LOAD 15 TO 20 PSF (73 to 98 kg/m^2)

- Manufacturing areas, combustible products, noncombustible building
- Congested combustible storage areas, noncombustible building

VERY SEVERE: TYPICAL FIRE LOAD GREATER THAN 20 PSF (98 kg/m^2)

- Flammable liquids
- Woodworking areas
- Office, combustible furniture and buildings
- Paper working, printing, etc.
- Furniture manufacturing and finishing
- Machine shop with combustible floors

Since the severity of an anticipated fire situation can be estimated on the basis of the fire load, occupancies are classified according to the fire load present. A low fire load occupancy includes those facilities or portions thereof that contain small quantities of various combustible materials. An occupancy is classified as a "moderate" fire load occupancy if it contains moderate quantities of combustible materials that are fairly evenly distributed. "High" fire load occupancies include facilities that contain hazardous operations or concentrated quantities of combustible materials or both. The fire load is sometimes expressed in terms of the equivalent weight of combustibles per square foot (m^2) of building area.

If fire inspectors can estimate the fire load when conducting inspections of a building, they can subsequently estimate the loss potential. This information can be extremely useful in preparing prefire plans, water requirements for automatic sprinkler and fire flow demands, and fire extinguisher placement. Some hazards contributing to high fire loads include hazardous materials, flammable and combustible liquids, high-rack storage, and large quantities of plastics.

7

Fire Protection and Water Supply Systems

This chapter provides information that addresses performance objectives described in NFPA 1031, *Standard for Professional Qualifications for Fire Inspector* (1987), particularly those referenced in the following sections:

3-16.1 Portable Fire Extinguishers.

3-16.1.4

3-16.1.5

3-16.1.6

3-16.1.7

3-16.1.8

3-16.1.9

3-16.2 Fixed Fire Extinguishing Systems.

3-16.2.1

3-16.2.2

3-16.2.3

3-16.3 Sprinkler Systems.

3-16.3.1

3-16.3.9

3-16.3.10

3-16.4 Standpipe and Hose Systems.

3-16.4.1

3-16.4.2

3-16.4.3

3-16.4.4

3-16.5 Water Supply Systems.

3-16.5.2

3-16.5.9

3-16.6 Heat, Smoke, and Flame Detection Systems.
(See NFPA 72E, *Standard on Automatic Fire Detectors,* **and NFPA 74,** *Standard for the Installation, Maintenance, and Use of Household Fire Warning Equipment.***)**

3-16.6.1

3-16.6.2

3-16.6.3

3-16.7 Fire Alarm Systems and Devices.
(See NFPA 72A, 72B, 72C, 72D, and 72E.)

3-16.7.1

3-16.7.2

3-16.7.3

3-16.7.4

3-16.7.5

3-16.7.6

3-16.7.7

10-90

Fire Inspector II

4-6.4 Water Supply Systems.

4-6.4.1

4-6.4.2

4-6.4.3

4-8.4 Building Equipment.

Chapter 7
Fire Protection and Water Supply Systems

This chapter covers five major areas of fire protection: portable extinguishers, standpipes, fixed fire extinguishing systems, detection equipment, and water supply systems. Included is an overview of each type of fire protection equipment or system, its uses, and procedures for testing and maintenance.

Extinguishing equipment provides a means of attacking fires during their incipient phase. Standpipe systems permit either building occupants and/or firefighters to operate fire streams on different floors or in different areas of a building. Properly installed and functioning fixed fire extinguishing systems can rapidly extinguish or control a fire before it reaches serious proportions. Fire detection systems are designed to alert building occupants during the early stages of a fire so they can quickly exit the structure. Many detection systems also send an alarm signal to fire suppression forces at the same time. Water supply systems provide the water necessary to supply standpipe and sprinkler systems and fire fighting operations. It is important that water supply systems be tested to be sure that an adequate supply is available during an emergency.

There are many types of extinguishing equipment and fire protection systems. A short summary includes the following:

- *Automatic Sprinkler Systems* — Sprinkler systems were designed in their basic form over 100 years ago. They have an outstanding record of achievement: records reveal that in buildings where automatic sprinklers were installed, 98 percent of all fires were controlled or extinguished by

these systems. The remaining 2 percent of fires that were not controlled were due mainly to inadequate or shutoff water supply, incorrect design, partial protection, obstructions, and improper maintenance.

- *Special-Agent, Fixed-Extinguishing Systems* — Special-agent, fixed-extinguishing systems are used in those situations where automatic sprinkler systems, which apply water, are not appropriate. Such systems use carbon dioxide (CO_2), dry or wet chemicals, foam, or halogenated agents.

- *Standpipe Systems* — Standpipe systems provide a quick and convenient means for deploying fire streams within a building, particularly in large area or multiple-story buildings. Depending upon the type installed, standpipe systems may be used by occupants, firefighters, or both.

- *Portable Fire Extinguishers* — The most common type of private fire protection appliance is the portable fire extinguisher. Fire prevention and suppression personnel and maintenance personnel must thoroughly understand the characteristics and applicability of each type, as well as the classification and distribution requirements.

- *Fire Detection and Alarm Systems* — Fire detection and alarm systems, otherwise known as protective signaling systems, provide visual and/or audible signals to alert building occupants and/or organized fire protection units. These systems are also designed to operate fire protection systems components. All detection systems use some type of device that is sensitive to heat, flame, or products of combustion. With greater emphasis being placed on early detection and with regulations that require detectors to be installed in single- and multi-family dwellings, it becomes increasingly necessary for all fire service personnel to understand these devices. Automatic fire detection systems are not an acceptable substitute for automatic sprinkler systems.

There are many types, brands, and models of fire detection and signaling equipment; however, all types of fire protection equipment must be installed, maintained, and tested in accordance with local, state, and/or federal codes, ordinances, and standards. Typically, the installation must follow the specific guidelines incorporated in both NFPA standards and UL or FM guidelines.

It is important that the fire inspector check to see that the equipment has been tested, listed, and labeled by UL, FM, or another recognized national testing laboratory. Generally, the more complex or sophisticated the equipment or system, the more important skilled maintenance becomes. The information in this chapter is descriptive in nature and should not be used by fire inspectors as the authority when they are inspecting extinguishers or fire protection systems. It is recommended that fire inspectors consult manufacturers' technical data sheets for answers to questions concerning design, installation, operation, and maintenance of the system components.

PORTABLE FIRE EXTINGUISHERS

The purpose of a portable fire extinguisher is to enable an individual with minimum training to extinguish an incipient fire. Portable fire extinguishers are not intended to substitute for other fire extinguishing systems. Their primary use is as a first-line defense against incipient or limited fires. Therefore, extinguishers are considered necessary even when a property is equipped with automatic, fixed fire protection systems.

The United States Department of Labor, Occupational Safety and Health Administration (OSHA) in cooperation with the National Associa-

tion of Fire Equipment Distributors performed a study to evaluate the effectiveness of portable fire extinguishers in certain occupancies. Of the 5,400 fires reported during the study, 5,073 fires, or 93.9 percent, were extinguished by personnel instructed in the use of portable fire extinguishers. In addition, 1,319 of the occupancies equipped with portable fire extinguishers also had automatic sprinkler protection. In 32 of these cases (2.4 percent) the sprinkler system activated to control or extinguish the fire.

Portable fire extinguishers are used, installed, and maintained in accordance with NFPA 10, *Portable Fire Extinguishers*. Further information can be found in IFSTA **Private Fire Protection and Detection.**

Types of Portable Fire Extinguishers

There are basically four common types of portable fire extinguishers:

- *Stored Pressure* — These models contain both the extinguishing agent and expellent gas in a single chamber; both are discharged directly from the control valve (Figure 7.1). Nearly all of these units have a pressure gauge located on or near the control valve. Stored pressure extinguishers

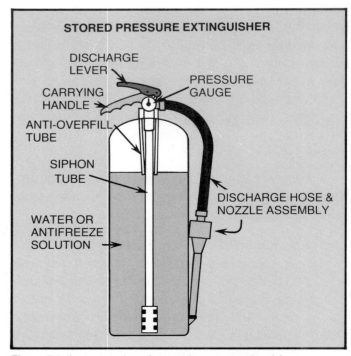

Figure 7.1 A cutaway view of a stored pressure extinguisher.

must be recharged by a qualified repair firm. These extinguishers are usually found in low-risk areas such as schools, department stores, and office buildings.

- *Cartridge Pressure* — These models contain the expellent gas in one cartridge and the extinguishing agent in another cartridge (Figure 7.2). During operation, the expellent gas must be released into the extinguishing agent cylinder. This type of extinguisher does not use a pressure gauge, but may have an indicator to show that the unit has been charged. Recharging is accomplished by replacing the expellent gas cylinder and refilling the extinguishing agent. This procedure can be an in-house operation since no special equipment is required. These types of extinguishers are commonly found in such industries as oil refineries, solvent manufacturing, and paper manufacturing.

- *Mechanically Pumped* — These models feature an internal pump (Figure 7.3). The operator provides the expelling energy by manually pumping pressure into the cylinder. Recharging is accomplished by refilling the cylinder with water. This type of extinguisher is limited to Class A fires.

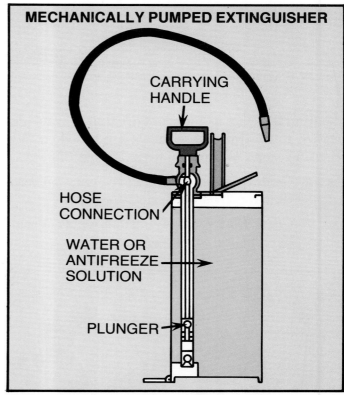

MECHANICALLY PUMPED EXTINGUISHER

CARRYING HANDLE

HOSE CONNECTION

WATER OR ANTIFREEZE SOLUTION

PLUNGER

Figure 7.3 A cutaway view of a mechanical pump extinguisher.

- *Hand Propelled* — This method of fire extinguishment includes 10- to 12-quart (9.5 L to 11.4 L) covered buckets of water, a 55-gallon (298 L) barrel filled with water with bails hung around or above the barrel, buckets filled with sand, and fire blankets (Figure 7.4 on pg. 114). All of these containers should be clearly stenciled with the word "Fire" across them. Hand-propelled extinguishers are valid when theft or vandalism of standard fire extinguishers is a problem.

Rating of Portable Fire Extinguishers

No portable fire extinguisher is suitable for use on all fires. Therefore, portable fire extinguishers are designated with a letter or letters in-

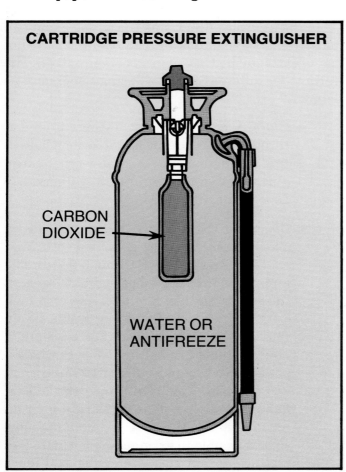

CARTRIDGE PRESSURE EXTINGUISHER

CARBON DIOXIDE

WATER OR ANTIFREEZE

Figure 7.2 A cutaway view of a cartridge-operated extinguisher.

HAND PROPELLED EXTINGUISHER

Figure 7.4 Buckets of sand can be used to extinguish fires, particularly Class D fires.

dicating the class or classes of fire that they are designed to control. These labels are based on four classifications, or types, of fires (Figure 7.5).

- *Class A:* Fires involving ordinary combustibles such as wood, cloth, or paper.

- *Class B:* Fires involving flammable or combustible liquids, greases, or gases.

Figure 7.5 Shown are the basic symbols for the classes of fire.

- *Class C:* Fires involving energized electrical equipment where the electrical nonconductivity of the extinguishing agent is of first importance. The materials involved are either Class A or B and can be handled as such once the equipment is deenergized.

- *Class D:* Fires involving combustible metals such as magnesium, titanium, zirconium, and potassium. These fires may require special extinguishing agents or techniques. If there is doubt, the inspector should consult NFPA 325M, *Manual on the Fire Hazard Properties of Flammable Liquids, Gases and Volatile Solids.*

Multiple letters or numerical-letter ratings are used on portable fire extinguishers that are effective on more than one class of fire. Class A and Class B extinguishers also receive a numerical rating that precedes the letter. This rating designates the size fire the extinguisher can be expected to suppress. It is very important that the correct agent be used on a fire. Using the wrong agent can be extremely dangerous and can result in a fire not being extinguished, a violent reaction, or both. The ratings for different types of extinguishers are as follows:

- *Class A Extinguishers:* Ratings from 1-A through 40-A are designated for Class A fire extinguishers. A water-type extinguisher rated 1-A requires 1¼ gallons (4.73 L) of water (Figure 7.6).

- *Class B Extinguishers:* Extinguishers for use on Class B fires are classified with numerical ratings from 1-B through 640-B. The number indicates the approximate area, in square feet (m^2), of fire involving an 8-inch (203 mm) deep layer of flammable liquid that can be extinguished. For example, a 10-B portable fire extinguisher can be expected to extinguish a fire of 10 square feet (0.9 m^2) of an 8-inch (203 mm) deep layer of flammable liquid. This is the rating for an untrained operator. A trained operator should be able to extinguish 25 square feet (2.3 m^2) with a 10-B extinguisher (Figure 7.7).

- *Class C Extinguishers:* Extinguishers for Class C fires have no numerical rating and are tested only for electrical nonconductivity. No Class C rating is provided in conjunction with a rating previously established for Class A and/or Class B fires. The size of the portable fire extinguisher should be appropriate for the size and extent of Class A and/or B materials in the electrical equipment or around the electrical hazard (Figure 7.8).

- *Class D Extinguishers:* Extinguishers for Class D fires have no numerical rating and the type of tests conducted vary depending upon the metals for which the extinguisher is intended. The faceplate of the extinguisher details the specific materials it should be used on and how to use it.

EXAMPLES OF EXTINGUISHER RATING LABELS

Foam Extinguisher Rated 4-A, 6-B — This extinguisher will extinguish four times the amount of fire a 1-A fire extinguisher will extinguish, and six times as much Class B fire as a 1-B extinguisher.

Dry Chemical Extinguisher Rated 5-10-B:C — This extinguisher will extinguish approximately five to ten times as much Class B fire as a 1-B unit and should extinguish a deep layer flammable liquid fire of a 5 to 10 square foot (2 m^2 to 3 m^2) area. It is also safe to use on fires involving energized electrical equipment.

Multipurpose Extinguisher Rated 4-A, 20-B:C — This extinguisher should extinguish a certain size Class A fire and approximately 20 times as much Class B fire as a 1-B extinguisher, and a deep-layer flammable liquid fire of 20 square foot (6 m^2) area. It is also safe to use on fires involving energized electrical equipment.

Extinguishing Agents

Portable fire extinguishers use several different types of extinguishing agents. Some agents

Figure 7.6 A stored-pressure water type extinguisher for Class A fires. *Courtesy of Ansul, Inc., Marinette, Wisconsin.*

Figure 7.7 A stored-pressure AFFF extinguisher for Class B fires.

Figure 7.8 A liquefied gas CO_2 extinguisher for Class B or Class C fires. *Courtesy of Ansul, Inc., Marinette, Wisconsin.*

control only one class of fire, although some are applicable to several classes of fire. (**NOTE:** There are no agents that can control all four classes of fire.)

There are six major extinguishing agents: water, carbon dioxide, halogenated agent, dry chemical, dry powder, and foam. Although there are a number of methods used for propelling the extinguishing agent, all portable fire extinguishers manufactured today are designed to be used in an upright position. Inverting-type extinguishers are obsolete. For more information on obsolete extinguishers, see Obsolete Extinguishers later in this section.

Water-based extinguishers are suitable for Class A fires; they are available in pump tank and stored-pressure models.

Foam extinguishers are rated and designed for Class B fires, but may be used with some success on Class A fires as well. Foam fire extinguishers have been available for many years. The earliest type utilized a chemical foam that was formed when two different chemicals were mixed together and reacted. These extinguishers are now considered obsolete. The primary type of foam extinguisher available today is the aqueous film forming foam (AFFF) extinguisher. This extinguisher contains a premixed solution of foam concentrate and water and is pressurized in a manner similar to a loaded stream water extinguisher. A special nozzle aerates the foam to produce bubbles.

Carbon dioxide extinguishers are used for Class B and Class C fires. They operate on the principle of an inert gas that will not support combustion. The carbon dioxide is self-propelling: because it is stored under its own pressure, it is ready for release at any time. Carbon dioxide extinguishers may be small, hand-held units or large wheeled units used in industrial applications.

NOTE: Carbon dioxide extinguishers with metal horns are not considered safe for use on fires in energized electrical equipment.

Halogenated agent extinguishers (either Halon 1211 [bromochlorodifluoromethane] or Halon 1301 [bromotrifluoromethane]) operate on the principle of liquefied compressed gas, which acts to extinguish fire because it interrupts the chemical chain reaction needed for combustion. Halon is stored under pressure and is ready to be released at any time.

Dry chemical extinguishers are used for Class A, Class B, and Class C fires, depending upon the rating of the specific extinguishing agent. The chemical compound in dry chemical extinguishers consists principally of sodium bicarbonate, potassium bicarbonate, ammonium phosphate, or potassium chloride. Whichever chemical is used, it has been chemically processed to make it moisture resistant and free flowing. Dry chemical extinguishers may be either stored pressure or exterior gas cartridge operated. Like carbon dioxide extinguishers, dry chemical extinguishers are available in small units or as large wheeled units. These large units may be suitable for Class A, B, or C fires.

Dry powder extinguishers are used for Class D (combustible metal) fires (Figure 7.9). Different dry powders are used for different metals. Some agents are applied through portable extinguishers; others are applied with a hand shovel or scoop.

Table 7.1 gives summary of the characteristics of portable extinguishers.

Figure 7.9 Dry powder extinguishers are gas cartridge operated and contain an agent for particular metals.

TABLE 7.1
CHARACTERISTICS OF EXTINGUISHERS

Extinguishing Agent	Method of Operation	Capacity	UL or ULC Classification
Water	Stored Pressure	2½ gal	2-A
Water	Pump Tank	1½ gal	1-A
	Pump Tank	2½ gal	2-A
	Pump Tank	4 gal	3-A
	Pump Tank	5 gal	4-A
Water (Antifreeze Calcium Chloride)	Cartridge or Stored Pressure	1¼, 1½ gal	1-A
	Cartridge or Stored Pressure	2½ gal	2-A
	Cylinder	33 gal	20 A
Water (Wetting Agent)	Stored Pressure	1½ gal	2-A
	Carbon Dioxide Cylinder	25 gal (wheeled)	10-A
	Carbon Dioxide Cylinder	45 gal (wheeled)	30-A
	Carbon Dioxide Cylinder	60 gal (wheeled)	40-A
Water (Soda Acid)	Chemically Generated Expellant	1¼, 1½ gal	1-A
	Chemically Generated Expellant	2½ gal	2-A
	Chemically Generated Expellant	17 gal (wheeled)	10-A
	Chemically Generated Expellant	33 gal (wheeled)	20-A
Water (Loaded Stream)	Stored Pressure	2½ gal	2 to 3-A:1-B
	Cartridge or Stored Pressure	33 gal (wheeled)	20-A
AFFF	Stored Pressure	2½ gal	3-A:20-B
	Nitrogen Cylinder	33 gal (wheeled)	20-A:160-B
Carbon Dioxide	Self-Expellant	2 to 5 lb	1 to 5-B:C
	Self-Expellant	10 to 15 lb	2 to 10-B:C
	Self-Expellant	20 lb	10-B:C
	Self-Expellant	50 to 100 lb (wheeled)	10 to 20-B:C
Dry Chemical (Sodium Bicarbonate)	Stored Pressure	1 lb	1 to 2-B:C
	Stored Pressure	1½ to 2½ lb	2 to 10-B:C
	Cartridge or Stored Pressure	2¾ to 5 lb	5 to 20-B:C
	Cartridge or Stored Pressure	6 to 30 lb	10 to 160-B:C
	Nitrogen Cylinder or Stored Pressure	75 to 350 lb (wheeled)	40 to 320-B:C
Dry Chemical (Potassium Bicarbonate)	Stored Pressure	1 to 2 lb	1 to 5-B:C
	Cartridge or Stored Pressure	2¼ to 5 lb	5 to 20-B:C
	Cartridge or Stored Pressure	5½ to 10 lb	10 to 80-B:C
	Cartridge or Stored Pressure	16 to 30 lb	40 to 120-B:C
	Cartridge	48 lb	120-B:C
	Nitrogen Cylinder or Stored Pressure	125 to 315 lb (wheeled)	80 to 640-B:C

Extinguishing Agent	Method of Operation	Capacity	UL or ULC Classification
Dry Chemical (Potassium Chloride)	Stored Pressure	2 to 2½ lb	5 to 10-B:C
	Stored Pressure	5 to 9 lb	20 to 40-B:C
	Stored Pressure	10 to 20 lb	40 to 60-B:C
	Stored Pressure	135 lb	160-B:C
Dry Chemical (Ammonium Phosphate)	Stored Pressure	1 to 5 lb	1 to 2-A and 2 to 10-B:C
	Stored Pressure or Cartridge	2½ to 8½ lb	1 to 4-A and 10 to 40-B:C
	Stored Pressure or Cartridge	9 to 17 lb	2 to 20-A and 10 to 80-B:C
	Stored Pressure or Cartridge	17 to 30 lb	3 to 20-A and 30 to 120-B:C
	Cartridge	45 lb	20-A and 80-B:C
	Nitrogen Cylinder or Stored Pressure	110 to 315 lb (wheeled)	20 to 40-A and 60 to 320-B:C
Dry Chemical (Foam Compatible)	Cartridge or Stored Pressure	4¾ to 9 lb	10 to 20-B:C
	Cartridge or Stored Pressure	9 to 27 lb	20 to 30-B:C
	Cartridge or Stored Pressure	18 to 30 lb	40 to 60-B:C
	Nitrogen Cylinder or Stored Pressure	150 to 350 lb (wheeled)	80 to 240-B:C
Dry Chemical (Potassium Chloride)	Cartridge or Stored Pressure	2½ to 5 lb	10 to 20-B:C
	Cartridge or Stored Pressure	9½ to 20 lb	40 to 60-B:C
	Cartridge or Stored Pressure	19½ to 30 lb	60 to 80-B:C
	Stored Pressure	125 to 200 lb (wheeled)	160-B:C
Dry Chemical (Potassium Bicarbonate Urea Base)	Stored Pressure	5 to 11 lb	40 to 80-B:C
	Stored Pressure	9 to 23 lb	60 to 160-B:C
	Stored Pressure	175 lb	480-B:C
Bromotrifluoro-methane	Stored Pressure	2½ lb	2-B:C
Bromochlorodi-fluoromethane	Stored Pressure	2 to 4 lb	2 to 5-B:C
	Stored Pressure	5½ to 9 lb	1-A and 10-B:C
	Stored Pressure	13 to 22 lb	1 to 4-A and 20 to 80-B:C

This table with Metric equivalents is printed in the Appendix K.

Obsolete Extinguishers

In 1969, American manufacturers stopped making inverting-type extinguishers including soda-acid, foam, and cartridge-operated water and loaded stream extinguishers (Figure 7.10). The soda-acid extinguisher is the most common obsolete type. When this extinguisher is inverted, acid from a bottle mixes with a basic soda-and-water solution and produces a gas that expels the liquid. The pressure on an acid-corroded shell has exploded many soda-acid extinguishers.

Figure 7.10 Extinguishers that require inverting to operate are obsolete and should be replaced.

Inverting foam extinguishers look like soda-acid extinguishers. Inverting mixes solutions (usually called A and B chemicals) from two chambers, forming a foam and gas that expels the foam.

Users of cartridge-operated water extinguishers have to invert and bump the units to puncture a CO_2 cylinder. The pressure of the gas released from the cartridge expels the water.

Some people still have liquid carbon tetrachloride extinguishers, obsolete since the 1960s. When carbon tetrachloride comes in contact with heat, it releases highly toxic phosgene gas.

All obsolete extinguishers should be replaced.

Using Portable Fire Extinguishers

Once the fire extinguisher is at the fire site, it must be used as quickly as possible. Although there are simple instructions located on every extinguisher, knowing the location and operating procedures of fire extinguishers will enable the operator to be more effective during the critical early seconds of a fire. Inspectors should be familiar with all types of extinguishers (most operate in basically the same manner) and should be able to demonstrate the procedure for operating fire extinguishers.

To operate an extinguisher, follow the letters P-P-P-S:

Step 1: PULL the pin at the top of the extinguisher that keeps the handle from being pressed. Break the plastic or thin wire inspection band.

Step 2: POINT the nozzle or outlet toward the fire. Some hose assemblies are clipped to the extinguisher body. Release it and point. Stand about 8 to 12 feet (2 m to 4 m) from the base of the fire.

Step 3: PRESS the handle above the carrying handle to discharge the agent inside. The handle can be released to stop the discharge at any time.

Step 4: SWEEP the nozzle back and forth at the base of the flames to disperse the extinguishing agent. After the fire is out, probe for remaining smoldering hot spots or possible reflash of flammable liquids. Make sure the fire is out.

Selection of Fire Extinguishers

In order to determine the number of fire extinguishers needed to adequately protect a property, fire inspectors must take into consideration a number of factors. The type of extinguisher needed to protect the hazards, the location of the extinguisher, and personnel available to operate the extinguisher are three of the most important considerations. Even though the hazard or hazards present in the occupancy are probably the most essential element in the selection of fire extinguishers, the following elements should be considered during extinguisher selection:

- The chemical and physical characteristics of the combustibles that might be ignited

- The potential severity (size, intensity, and rate of travel) of any resulting fire

- The effectiveness of the extinguisher for the hazard in question

- The personnel available to operate the extinguisher, including their physical abilities, emotional characteristics, and any training they may have in the use of extinguishers

- Environmental conditions that may affect the use of the extinguisher (temperature, winds, presence of toxic gases or fumes)

- Any anticipated adverse chemical reactions between the extinguishing agent and the burning material

- Any health and occupational safety concerns, such as exposure of the extinguisher operator to heat and products of combustion during fire fighting efforts

- The inspection and maintenance required to maintain the extinguishers

The type, size, and number of extinguishers needed varies according to whether the occupancy is classified as light hazard, ordinary hazard, or extra hazard. The occupancy classifications are defined in NFPA 10, *Portable Fire Extinguishers,* as follows:

- *Light Hazard Occupancy:* An occupancy where the amount of combustibles or flammable liquids present is such that a fire of small size may be expected. Some examples are offices, school classrooms, churches, and assembly halls.

- *Ordinary Hazard Occupancy:* An occupancy where the amount of combustible or flammable liquids present is such that fires of moderate size may be expected. Some examples are mercantile storage and display, automobile showrooms, parking garages, light manufacturing, school shop or laboratory areas, and warehouses not classified as extra hazard.

- *Extra Hazard Occupancy:* An occupancy

where the amount of combustibles or flammable liquids present is such that a fire of large size may be expected. Some examples are manufacturing processes such as painting, dipping, coating, flammable liquids handling, auto repair garages, and aircraft and boat maintenance facilities.

For each hazard classification, NFPA 10 specifies the type of rated extinguisher needed, the maximum travel distance (the distance the operator has to travel to get the extinguisher and bring it to the fire before it can be used), and the maximum areas that can be protected by each extinguisher. These specifications can be found in Tables 7.2 and 7.3.

In ordinary or low hazard occupancies, the authority having jurisdiction may approve the use of several lower-rated extinguishers in place of one higher-rated extinguisher. For example, two or more extinguishers may be used to fulfill a 6-A rating if there are enough individuals trained to use the extinguishers. When the weight of the extinguisher causes problems for those who will be operating it, two extinguishers of lesser weight may be used to replace the heavier extinguisher. If the area to be protected is less than 3,000 square feet (288 m^2), at least one extinguisher of the minimum rating should be provided.

Table 7.2 on pg. 120 shows the maximum area that each size Class A extinguisher can protect. These numbers can be used to determine the minimum number of extinguishers required to protect a particular area or building. For example, a building owner proposes to protect a 120,000 square foot (11 148 m^2) light hazard occupancy with 2-A rated extinguishers. From the table we see that each 2-A extinguisher can cover an area of 6,000 square feet (557 m^2). Thus, by division the minimum number of extinguishers that will be required can be determined:

$$\frac{120{,}000 \text{ square feet } (11\ 148^2)}{6{,}000 \text{ square feet } (557 \text{ m}^2)} = 20 \text{ extinguishers required}$$

Twenty extinguishers will be required, assuming that the 75-foot (23 m) maximum travel distance can be maintained.

TABLE 7.2
FIRE EXTINGUISHER SIZE AND PLACEMENT FOR CLASS A HAZARDS

Basic Minimum Extinguisher Rating for Area Specified	Maximum Travel Distances to Extinguishers ft (m)	Areas to be Protected per Extinguisher		
		Light Hazard Occupancy sq ft (m²)	Ordinary Hazard Occupancy sq ft (m²)	Extra Hazard Occupancy sq ft (m²)
1-A	75 (23)	3,000 (279)	—	—
2-A	75 (23)	6,000 (557)	3,000 (279)	2,000 (186)
3-A	75 (23)	9,000 (836)	4,500 (418)	3,000 (279)
4-A	75 (23)	11,250 (1 045)	6,000 (557)	4,000 (372)
6-A	75 (23)	11,250 (1 045)	9,000 (836)	6,000 (557)
10-A	75 (23)	11,250 (1 045)	11,250 (1 045)	9,000 (836)
20-A	75 (23)	11,250 (1 045)	11,250 (1 045)	11,250 (1 045)
40-A	75 (23)	11,250 (1 045)	11,250 (1 045)	11,250 (1 045)

Note: 11,250 sq ft (1 045 m²) is considered a practical limit.

Reprinted with permission from the *Fire Protection Handbook*, 16th Edition, Copyright©, 1986, National Fire Protection Association, Quincy, MA 02269.

For Class B (flammable liquid) hazards, extinguisher placement must be based upon the depth of the fuel. In most cases, however, flammable liquids are used and stored at depths no greater than ¼-inch (6.4 mm). (Consult NFPA 10 for more information regarding flammable liquids stored at greater depths.) Since flammable liquids burn so rapidly, maximum travel distance is of major importance. Because of the higher level of hazard, two or more extinguishers of lower rating should not replace one higher-rated extinguisher for protection of Class B hazards. The exception to this rule is that up to three AFFF (aqueous film forming foam) extinguishers can be used to provide an adequate level of protection. A higher-rated portable fire extinguisher can be used as long as the maximum travel distance is not exceeded. The distribution of Class B extinguishers is based solely on the size of the extinguisher and the maximum travel distance allowed. Table 7.3 lists the requirements for Class B extinguishers.

Extinguishers with Class C ratings are required where energized electrical equipment may be encountered. The fire itself is a Class A or B hazard, and the size and location of portable fire extin-guishers are based on the anticipated Class A or B hazard once the energy supply has been cut off.

TABLE 7.3
FIRE EXTINGUISHER SIZE AND PLACEMENT FOR CLASS B HAZARD EXCLUDING PROTECTION OF DEEP LAYER FLAMMABLE LIQUID TANKS

Type of Hazard	Basic Minimum Extinguisher Rating	Maximum Travel Distance to Extinguishers ft (m)
Low	5-B	30 (9)
	10-B	50 (15)
Moderate	10-B	30 (9)
	20-B	50 (15)
High	40-B	30 (9)
	80-B	50 (15)

Reprinted with permission from the *Fire Protection Handbook*, 16th Edition, Copyright©, 1986, National Fire Protection Association, Quincy, MA 02269.

Placement of Class D fire extinguishers should be no more than 75 feet (23 m) from the Class D fire hazard.

Placement of Fire Extinguishers

Extinguishers should be placed where they are

- Readily visible
- Uniformly distributed
- Easily accessible
- Relatively free from blockage by storage and equipment
- Near normal paths of travel
- Near points of ingress and egress
- Protected from potential accidental or malicious damage

Where visual obstructions cannot be avoided, a means must be provided to conspicuously indicate the extinguisher's location and intended use.

The height of extinguisher placement affects its accessibility. Extinguishers weighing less than 40 pounds (18.1 kg) should be installed with the top not more than 5 feet (1.5 m) from the floor. Those weighing more than 40 pounds (18.1 kg), except wheeled units, should be installed with the top not more than 3½ feet (1.1 m) from the floor. A clearance of at least 4 inches (102 mm) must be maintained between the floor and the bottom of the extinguisher.

Inspecting Fire Extinguishers

Fire extinguishers must be inspected regularly to ensure that they are accessible and operable. This is done by verifying that the extinguisher is in its designated location, it has not been actuated or tampered with, and there is no obvious physical damage or condition present that will prevent its operation. Inspection of portable fire extinguishers (or any other privately owned fire suppression or detection equipment) is the responsibility of the property owner or building occupant.

Though it is usually performed by the building owner or the owner's designate, fire inspectors should include extinguisher inspections in their building inspection and prefire planning pro-

grams. During the inspection, the inspector should remember there are three important factors that determine the value of a fire extinguisher: its serviceability, its accessibility, and the user's ability to operate it.

The following steps should be part of every fire extinguisher inspection:

- Check to ensure that the extinguisher is in a proper location and that it is accessible.
- Inspect the discharge nozzle or horn for obstructions. Check for cracks and dirt or grease accumulations.
- Check to see if the operating instructions on the extinguisher nameplate are legible.
- Check the lockpins and tamper seals to ensure that the extinguisher has not been tampered with (Figure 7.11).
- Determine if the extinguisher is full of agent and/or fully pressurized by checking the pressure gauge, weighing the extinguisher, or inspecting the agent level (Figure 7.12 on pg. 122). If an extinguisher is found to be deficient in weight by 10 percent, it should be removed from service and replaced.
- Check the inspection tag for the date of the previous inspection, maintenance, or recharging (Figure 7.13 on pg. 122).

Figure 7.11 A properly sealed extinguisher should have a breakaway plastic or wire tie that holds the trigger pin in place. The tie also ensures that the pin has not been removed.

Figure 7.12 Most extinguisher gauges have a marked zone where the needle should rest when the extinguisher is properly charged.

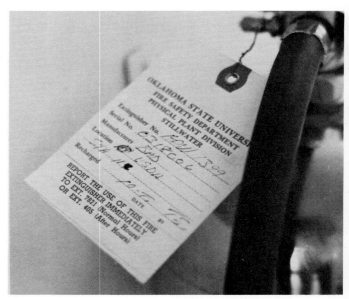

Figure 7.13 An extinguisher that has not been inspected for several years may be unsafe or may fail to operate. Every check or recharge should be recorded.

- Examine the condition of the hose and its associated fittings.

If any of the items listed are deficient, the extinguisher should be removed from service and repaired as required. The extinguisher should be replaced with an extinguisher that has an equal or greater rating.

Extinguisher Maintenance Requirements and Procedures

All maintenance procedures should include a thorough examination of the three basic parts of an extinguisher: 1) mechanical parts, 2) extinguishing agents, and 3) expelling means. Building owners should keep accurate and complete records of all maintenance and inspections including the month, year, type of maintenance, and date of the last recharge. It is the inspector's responsibility to review these records.

Fire extinguishers should be thoroughly inspected at least once a year. Such an inspection is designed to provide maximum assurance that the extinguisher will operate effectively and safely. A thorough examination of the extinguisher will also determine if any repairs are necessary or if the extinguisher should be replaced.

Stored-pressure extinguishers containing a loaded stream agent should be disassembled for complete maintenance. Prior to disassembly, the extinguisher should be discharged to check the operation of the discharge valve and pressure gauge.

Stored-pressure extinguishers that require a 12-year hydrostatic test must be emptied every 6 years for complete maintenance. Extinguishers having nonrefillable disposable containers are exempt.

All carbon dioxide hose assemblies should have a conductivity test. Hoses found to be nonconductive must be replaced. (Hoses must be conductive because they act as bonding devices to prevent the generation of static electricity.)

For additional information regarding fire extinguishers, extinguishing agents, distribution, and applications consult IFSTA **Private Fire Protection and Detection, Essentials of Fire Fighting,** and NFPA 10, *Portable Fire Extinguishers.*

STANDPIPE SYSTEMS

Standpipe systems provide a fire hose outlet at each floor level and sometimes on the roof of multistory buildings. Standpipe systems may also be used in various areas of large, one-story buildings. Their purpose is to provide a quick and convenient means for operating fire streams on different floors or in different areas of a building. To be effective, standpipes must be supplied with adequate water and pressures. Although standpipes are required in buildings of large area and height, they do not take the place of automatic extinguishing systems. NFPA 14, *Installation of Standpipe and Hose Systems,* governs the design and installation of standpipes.

Classes of Standpipe Systems

NFPA 14 classifies standpipe systems as Class I, Class II, or Class III, depending upon their intended use.

Class I standpipes are intended for use by fire departments and individuals trained in handling heavy fire streams (2½-inch [63.5 mm] hose) (Figure 7.14). These systems must be capable of furnishing the effective fire streams required during the more advanced stages of interior fires or for exposure protection from fires nearby.

Figure 7.14 A Class I standpipe is for fire department use in developing heavy fire streams.

Class II standpipes are intended for use primarily by building occupants until the arrival of the fire department. It is assumed that occupants have no specialized training. Class II systems fea-

ture smaller hose, either 1½-inch or 1-inch (38.1 mm or 25.4 mm) (Figure 7.15). If 1-inch (25.4 mm) hose is used, it must be listed and labeled for standpipe service and must be approved by the authority having jurisdiction. This size hose must also be kept on continuous-flow hose reels.

Figure 7.15 A Class II standpipe is for occupant use until the fire department arrives.

Some areas restrict the installation of Class II standpipe systems to industrial occupancies where personnel receive fire fighting training. Some authorities also feel that the use of hoselines in combination with automatic sprinkler systems will tax the sprinkler system's water supply. A further area of concern is that use of hoselines may delay the escape of occupants because they feel that they should fight the fire instead of exiting the building. Occupancies populated extensively by the general public, such as office buildings and auditoriums, can create a risk of occupant injury from the operation of hoselines. Therefore, the assumed benefits of fire control or suppression from the operation of the hose on a Class II standpipe are neither dependable nor guaranteed.

Class III standpipe systems combine the features of Class I and II systems. Because Class III systems feature both the smaller 1½-inch (38.1 mm) hose and the larger 2½-inch (63.5 mm) outlet, they are intended for use either by fire department personnel and those trained in handling heavy streams or by building occupants. The smaller hose enables occupants to attack the fire, and the larger hose outlet provides the fire department with a means of deploying 2½-inch (63.5 mm) hose streams (Figure 7.16).

Figure 7.16 A Class III standpipe can be used by either the occupants or the fire department.

Water Supply for Standpipe Systems

The flow rate and amount of water required for standpipe systems is determined by the size and number of fire streams that will be needed and the probable length of time they will be used. Both of these factors are largely influenced by the condition of the building. Consult NFPA 14 for further information regarding character of water supply, fire pumps, storage tanks, and automatic sprinkler systems.

Class I standpipe systems must provide a minimum of 500 gpm (1 890 L/min) for at least 30 minutes. When more than one standpipe is required, the system must be capable of flowing 500 gpm (1 890 L/min) from the first standpipe and 250 gpm (945 L/min) for each additional standpipe for at least 30 minutes. A minimum residual pressure of 65 psi (448 kPa) must be maintained at the topmost outlet with 500 gpm (1 890 L/min) flowing.

A Class II standpipe system must provide 100 gpm (378 L/min) for at least 30 minutes. The system must be capable of maintaining a residual pressure of 65 psi (448 kPa) at the topmost outlet with 100 gpm (378 L/min) flowing.

A Class III standpipe system must have the same minimum flow, pressure, and duration as a Class I system.

Types of Standpipe Systems

There are four types of standpipe systems:

- Wet standpipe systems that have the water supply valve open and water maintained throughout the system at all times.

- Dry standpipe systems that admit water to the system through the operation of a manually activated, approved remote control device located at each hose station.

- Dry standpipe systems that admit water to the system automatically by the use of a device similar to a dry-pipe valve in an automatic sprinkler system. When a discharge outlet valve is opened, the air pressure in the system is depleted and the dry-pipe valve opened to admit water into the system. This type of system is suitable for use in unheated structures.

- Dry standpipe systems that have no permanent water supply and must be charged by the fire department pumping into a connection outside the building.

A variation of the wet standpipe system is known as a primed system. Water is maintained in the system using a ¾- to 1-inch (19.1 mm to 25.4 mm) water line connection. The system cannot be used until the fire department pumps into the connection on the outside of the building. The advantages of a primed system are that it is usable as soon as the fire department starts pumping into it, it does not have to be filled with water, and air does not evacuate from the system through the hoseline. Corrosion of the internal surface of the piping system is also reduced.

Pressure Reducing Valves

Where pressure at the standpipe outlet exceeds 100 psi (684 kPa), an approved pressure reduction device may be installed at each outlet to reduce the pressure with the required flow to 100 psi

(684 kPa). Pressure reducing valves used on Class II and Class III systems normally cannot be adjusted to provide a pressure higher than 100 psi (684 kPa). Where the pressure is greater than 150 psi (1 034 kPa), an appropriate warning sign should be installed at each outlet.

Standpipe Systems in High-Rise Buildings

Standpipe systems are generally limited to 275 feet (84 m) in height. Buildings in excess of 275 feet (84 m) are zoned so that this requirement can be satisfied. The term "zone" refers to those upper and lower levels of a building that are serviced by standpipes. Although 275 feet (84 m) is a maximum, zones may be less than this limit. For example, a 25-story building may be approximately 325 feet (99 m) in height. Zones may be established that are less than 275 feet (84 m). There-

fore, the low-level zone might be 200 feet (61 m) and the high-level zone 125 feet (38.1 m).

NOTE: Standpipe zone heights up to 400 feet (122 m) are permitted when all system components are rated for the maximum system pressure.

Fire Department Connections

Each Class I or Class III standpipe system must have one or more fire department connections (FDC) through which a fire department pumper can supply water. High-rise buildings having two or more zones require an FDC for each zone. There must be no shutoff between the FDC and the standpipe riser; however, an approved straightway check valve must be installed on the main water supply line and located as near as practical, but below the point where the FDC joins the standpipe system (Figure 7.17). This valve prevents the

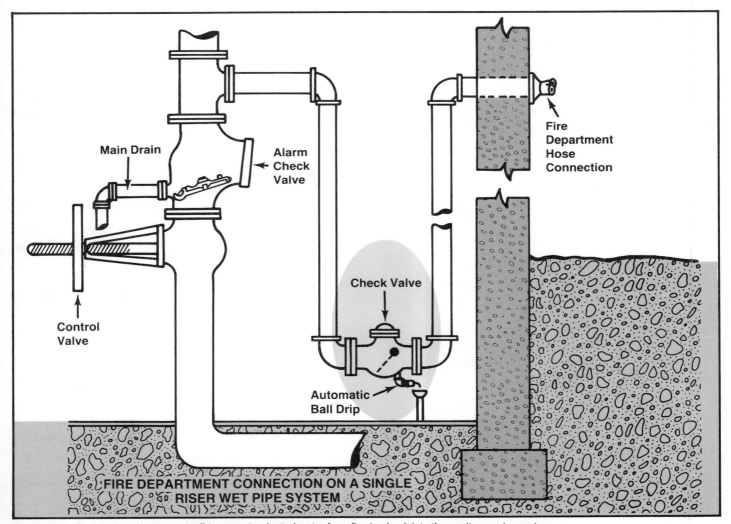

Figure 7.17 The check valve prevents dirty or contaminated water from flowing back into the sanitary water system.

water in the standpipe system from backflowing into the public water supply piping.

The hose connections to the FDC must be of the female or sexless type and equipped with standard caps. It is important that the hose coupling threads or connection conform to those used by the local fire department. The fire department hose connection must be designated by a raised-letter sign on a plate or fitting reading "STANDPIPE." If the FDC does not service all of the building, an appropriate sign must be attached indicating the portion of the building served.

Inspecting and Testing Standpipes

There are two types of tests performed on standpipe systems: initial installation tests and periodic inspections.

INITIAL INSTALLATION

A new standpipe installation should be inspected to ensure that

- All devices (valves, racks, cabinets, and so on) are listed and labeled by a nationally recognized testing laboratory.

- Hose stations are not over 6 feet (1.83 m) from the floor and are within easy reach when standing on the floor.

- Each hose cabinet or closet for Class II or Class III systems is provided with a conspicuous sign that reads "FIRE HOSE" and/or "FIRE HOSE FOR USE BY OCCUPANTS OF BUILDING."

- Fire department connections have the proper fire department hose threads or connection and are designated with a sign reading "STANDPIPE."

- Dry standpipes are posted with a sign reading "DRY STANDPIPE FOR FIRE DEPARTMENT USE ONLY."

- Piping, feed mains, and connections are flushed to remove all debris.

- System components are tested hydrostatically at not less than 200 psi (1 379 kPa) for two hours.

PERIODIC INSPECTION

Wet standpipe systems should be inspected at

regular intervals, usually every six months, to ensure the following:

- Proper water level is maintained in water supply tanks, either the pressure or gravity type. If the tank is pressurized, at least 75 psi (517 kPa) must be maintained. Precautions against freezing must be taken when necessary.

- Fire pumps start and operate. NFPA 20, *Centrifugal Fire Pumps*, specifies that all engine- and electric-driven pumps be started and run no less than 30 minutes each week by the owner or a designated agent. The inspector should ensure that this test is being performed. In addition, transfer switches for electric-driven pumps should be inspected for switching time from the primary to secondary pump power supply source. Fuel supplies for engine-driven pumps should be able to provide 8 hours of service to the driver.

- Fire hose cabinets and closets are used for the storage of fire fighting equipment only.

- Water control valves are open and supervised.

- Individual discharge valves operate properly, gaskets are in good condition, and there is no corrosion or evidence of leaks. (**NOTE:** Wet standpipe systems will have to be shut down and drained prior to inspection to determine if the valves operate properly.)

- Discharge outlet threads are compatible with fire department threads and are not damaged.

- The fire department connection is inspected for access, protective caps, compatibility with fire department threads or connections, thread or connector damage, proper operation of the check valve (DO NOT HAMMER CHECK VALVES), and that no debris has been placed in the connection.

- Drains are free of dirt and/or sediment.

In addition, Class II and Class III systems are inspected to ensure that

- The hose is in good condition, dry and properly racked. (**NOTE:** Hose should be removed and re-racked periodically so that it will not deteriorate at the bends. Hose folds should not occur in the same place.) NFPA 1962, *Care, Use, and Maintenance of Fire Hose,* provides the following requirements for pressure service testing of 1½-inch (63.8 mm) fire hose: For single-jacket 1½-inch (63.8 mm) hose, the pressure shall be held at 150 psi (1 034 kPa) for 5 minutes. For double-jacket 1½-inch (63.8 mm) hose, the pressure shall be held at 250 psi (1 724 kPa) for 5 minutes.

- The swingout rack should be checked for ease of operation.

- Threads on the combination nozzles and hose couplings are not damaged and the gaskets are in good condition.

- The nozzle is not obstructed and the shutoff valve, if there is one, is working properly.

Dry standpipe systems should be hydrostatically tested every five years. Any standpipe out of service for a period of time should be tested with air compressed at 25 psi (172 kPa) to ensure pipe connection tightness. See Appendix G for more information on standpipe testing and inspection.

FIXED FIRE EXTINGUISHING SYSTEMS

Fixed fire extinguishing systems are designed to respond rapidly to fire, thus increasing life safety and reducing property damage. Other advantages of installing fixed fire extinguishing systems are

- Building and fire code concessions granted during construction that allow for longer allowable exit travel distances

- Reduced construction costs by using less fire-resistive materials

- Greater architectural design freedom

- Reduced insurance premiums

- Less water damage during a fire than that caused by firefighters' handlines

Two very important features of automatic sprinkler systems are that they automatically initiate a fire alarm and that they begin to combat the fire while it is in its incipient phase.

This section discusses the following types of fixed fire extinguishing systems: automatic sprinkler systems, foam systems, carbon dioxide systems, halogenated agent systems, and chemical systems.

Carbon dioxide systems can be used to extinguish or control fires both in flammable gases or liquids and in electrical equipment. Typical installations include flammable liquid dip tanks, engine test cells, and electrical switch gear rooms. Since carbon dioxide is a gas, it will leave no residue, thus cleanup will be unnecessary and equipment is not damaged by the use of the agent. In some instances, large carbon dioxide concentrations are needed to extinguish fires, and in these cases, a health hazard does exist. For this reason, a predischarge alarm is incorporated into total flooding systems so occupants can evacuate the area before the agent discharges.

Dry-chemical fixed systems are used primarily on flammable liquid fires and electrical equipment, especially where flammable liquids are contained in electrical equipment, such as an oil-filled transformer. Dry chemical systems are **NOT RECOMMENDED**, however, for delicate electrical equipment because of the residue and corrosion problem created. Wet-chemical systems are used primarily in restaurant hood and duct systems.

Foam fixed systems are used mainly for flammable liquids storage. High expansion foam flooding systems may be used to protect high-piled stock such as rolled paper.

Halogenated-agent fixed systems can be used on fires involving Class A, B, or C materials. Halogenated agents leave no residue and are sometimes referred to as a "clean extinguishing agent." The concentration of agent needed to extinguish fires under normal circumstances is between 3 to 7 percent. Although these concentrations would allow people to remain in the area of discharge for a short period of time with no ill effects, it is not recommended they do so. Therefore, pre-activation alarms are usually used if the area protected is, or may be, occupied. Halogenated agents are highly recommended for areas containing high-value materials that are easily damaged such as computer rooms, fur vaults, and archives.

Automatic Sprinkler Systems

The value to life safety of automatic sprinkler systems has been proven again and again. According to the National Fire Sprinkler Association, there has never been a multiple loss of life in a fully sprinklered building due to fire or smoke. If loss of life has occurred, it is a result of major explosions or strong collapse. Individual fatalities have occurred when victims or their clothing became the source of the fire.

"Rapid Response" sprinkler technology, without a doubt, has increased the level of occupant safety both in residential and in industrial occupancies. Automatic sprinkler systems have been particularly beneficial in high-rise buildings, where providing fire protection continues to be a challenge.

Automatic sprinkler systems consist of a series of nozzle-like devices so arranged that the system will automatically distribute sufficient quantities of water to either extinguish a fire or to hold it in check until firefighters arrive. Water is supplied to the sprinklers through a system of piping. Most sprinklers (except deluge system sprinklers) are kept closed by fusible links or other heat-sensitive devices (Figure 7.18). Heat from a fire causes affected sprinklers to open automatically. Sprinklers can either extend upward from exposed pipe (upright sprinklers), protrude downward through the ceiling from hidden pipes (pendant sprinklers), or extend from the piping along a wall (sidewall sprinklers) (Figure 7.19).

Figure 7.19 Sprinklers may be of the upright, pendant, or sidewall type.

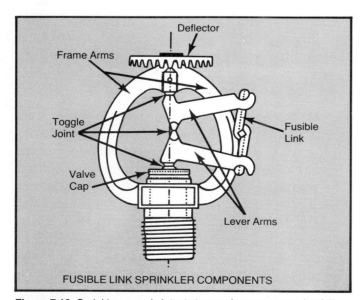

Figure 7.18 Sprinklers remain intact at normal temperatures, but fall away when the temperature exceeds the rated temperature of the sprinkler.

Deflector

Frame Arms

Toggle Joint

Valve Cap

Fusible Link

Lever Arms

FUSIBLE LINK SPRINKLER COMPONENTS

A sprinkler system layout, as shown Figure 7.20, consists of different sizes of pipe. The system starts with a feeder main that originates from a city or private water supply. The feeder main contains a check valve to prevent sprinkler water from backflowing into the potable water supply. The feeder main also has a pipe tee to allow the fire department to augment the system through a fire department connection. Risers are vertical sections of pipe that connect to the feeder main. The riser has the system control valve and associated hardware, or "trim," which is used for testing and system operation. Risers supply the cross main. The cross main directly serves a number of branch lines on which sprinklers are installed. The entire system is supported by hangers and clamps and may be pitched to facilitate drainage.

10-90

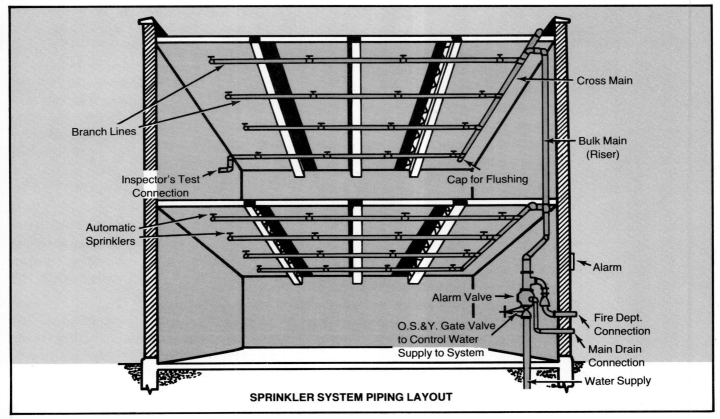

Branch Lines

Cross Main

Inspector's Test Connection

Cap for Flushing

Bulk Main (Riser)

Automatic Sprinklers

Alarm

Alarm Valve →

O.S.&Y. Gate Valve to Control Water Supply to System

Fire Dept. Connection

Main Drain Connection

Water Supply

SPRINKLER SYSTEM PIPING LAYOUT

Figure 7.20 Shown are the components of a typical wet-pipe automatic sprinkler system.

Automatic sprinklers, often called "sprinkler heads," discharge water after the release of a cap or plug that is activated by a heat-responsive element. Sprinklers are kept closed by a number of devices. Some of the most commonly used release mechanisms are fusible links, glass bulbs, and chemical pellets, all of which fuse or open in response to heat (Figure 7.21).

The sprinkler used for a specific application should be based on the maximum temperature expected at the level of the sprinkler under normal conditions. Another important consideration is the anticipated rate of heat release that would be produced by a fire in the particular area. These temperature ratings are given in Table 7.4 on pg. 130.

There are four basic types of sprinkler systems: wet-pipe, dry-pipe, deluge, and pre-action systems. Wet-pipe systems contain water in the system at all times. Dry-pipe systems maintain air under pressure in the sprinkler piping. When sprinklers fuse, air escapes and water is automatically admitted into the system. A pre-action system is a type of dry system that employs a deluge-

Figure 7.21 Release mechanisms for sprinklers include fusible links, glass bulbs, and chemical pellets.

TABLE 7.4
SPRINKLER TEMPERATURE RATINGS

TEMPERATURE RATINGS	TEMPERATURE CLASSIFICATION	FRAME COLOR
135 to 170°F 57°C to 77°C	Ordinary	Unpainted (or partly black or chrome)*
175 to 225°F 79°C to 107°C	Intermediate	White*
250 to 300°F 121°C to 149°C	High	Blue
325 to 375°F 163°C to 191°C	Extra High	Red
400 to 475°F 204°C to 246°C	Very Extra High	Green
500 to 575°F 260°C to 302°C	Ultra High	Orange

*The 135°F (57°C) sprinklers of some manufacturers are half black and half painted.

The 175°F (79°C) sprinklers of these same manufacturers are yellow.

type valve, fire detection devices, and closed sprinklers. This system only discharges water into the piping in response to the detection system. The deluge system is another type of dry-pipe system. It is usually equipped with open sprinklers and a deluge valve. Upon fire detection, the deluge valve opens, permitting water to flow out of all the sprinklers at once. Each type of system and its inspection and testing procedures is discussed more fully in the sections that follow.

CONTROL AND OPERATING VALVES

Every sprinkler system is equipped with a main water control valve and various test and drain valves. Control valves are used to cut off the water supply to the system when it is necessary to perform maintenance, change sprinklers, or interrupt operation. These valves are located between the source of water supply and the sprinkler system (Figure 7.22).

Control valves are indicating valves, that is, they indicate at a glance whether the valves are open or closed. There are several common types of

indicator control valves used in sprinkler systems: outside screw and yoke (OS&Y), post indicator valve (PIV), wall post indicator valve (WPIV), and the post indicator valve assembly (PIVA). The OS&Y valve has a yoke on the outside with a threaded stem; the threaded portion of the stem is out of the yoke when the valve is open and inside the yoke when the valve is closed. The post indicator valve is a hollow metal post that is attached to the valve housing. The valve stem inside the post has a target on which the words "OPEN" or "CLOSED" appear (Figure 7.23). The WPIV is similar to a PIV except that it extends through the wall with the target and valve operating nut on the outside of the building. A PIVA is similar to the PIV except that it uses a butterfly valve, while the PIV uses a gate valve.

In addition to the main water control valves, sprinkler systems employ various operating valves such as globe valves, stop or cock valves, check valves, and automatic drain valves. The alarm test valve is located on a pipe that connects the supply side of the alarm check valve to the retard chamber. This valve is provided to stimulate actuation of the system by allowing water to flow into the retard chamber and operate the waterflow alarm devices.

PRIVATE WATER SUPPLY SYSTEMS AND FIRE PUMPS

Every sprinkler system must have an automatic water supply of adequate volume, pressure, and reliability. A water supply must be able to deliver the required volume of water to the highest sprinkler in a building at a residual pressure of 15 psi (103 kPa). The minimum water flow required is determined by the hazard to be protected, the occupancy classification, and fire loading conditions.

A connection to a public water supply system that has adequate volume, pressure, and reliability is a good source of water for automatic sprinklers. This type of connection is often the only water supply. A gravity tank of the proper size also makes a reliable primary water supply. In order to give the minimum required pressure, the bottom of the tank should be at least 35 feet (11 m) above the highest sprinkler in the building. In some instances, a second independent water supply is not only desirable but also required. Pressure tanks,

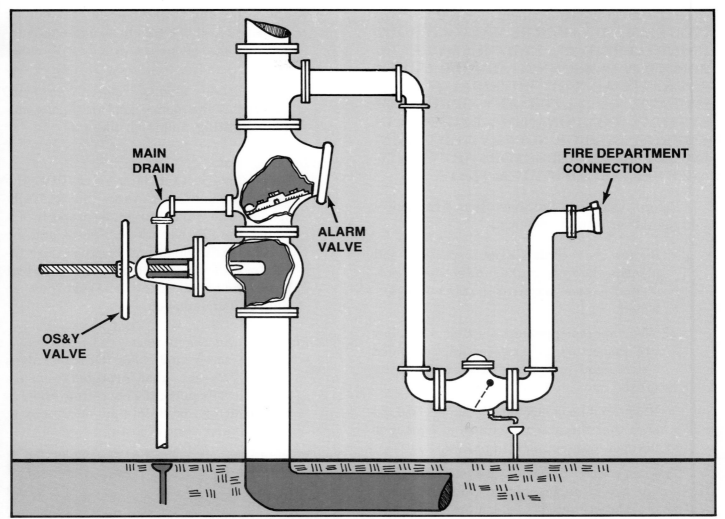

MAIN DRAIN

FIRE DEPARTMENT CONNECTION

ALARM VALVE

OS&Y VALVE

Figure 7.22 The main control valve is between the water supply and the sprinkler system.

Figure 7.23 Post indicator valves clearly show whether the water supply is turned on or off.

another source of water supply, are usually used in connection with a secondary supply. This type of tank is filled two-thirds full with water and has an air pressure of at least 75 psi (517 kPa). Adequate fire pumps that take suction from large reservoirs are used as a secondary source of water supply. When properly powered and supervised, these pumps may be used as a primary source of water supply.

INSPECTING SPRINKLER SYSTEMS

Sprinkler systems require periodic inspection, testing, and maintenance to ensure that they will perform properly during a fire. It is the facility management's responsibility to provide or contract for personnel to perform these functions during normal conditions. Before making an inspection, fire inspectors should obtain permission to do so from the facility management. **THE INSPECTION SHOULD NEVER BE MADE WITHOUT PRIOR APPROVAL. FURTHERMORE, INSPECTION PERSONNEL SHOULD NEVER OPERATE, ADJUST, PHYSICALLY MANIPULATE, OR ALTER ANY SPRINKLER SYSTEM COMPONENTS EXCEPT IN EMERGENCIES OR SUPERVISED TRAINING SESSIONS. INSPECTORS ARE TO WITNESS SPRINKLER SYSTEM TESTS.**

Before beginning the inspection, fire inspectors should perform the following:

- Review the records of prior inspections and identify the type, make, and model of protection system as well as the area to be protected.

- Wear appropriate clothing that will allow fire inspectors to enter attics, concealed spaces, and processing or manufacturing areas.

- If equipment is electrically supervised, be sure that the alarm monitoring agency is notified that testing is in process.

The inspection of a sprinkler system involves three main areas: valves, piping, and sprinklers. Valves must be inspected to determine if all valves controlling water supplies to and within the sprinkler system (sectional or "zone" control valves) are kept open. Inspectors should further determine if a procedure exists for notifying the appropriate person when valves do have to be turned off. Each valve should be examined for the following:

Control Valves
- The valve is fully open and supervised in an approved manner. Supervision can be achieved with lock and chain, electrically, or with the use of wire valve seals. If wire valve seals are used, the owner should inspect those valves once a week to ensure they are open.

- The valve is accessible at all times. If a permanent ladder is provided, inspectors should check the ladder to be sure it is in good condition.

- The valve and its associated components are not subject to mechanical damage. Adequate guards should be called for if necessary.

- If the control valve is enclosed in a room, ensure that the room is kept clean, has adequate lighting, and is sprinklered.

Post Indicator Valve
- If there is a post indicator valve (PIV), the inspector should check to see that it is fully opened, the valve target is properly adjusted, the PIV wrench is in place, and the cover glass is clean and in place. Fire inspectors should also check the PIV bolt heads and barrel casing to be sure they are tight and undamaged.

Outside Screw and Yoke Valves
- Outside screw and yoke (OS&Y) valves (Figure 7.24) should be inspected to ensure they are fully open, the operating wheel is backed off approximately one-quarter turn, and the valve stem is clean.

Figure 7.24 The outside screw and yoke valve is an indicating control valve; stem out is open, stem in is closed.

Other Valves
- The alarm line shutoff cock valve is open.
- The pressure gauge valves are open.

- The static pressure above the clapper is equal to or greater than the static pressure below the clapper. (**NOTE:** Systems without check valves may have only one pressure gauge on the riser.)

- The main drain, auxiliary drain(s), and the inspector's test valve are closed.

- The ball drip valve in the fire department connection moves freely and allows trapped water to drain (Figure 7.25).

- The velocity drip valve will move freely and allow trapped water to drain into the retard chamber (Figure 7.26).

Figure 7.26 Check the velocity drip on the retard chamber to ensure it is working properly.

Inspect all sprinkler piping and hangers to determine that they are in good condition. Check for corrosion and areas that could be subject to physical damage. Sprinkler piping or its hangers are not to be used as a support for ladders, stock, banners, or other material. Report all damaged pipe and missing hangers. Report all loose hangers, as they may put an unnecessary strain on the piping and fittings and interfere with proper drainage (Figure 7.27).

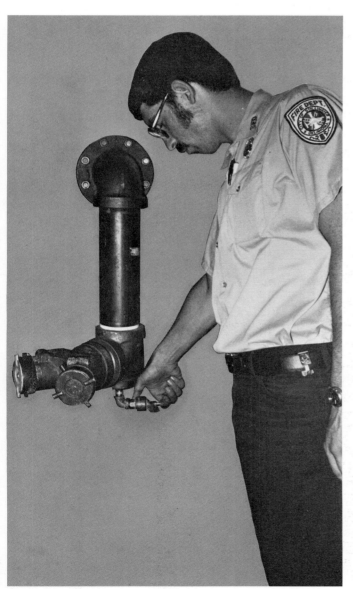

Figure 7.25 An inspection should include a check of the operation of the ball drip valve on a fire department connection.

Figure 7.27 Hangers for sprinkler systems may be attached to the wall or the ceiling. *Courtesy of Edward Prendergast.*

Piping in wet systems must be protected against freezing. Freezing can stop the flow of water to sprinklers or cause the failure of control and alarm devices. Frozen pipes can also rupture, resulting in an interruption of sprinkler protection as well as costly water and restoration damages.

Fire inspectors should inspect all sprinklers to make sure that they are clean, in good condition, and free of corrosion. Sprinklers that need guards to protect them from mechanical damage should be reported. Sprinklers in buildings subject to high temperatures, such as industrial clothing cleaning facilities, can develop cold flow. Cold flow occurs when a sprinkler is repeatedly heated near its operating temperature. Cold flow is evidenced by a creeping or sliding apart of the fused parts (which is known as cold flow) or water leakage around the sprinkler. Cold flow problems can be eliminated by using frangible-bulb sprinklers with the same temperature rating or by using fusible-link sprinklers with a higher temperature rating. Sprinklers exposed to a corrosive atmosphere should have a special, manufacturer-applied protective coating. Any sprinklers that are corroded, painted, or heavily loaded with foreign material should be replaced.

Sprinklers must have adequate room in which to operate if they are to be effective. Partitions or stock must not be arranged so that they can obstruct the distribution of water from sprinklers, and sprinklers must be kept free of hanging displays. A clearance of at least 18 inches (457 mm), measured from the deflector, should be maintained under sprinklers. High-piled storage requires 36 inches (914 mm) of clearance. An adequate supply of spare sprinklers, together with a sprinkler wrench, should be kept in an approved cabinet near the main system control valve. Sprinklers of various designs and temperature ratings in service should be included so used or damaged sprinklers can be promptly replaced and full protection restored. Sprinklers for hallways, shafts, and special rooms may have special deflectors or orifices.

It is important for inspectors to examine sprinklers when changes occur in occupancy. The presence of different fire hazards, as well as changes in heating, lighting, or mechanical equipment, may require different types or styles of sprinklers to be installed.

WET-PIPE SYSTEMS

A wet-pipe sprinkler system, as the name suggests, contains water within the piping at all times. Wet-pipe sprinkler systems are the most reliable of all sprinkler systems. Water under pressure is maintained throughout the system, with the exception of the piping from the fire department connection. The system is connected to a water source such as the municipal water system. When a sprinkler fuses, therefore, it will immediately discharge a continuous flow of water. Systems are designed so the water flow actuates an alarm. This is accomplished by installing an alarm check valve or paddle-flow indicator in the main riser. Water flow lifts a clapper valve and allows water to flow into piping leading to a water motor gong. The water motor gong is equipped with a false alarm training device called a retard chamber. The water chamber of the retard chamber must be filled with water before water will flow into the water gong. The retard chamber is equipped with a ball check valve leading to a drain. Water surges will partially fill the chamber but will subsequently drain, preventing false alarms.

Acceptance Tests

In many jurisdictions, fire inspectors witness sprinkler system acceptance tests. These tests are conducted by a representative of the installation firm. This releases the fire department from liability resulting from damaged equipment due to improper operation or installation. The tests performed and the procedures for each are as follows:

- Flushing of Underground Connections. Underground mains and lead-in connections should be flushed *BEFORE* connection is made to the sprinkler piping. A flow rate of not less than 750 gpm (2 838 L/min) is required for 6-inch (152.4 mm) pipe, 1,000 gpm (3 785 L/min) for 8-inch (203 mm) pipe, 1,500 gpm (5 677 L/min) for 10-inch (254 mm) pipe, and 2,000 gpm (7 570 L/min) for 12-inch (305 mm) pipe. Flushing is continued until the water is clear.

- Hydrostatic Tests. All piping is required to be hydrostatically tested at not less than 200 psi (1 379 kPa) for two hours. There should be no visible leakage while the system is pressurized.

Wet-Pipe Systems Testing

On a wet-pipe system with an alarm check valve, the alarm bypass valve should be opened so the alarm can be tested without unseating the valve (Figure 7.28). The pressure gauge readings should not change significantly, but the water should flow to the retard chamber (if so equipped) and then to the alarm line. The water motor gong or electric alarm should sound. The retard chamber drain should empty the chamber after the alarm bypass valve is closed. If there is no retard chamber, the alarm line should be drained at the conclusion of the test.

Figure 7.28 To test the alarm devices without unseating the alarm check valve, have the alarm bypass valve opened.

Waterflow Alarm Test

Inspectors should witness an alarm test where the alarm is tripped by the use of the inspector's test connection (Figure 7.29). Two persons are required: one person is positioned at the main valve and the other at the inspector's test connection valve. The inspector's test valve is opened, causing water to flow from the exterior orifice at the test connection. The alarm should sound. The inspector's test valve may then be closed. The person at

the main valve should observe the pressure drop, which should be slight. Of course, the alarm company should be notified of the test.

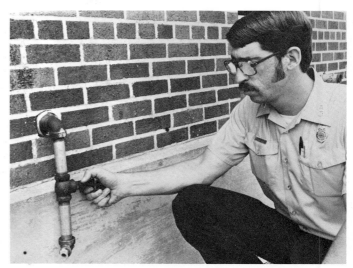

Figure 7.29 To test the devices through the alarm check valve, have someone open the inspector's test valve until the alarm operates.

Main Drain Test

To conduct a main drain test:

Step 1: Record the pressure indicated on the riser gauge. If the system has two gauges on the riser (one below the alarm valve and one above it) read the lower gauge. This gauge monitors the incoming water supply pressure (Figure 7.30).

Figure 7.30 First, the inspector records the incoming water supply pressure indicated on the riser gauge.

Step 2: Have the 2-inch (50 mm) main drain valve opened fully. When the alarm sounds, record the pressure indicated on the gauge (lower gauge if there are two) (Figure 7.31).

Figure 7.31 Have the main drain opened fully and record the pressure once the alarm sounds.

Step 3: Have the 2-inch (50 mm) main drain valve closed and calculate the amount of pressure drop (Figure 7.32).

Figure 7.32 Have the main drain valve closed and record the amount of pressure drop.

If there has been a significant pressure drop, a supply main valve may be partially closed or the supply main may be obstructed.

Frequency of Inspecting and Testing Wet-Pipe Sprinkler Systems

Inspectors should determine if competent personnel are making a visual check of the system at least once a month. Weekly visual inspections are recommended since they increase the chances of discovering deficiencies.

Alarm tests and waterflow alarm tests should be performed at least once a year, but monthly or even weekly tests are recommended. The 2-inch (50 mm) main drain test should be performed at least once a year, although a quarterly test is suggested.

Alarm tests, waterflow alarm tests, and 2-inch (50 mm) main drain tests should only be performed during nonfreezing weather so that ice does not form on sidewalks and roadways. An additional hazard is that piping to the alarm can be damaged by freezing resulting from cold weather testing.

After all testing is complete, the inspector should be sure the alarm monitoring organization is informed that all testing has been completed. At that time, the alarm monitoring organization should confirm that the alarm equipment did function when testing was performed. Written records should be kept of all testing.

DRY-PIPE SPRINKLER SYSTEMS

Dry-pipe sprinkler systems are necessary where piping is subject to freezing. In a dry-pipe system, air under pressure replaces the water in the sprinkler piping above a device called the dry-pipe valve. The dry-pipe valve is located near the base of the riser in a heated area at the point where the supply main enters the premises. Dry-pipe valves are designed so that a small amount of air pressure in the sprinkler system piping will hold back a much greater amount of water pressure. The dry-pipe valve balances the air pressure in the system against the incoming water pressure. This keeps the valve closed and prevents water from entering the sprinkler system piping. When sprinkler(s) are actuated, air escapes which lowers the air pressure. Water pressure on the incoming side of the valve then causes the valve to open, allowing water to enter the system piping.

INSPECTING DRY-PIPE SPRINKLER SYSTEMS

Acceptance Tests

As with wet-pipe sprinkler systems, acceptance tests for dry-pipe systems are conducted by a representative of the installation firm. The tests performed and the procedures for each are as follows:

- Flushing Underground Connections. Underground connections should be flushed following the same procedures described for wet-pipe sprinkler systems.

- Hydrostatic Testing. Hydrostatic testing of dry-pipe systems is performed in the same manner as described for wet-pipe sprinkler systems. In freezing weather, however, dry-pipe systems are tested for two hours with not less than 50 psi (348 kPa) air pressure.

- Air Test. In this test, the entire system is placed under an air pressure of 40 psi (276 kPa). After 24 hours the air pressure is measured again. If there is a loss of more than 1½ psi (10 kPa), the leaks should be located and corrected.

Operations Tests

During an inspection of an operating sprinkler system, inspectors should ensure that

- All indicating control valves are open and properly supervised in the open position.

- Air pressure readings correspond to previously recorded readings.

- The ball drip valve moves freely and allows trapped water to seep out of the fire department connection.

- The velocity drip valve located beneath the intermediate chamber is free to move and allow trapped water to seep out. Inspectors can check this valve by instructing someone to lift a push rod that extends through the drip valve opening. Where an automatic drip valve is installed, the velocity drip valve can be checked by moving the push rod located in the valve opening.

- The fire department connection threads are in good condition and the caps are in place.

- Any drum drips are drained to eliminate the moisture trapped in the low areas of the system (Figure 7.33).

- During freezing weather, the dry-pipe valve enclosure heating device keeps the temperature of the dry-pipe valve at or above 40°F (4.4°C).

- The priming water is at the correct level. If necessary, personnel can drain water by

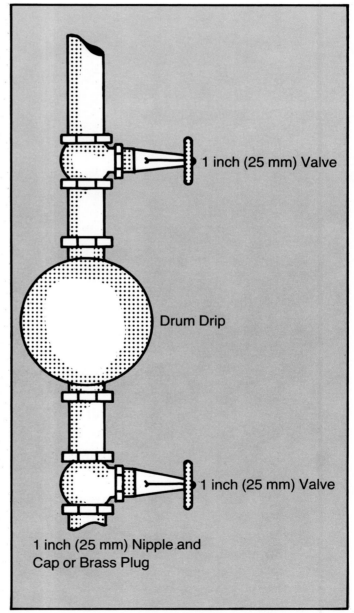

1 inch (25 mm) Valve

Drum Drip

1 inch (25 mm) Valve

1 inch (25 mm) Nipple and Cap or Brass Plug

Figure 7.33 A drum drip valve for removing moisture from the system piping.

opening the priming water test-level valve until air begins to escape (Figure 7.34).

(**NOTE:** If the system is equipped with a quick-opening device, opening the priming water test line could trip this system.)

• The system's air pressure is maintained at 15 to 20 psi (103 kPa to 138 kPa) above the trip point and no air leaks are indicated by a rapid or steady air loss. If inspectors note excessive air pressure, they should have the system drained down.

Figure 7.34 Have priming water drained, if necessary.

Main Drain and Alarm Test

To conduct a main drain and alarm test:

Step 1: Record the static pressure showing on the lower gauge (Figure 7.35).

Step 2: Have the 2-inch (50 mm) drain opened *slowly* (Figure 7.36).

Step 3: Record the residual pressure (Figure 7.37).

Step 4: Have the 2-inch (50 mm) main drain valve closed slowly. Calculate the difference in readings (Figure 7.38). If the pressure has dropped significantly, try to determine if a supply line valve is partly closed or supply line piping is obstructed.

Step 5: Test the alarm by having the alarm test bypass valve opened (Figure 7.39).

Figure 7.35 Record the static pressure, which will be shown on the lower gauge.

Figure 7.36 Have the 2-inch (50 mm) main drain opened slowly.

Figure 7.37 Record the residual pressure, again from the lower gauge.

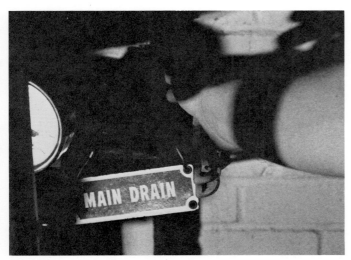

Figure 7.38 Have the main drain closed slowly and calculate the difference in the pressure readings.

Figure 7.39 Have the alarm bypass valve opened to determine whether or not the alarm is operating properly.

System Trip Test

To conduct a dry-pipe valve trip test:

Step 1: Have the water supply control valve checked to see that it operates freely. This can be done by having the 2-inch (50 mm) main valve opened slightly while the water supply control valve is closed and re-opened (Figure 7.40).

Step 2: Have the 2-inch (50 mm) main drain valve closed (Figure 7.41).

NOTE: The 2-inch (50 mm) main drain valve is used to prevent accidental tripping. Closing the main valve without opening the 2-inch (50 mm) main drain valve can cause water trapped between

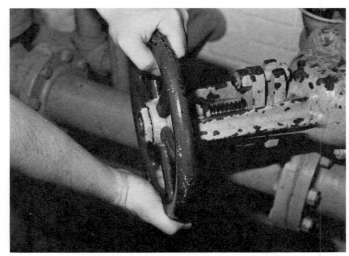

Figure 7.40 Have the water supply control valve closed, then re-opened, to ensure it is operating properly.

Figure 7.41 Have the 2-inch (50 mm) main drain closed.

the water supply control valve and the dry-pipe clapper to push the clapper off its seat.

Step 3: Assign someone to open the inspector's test valve. This person should record the time the test valve is opened and the time when water discharges (Figure 7.42).

NOTE: This should not take longer than 60 seconds.

Step 4: Watch the air pressure gauge. When the dry valve trips, note the air pressure (Figure 7.43).

Step 5: Check the water and the air pressure gauges to ensure the pressure has equalized after tripping (Figure 7.44).

Step 6: Have the water supply control valve closed, then have the main drain valve opened (Figure 7.45). *Be sure that the system is completely drained before proceeding.*

Step 7: Have the inspector's test valve closed after the system is drained (Figure 7.46).

Step 8: Have the dry-pipe valve cover removed and check to make sure that the clapper is latched in the open position (Figure 7.47). The system is now ready to be restored to service.

Figure 7.43 Note the air pressure at which the dry pipe valve trips.

Figure 7.44 Check the air pressure and water pressure gauges after the valve has tripped to make sure that they are indicating equal pressures.

Figure 7.42 Have the inspector's test valve opened and record the amount of time it takes for water to flow from the discharge.

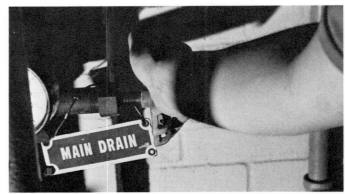

Figure 7.45 After the water supply valve has been closed, have the main drain opened.

Figure 7.46 After the system has drained, have someone close the inspector's test valve.

Figure 7.48 Inspect the clapper seat area and have it cleaned to ensure that the clapper will reseat properly.

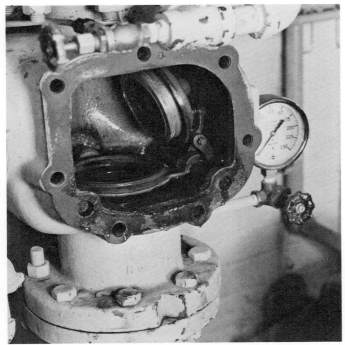

Figure 7.47 Have the dry-pipe valve cover removed and check to see that the clapper is latched in the open position.

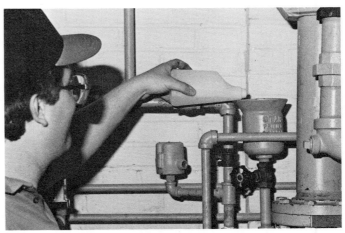

Figure 7.49 See that the priming chamber is refilled to the proper level.

Step 9: Have the clapper seat area cleaned with a cloth and the valve cover replaced (Figure 7.48).

Step 10: See that the priming chamber is refilled to the proper level (Figure 7.49).

Step 11: Observe that the system is pressurized with air to the recommended air pressure (Figure 7.50).

Figure 7.50 Note that system is pressurized with air to the recommended pressure.

Step 12: Observe that the water supply valve is opened (Figure 7.51).

Step 13: See that the main drain is closed (Figure 7.52).

Figure 7.51 Have the water supply valve opened slowly.

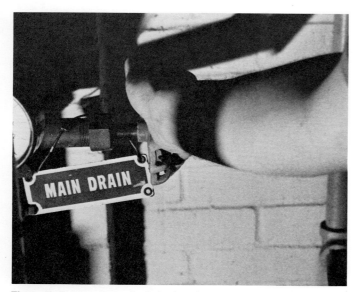

Figure 7.52 Have the main drain closed.

Frequency of Inspecting and Testing Dry-Pipe Systems

Dry-pipe sprinkler systems should be visually inspected by the owner or owner's representative at least weekly. A weekly visual inspection is recommended because it keeps the in-plant inspector familiar with the system and provides an opportunity to discover system deficiencies soon after they develop.

The 2-inch (50 mm) main drain test should be performed at least annually; however, a quarterly test is recommended.

An alarm test should be performed at least monthly, although a weekly test is recommended.

The trip test should be performed at least annually. (A visual inspection of the system and 2-inch (50 mm) main drain test can be performed along with the trip test).

Of course, written records of all tests and inspections should be kept on file. More information about automatic sprinkler systems may be found in NFPA 13, *Installation of Sprinkler Systems* and in NFPA 13A, *Inspections, Testing, and Maintenance of Sprinkler Systems.*

DELUGE AND PRE-ACTION SPRINKLER SYSTEMS

A pre-action sprinkler system contains piping with *closed* sprinklers. Air is maintained under pressure within the system piping. Water is held back by a deluge valve, which is tripped when fire detection devices activate. This type of system is used when it is especially important that water damage be prevented: if a sprinkler is broken or piping is broken, no water will flow. If a detector malfunctions, water will enter the system but none will be discharged unless a sprinkler actuates.

A deluge sprinkler system contains piping with *open* sprinklers. No water enters the piping until a device called a deluge valve opens. The deluge valve is tripped automatically by activation of fire detection devices that are installed in the same areas as the sprinklers. When the deluge valve operates, water flows from ALL the sprinklers.

A combined dry-pipe and pre-action sprinkler system may be used where it is important that failure of the fire detection system not prevent operation of the sprinkler system or vice versa. A combined system features closed sprinklers, air under pressure within the piping, and a dry-pipe valve. Operation of the fire detection system activates tripping devices that open the dry-pipe valve WITHOUT causing loss of air pressure in the system. Air exhaust valves at the end of the feed main act to speed the filling of the system and usually precede opening of sprinklers.

Inspecting and Testing Deluge or Pre-action Sprinkler Systems

Testing and inspecting of deluge and pre-action sprinkler systems generally follow the same guidelines already detailed for dry-pipe systems. In addition, the fire detection devices must be inspected in these systems to ensure they are functioning.

Trip tests for deluge systems require a variation of procedures previously outlined for trip tests:

- For a small deluge system, open sprinklers are replaced with standard sprinklers, plugs, and caps.

- For large deluge systems, where it is not practical to replace or plug sprinklers, the main water supply valve (OS&Y) is closed to within two turns of being completely shut (Figure 7.53). The deluge valve is then tripped and the main water supply valve (OS&Y) is closed immediately (Figure 7.54). (**NOTE:** The system may be activated by using a heating device on a detector or by manual means.)

Figure 7.54 The deluge valve is tripped and the main water supply valve closed immediately.

Fire inspectors should ensure that plant personnel are inspecting deluge and pre-action systems at least monthly, although a weekly inspection is recommended. They should also determine if a main drain test is being performed at least annually, although a quarterly test is recommended. If possible, inspectors should witness the annual trip test, at the same time making a visual inspection of the system. As with all other types of sprinkler systems, written records should be kept of all tests and inspections.

RESIDENTIAL SPRINKLER SYSTEMS

Like other automatic sprinkler systems, residential sprinklers are designed to extinguish an incipient fire before it progresses to the flashover stage. In this way, occupants will have enough time to escape a potentially life-threatening fire. This area of sprinkler protection has gained the most from advances in rapid-response sprinkler technology. NFPA 13D, *Sprinkler Systems in One- and Two-Family Dwellings and Mobile Homes* is becoming widely accepted and adopted by many communities.

The three primary components of a residential sprinkler system are the sprinklers, the piping,

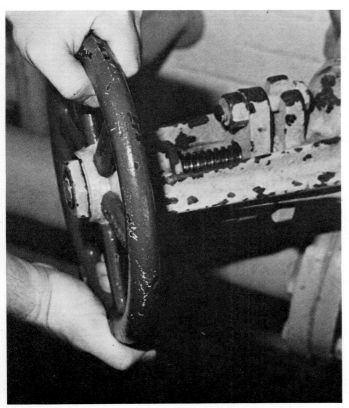

Figure 7.53 See that the water supply valve is closed to within two turns of being completely shut.

and the valves. Each component in a residential system is different in design, construction, and material from its industrial counterpart.

Residential sprinklers must be able to respond quickly to incipient fires, operate at low water pressure, and provide an adequate density to suppress fires. Sprinklers have been developed that are 5 to 15 times more sensitive than commercial sprinklers. Pendant sprinklers and sidewall sprinklers respond quickly, have low pressure demands, and can control the distribution of water.

Most residential sprinkler systems are wet-pipe systems. The recommended piping for residential systems is constructed of polybutylene plastic. Polybutylene plastic pipe has advantages over steel or copper because it is flexible, lightweight, resists corrosion, has high flow rates, and costs less to install.

The valve package provides the system with components that monitor water pressure, give an audible signal when there is water flow, and prevent the flow of sprinkler water back into the domestic water supply. To perform these functions, the following devices are required:

- Waterflow detector
- Anti-water hammer device
- Pressure gauge
- Check valve
- Main control valve

Other components and devices that may be needed are a water pump to boost water pressure, additional signaling devices to a fire department, a drain valve, and a test connection. Extra sprinklers and a sprinkler wrench must also be stored near the riser.

Inspecting and maintaining a residential sprinkler system is the responsibility of the homeowner. The monthly maintenance program should include the following:

- Visually inspecting all sprinklers for obstructions, damage, or paint
- Visually inspecting all control valves to ensure they are open

- Testing all waterflow devices
- Testing the alarm (notify the fire department that it is being tested)
- Ensuring pumps are operational according to NFPA 20, *Installation of Centrifugal Fire Pumps*
- Checking air pressure in dry-pipe systems
- Checking water levels in tanks

For further information regarding sprinkler system maintenance, consult NFPA 13A, *Inspection, Testing, and Maintenance of Sprinkler Systems*. Further information about the history of residential sprinkler systems can be obtained in *ALPHA TO OMEGA, The Evolution in Residential Fire Protection* by Ronald J. Coleman.

Foam Extinguishing Systems

Foam extinguishing systems are most frequently used to protect flammable liquid hazards. High-expansion foam flooding systems may be used to protect high-piled stock such as rolled paper. Fire extinguishing foams are lightweight, have a high water content, have blanketing tendencies, and are resistant to rapid breakdown. These properties cause the foam to float on top of the flammable liquid, excluding air needed for combustion and thus extinguishing the fire. The high water content of the foam also cools heated liquids and surrounding objects. Foams are not suitable for water-reactive or three-dimensional fires, such as occurs when there is a fire at a leaking flange.

Most foams used today are mechanical foams. They are made by mixing air with a solution of water and foam concentrate. These finished foams can be classified as low- medium-, or high-expansion foams, depending upon the size of the air/water ratio. The foam concentrates used to make the foam solution are divided into three categories: protein, fluoroprotein, and synthetic.

- Protein Foam Concentrate — primarily a protein hydrolysate with stabilizing additives and inhibitors diluted in water.
- Fluoroprotein Foam Concentrate — similar to protein foam but contains a synthetic

fluoronated surfactant additive. This type of foam forms a foam blanket and a vaporization film to help cool the fuel surface.

- Synthetic Foam Concentrate — synthetic foams contain no organic bases such as hydrolysed proteins. They include:

  — Aqueous Film Forming Foam — fluoronated surfactants plus foam stabilizers in solution with water. AFFF acts as a barrier to exclude air and create a film on the fuel surface that suppresses the emission of fuel vapors.

  — High-Expansion Foam Concentrates — usually a hydrocarbon surfactant mixed with water. A soapsuds-like foam is produced by air aspiration.

  — Other Synthetic Concentrates — hydrocarbon active agents (typically named "wetting agents") that are usually applied with a portable nozzle.

There are also special alcohol-type foam concentrates that have an insoluble barrier in the bubble structure that resists breakdown when used to attack polar solvent fires. Manufacturers typically designate these foams as "alcohol-type concentrate" or "ATC."

Usually, fire protection engineers determine the type of foam to be used for fixed systems. Some typical uses of various foam types include:

- Protein or fluoroprotein foams used for hydrocarbon hazards, particularly subsurface injection systems

- High-expansion foams used for protection of Class A combustibles or flammable liquids with total flooding systems

- Alcohol-type foams used for the protection of polar solvents

There are two types of discharge devices used on foam extinguishing systems protecting indoor hazards: those producing foam in a spray or dispersed pattern, and those producing foam in compact, low velocity streams. NFPA 11, *Foam Extinguishing Systems*, should be consulted for more detailed information.

INSPECTING AND TESTING FIXED FOAM SYSTEMS

Inspectors should compare the extinguishing system installation to the engineering plans to determine that the system has been properly installed. In addition, systems should be checked for continuity of piping; removal of temporary pipe blinds; accessibility of valves, controls, and gauges; and the proper installation of vapor seals, where applicable. Inspectors should also check to be sure that proper operating instructions are posted with the system.

Piping should be subjected to a 2-hour hydrostatic test at 200 psi (1 379 kPa) or 50 psi (345 kPa) in excess of the maximum pressure anticipated, whichever is greater. Inspectors should also determine if systems are being thoroughly inspected by competent personnel at least once a year. Periodic inspections must also be made of air-foam concentrates and their tanks or storage containers for evidence of excessive sludging or deterioration.

Carbon Dioxide Extinguishing Systems

Carbon dioxide extinguishing systems can be effectively used on practically all combustibles or flammable materials, except for active metals; metal hydrides; materials such as cellulose nitrate, which polymerizes its own oxygen; and deep-seated Class A fires. Carbon dioxide is a nonconductor of electricity, making it suitable for use on energized electrical equipment. It is noncorrosive and leaves no residue, making it suitable for use on electrical equipment and other items that would be damaged if water, dry chemical, or dry powder-type extinguishing agents were used.

There are four types of fixed carbon dioxide systems:

- *Total Flooding.* A total flooding system has nozzles arranged to discharge carbon dioxide into an enclosed space or an enclosure around the fire hazard. Piping connects the nozzles to a fixed supply of carbon dioxide.

- *Local Application.* A local application system has nozzles arranged to discharge carbon dioxide directly on burning material. Piping connects the nozzles to a fixed supply of carbon dioxide.

- *Hand Hoselines.* A hand hoseline system has hoselines located next to hazards. This system is designed for manual local application. Piping connects the hoselines to a fixed supply of carbon dioxide.

- *Standpipe with Mobile Supply.* A standpipe with a mobile supply system has fixed nozzles or hoselines, or both, connected to fixed piping. These systems may be used for either total flooding or local applications. A mobile supply of carbon dioxide is maintained where it can be quickly moved into position and manually connected to the fixed piping.

There are two kinds of carbon dioxide storage (supply) systems: low pressure and high pressure. Both store carbon dioxide as a liquefied compressed gas.

Low-pressure systems utilize refrigerated pressure vessels ranging in capacity from 500 pounds to 125 tons (227 kg to 113.5 metric tons) (Figure 7.55). Vessels are designed to maintain a normal pressure of 300 psi (2 069 kPa) corresponding to a temperature of approximately 0°F (-18°C).

High-pressure storage systems have rechargeable cylinders. Carbon dioxide is stored at atmospheric temperatures corresponding to a nominal pressure of 850 psi (5 861 kPa) at 70°F (21°C) (Figure 7.56).

NFPA 12, *Carbon Dioxide Extinguishing Systems,* classifies carbon dioxide systems as automatic or manual according to the following methods of activation:

- Automatic Operation — No human action required.

- Normal Manual Operation — A control must be operated to activate the system. The single control device must be easily accessible from the fire hazard at all times.

- Emergency Manual Operation — A fully mechanical device is used to cause operation if normal manual operation fails. It is located along the emergency egress route.

Carbon dioxide systems must be equipped with an indicator to show that the system has oper-

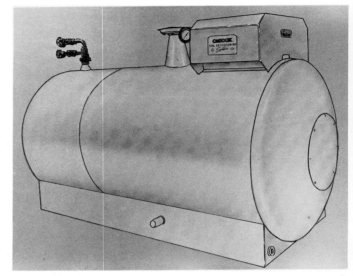

Figure 7.55 Low-pressure storage tanks for carbon dioxide systems are of steel or fiber glass construction. *Courtesy of Cardox Division of Chemtron Corp.*

ated and needs recharging. An audible alarm should also be provided to indicate that the system has operated. Predischarge alarms are required where hazards to personnel may exist. Further, NFPA 12 requires that signs warning of actions to be taken upon activation of a carbon dioxide fixed system must be located in and near the area of protection.

INSPECTING AND TESTING CARBON DIOXIDE EXTINGUISHING SYSTEMS

NFPA 12 requires that the manufacturer provide the owner with a testing and maintenance procedure. Inspectors should review this material before conducting an inspection or witnessing any tests. Systems should be thoroughly inspected and

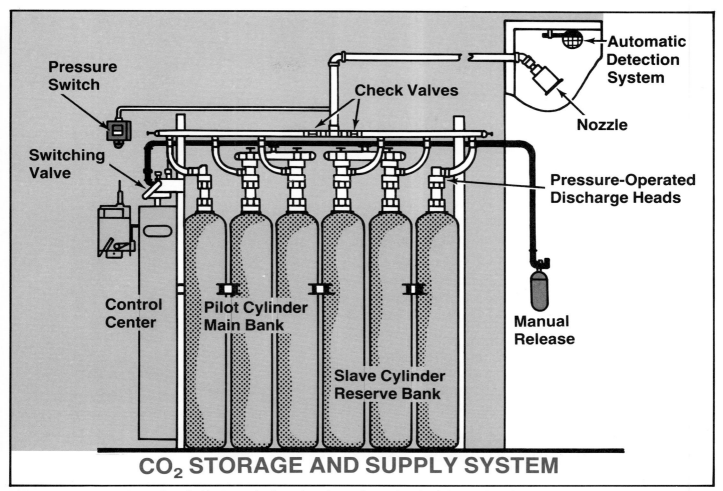

Figure 7.56 A high-pressure carbon dioxide storage bank consists of a set of controls and switches, main and reserve supply cylinders, and a manual operation station. *Courtesy of Cardox Division of Chemtron Corp.*

tested for operation at least annually. This will normally require specially trained personnel. The fire inspector's role is to ensure the inspection and testing is being performed. Inspectors should be sure that all persons who may be required to inspect, test, maintain, or operate carbon dioxide fire extinguishing systems are thoroughly trained in the functions they are expected to perform.

All high-pressure cylinders should be weighed at least semiannually and the date of the last hydrostatic test noted (to be current, the cylinder must have been tested within the past five years). If a container shows a loss in net content of more than 10 percent, it should be refilled or replaced.

Low-pressure containers have liquid level gauges which should be inspected on a weekly basis. If the gauge shows a loss of more than 10 percent, the container should be refilled.

Halogenated Agent Extinguishing Systems

Halon is a generic term for halogenated hydrocarbons. These are chemical compounds that contain carbon plus one or more elements from the halogen series (fluorine, chlorine, bromine, and iodine). A large number of halogenated compounds exist, but only a few are used as fire extinguishing agents. The two most common are Halon 1211 (bromochlorodifluoromethane) and Halon 1301 (bromotrifluoromethane).

Halogenated agents are effective in extinguishing surface fires in flammable liquids, most solid combustible fires, and electrical fires. They work to extinguish fires chemically and also to stop the combustion process itself by breaking the chemical chain reaction. Halogenated agents are not effective on fires involving fuels, such as potassium or sodium, which contain their own oxidizing agent, or on deep-seated fires in Class A materials.

Both Halon 1211 and 1301 are expelled as a vapor. The agent itself leaves no residue. It is nonconductive and noncorrosive, thus it is classified as a clean agent.

Only a low concentration (the percent in air) of halogenated agent is required to extinguish most fires. Concentrations of up to 10 percent may be used in areas normally occupied by people. Halon 1211 is suitable for areas that are seldom occupied such as switchgear rooms, pump rooms, and vaults (Figure 7.57). Halon 1301 is a particularly effective extinguishing agent for areas containing electronic equipment such as computer enclosures.

Halogenated agents are used in both total flooding and local application systems. Both types of systems require automatic detection equipment for actuation (Figure 7.58).

Total flooding systems require an enclosed area that is sealed sufficiently to maintain the required concentration of agent for a specified length of time (Figure 7.59). In addition, NFPA 12A and 12B require predischarge alarms for total-flooding Halon 1301 and 1211 systems.

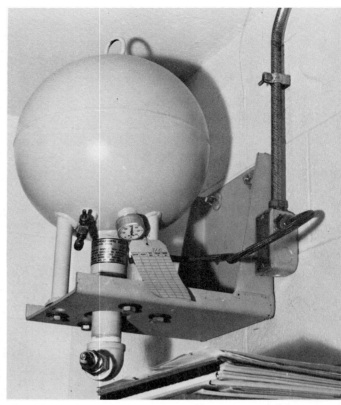

Figure 7.58 A typical halogenated system has automatic detection equipment and a liquefied Halon storage container.

Figure 7.57 A closed vault for record protection is an ideal application for a halogenated system.

INSPECTING HALON SYSTEMS

Inspectors should be sure that systems are being inspected and tested at least annually for proper operation by competent personnel and that a visual inspection is being made on a monthly basis. The inspector should review the approved schedule and procedure. The inspector should also determine that the agent quantity and pressure of refillable containers are checked at least semiannually. **NOTE:** If a container shows a loss in net weight of more than 5 percent or a loss in pressure (adjusted for temperature) of more than 10 percent, it should be refilled or replaced.

Chemical Extinguishing Systems

Chemical fire extinguishing systems utilize both dry and wet extinguishing agents. Dry chemical systems may have one of the following chemical agents as their base: monoammonium phosphate, sodium bicarbonate, potassium chloride, or potassium carbonate (a product of the fusion of potassium bicarbonate and urea). Dry chemicals are effective against flammable liquids, gases, greases,

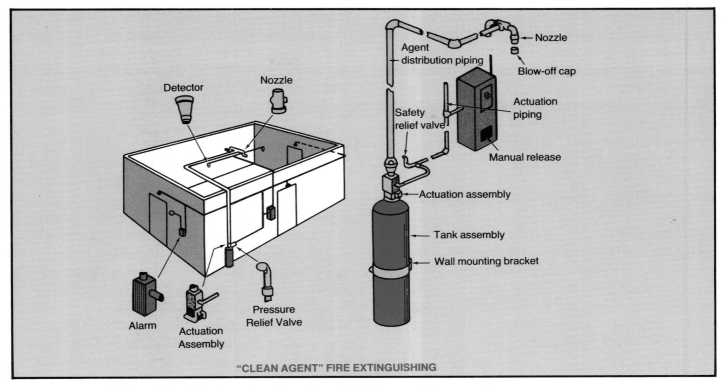

"CLEAN AGENT" FIRE EXTINGUISHING

Figure 7.59 A typical Halon 1301 total flooding system plus a schematic showing essential components.

and electrical fires (Figure 7.60). All of the dry chemical agents leave a residue, which must be vacuumed or brushed away after the fire is completely extinguished. (**NOTE:** Monoammonium phosphate residue may also adhere to hot surfaces and can etch heated metals.)

Because of problems caused by residue, dry chemicals are usually *not* recommended for use on electronic and computer equipment. Dry chemicals can be stored under a wide range of temperature conditions. NFPA 17, *Dry Chemical Extinguishing Systems,* provides detailed information concerning these systems and should be consulted prior to any inspections.

Wet chemical extinguishing systems contain an aqueous solution of potassium carbonate, which is discharged as a spray. Potassium carbonate is an excellent extinguishing agent for fires involving flammable liquids, gases, greases, and ordinary combustibles such as paper and wood. It is not suitable for electrical fires because the spray may act as a conductor. The design, installation, and maintenance of wet chemical extinguishing systems is governed by NFPA 17A, *Wet Chemical Extinguishing Systems.*

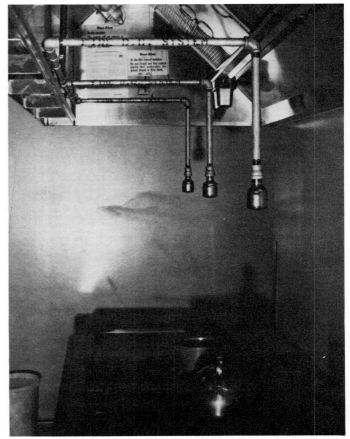

Figure 7.60 Dry chemical systems are popular for the protection of restaurants.

Wet chemical extinguishing systems are especially effective on fires involving cooking hazards. The nature of the chemical is such that it reacts with animal or vegetable oils to form a soap-like solution. As a result, grease or oil fires are attacked four ways: fuel removal, cooling, smothering, and flame inhibition. Because the agent is a liquid, NFPA 17 recommends that it be used only in environments ranging from 40°F to 120°F (4°C to 49°C).

INSPECTING AND TESTING CHEMICAL EXTINGUISHING SYSTEMS

Inspectors should check to ensure competent personnel are inspecting and testing chemical extinguishing systems following an approved schedule and procedure. These inspections should be made at least annually. A discharge test should be required if there is any question as to whether the system is fully functional.

The fire inspector should also determine if qualified personnel are inspecting agent quantity and pressure of refillable containers at least semiannually. (If a container shows a loss in net weight of more than 5 percent or a loss in pressure [adjusted for temperature] of more than 10 percent, it should be refilled or replaced.) Dry chemical agents must be checked for caking.

FIRE DETECTION AND ALARM SYSTEMS

Fire detection and alarm systems used for private fire protection are highly technical and include many types of equipment that are usually installed and maintained by specially trained persons.

During inspections, fire inspectors should note the functional aspects of fire detection and alarm systems. Observant inspectors should be able to recognize physical and environmental conditions that may render the system inoperative or unresponsive to a hostile fire. Inspectors should also recognize conditions that may trigger an unwanted alarm, and by recommending corrective action, reduce fire department responses to false alarms.

To ensure operational reliability, all the components in a fire detection and alarm system should be listed by a nationally recognized testing laboratory such as Underwriter's Laboratories, Inc. or Factory Mutual Engineering. The installation of the system should also conform to applicable provisions of NFPA 70, *National Electrical Code*, and the respective NFPA standard for the particular type of system.

Local Alarms

A local alarm system is a combination of alarm components designed to detect a fire and/or transmit an alarm on the immediate premises only. The primary purpose of a local alarm system is to alert the occupants in order to ensure their life safety. The secondary purpose of a local alarm system is to conserve property. A local alarm system does not retransmit an alarm to any agency or group away from the premises. Installation and maintenance of local alarm systems is addressed in NFPA 72A, *Installation, Maintenance and Use of Local Protective Signaling Systems for Guard's Tour, Fire Alarm and Supervisory Service*.

The basic components of a local alarm system (and, in fact, of all alarm systems) are as follows (Figure 7.61):

- Control panel
- Alarm-initiating devices
- Signaling devices
- Power sources
- Auxiliary devices

Additional equipment that may be required for use in high-rise occupancies:

- Voice Alarm Systems
- Public Address System
- Two-way Fire Department Communications System
- Door Unlocking Devices

The control panel is the central nervous system of the local alarm system and serves many purposes (Figure 7.62). All the components of the alarm system are connected to the control panel and are controlled from the panel. The hardware contained in the control panel determines the capabilities of the system. These capabilities range from simply receiving a detection signal and

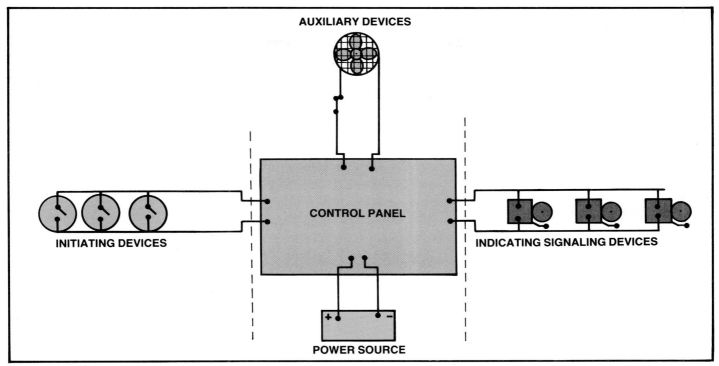

Figure 7.61 The basic components of all alarm systems are the same. The complexity of the system depends upon the needs of the area being protected.

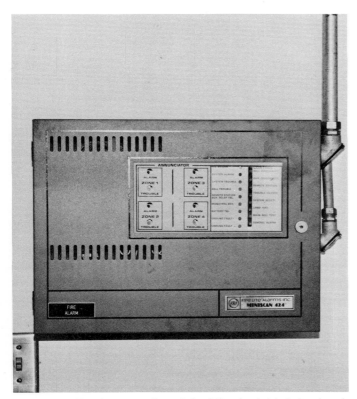

Figure 7.62 The alarm control panel should be clearly labeled and readily accessible to firefighters.

sounding an alarm to complicated zoned presignal systems. The control panel hardware also determines at what level the system can supervise itself.

The alarm-initiating devices are the nerve endings of the system. It is through these devices that the alarm signal is originated. The basic types of initiating devices are

- Manual
- Thermal sensitive
- Products of combustion
- Flame detectors
- Extinguisher system related

Except for the pull station alarm, which operates manually, initiating devices monitor the physical conditions in the areas around them. When fire conditions develop, the initiating device transmits a signal to the control panel, which then activates the signaling devices. Local alarm systems may be

- Noncoded
- Zone noncoded
- Zone coded
- Master coded
- Presignal

A noncoded system is the least complicated type of system. Every alarm initiating device will sound an identical alarm signal. Figure 7.63 on pg. 152 shows a schematic of a typical noncoded system.

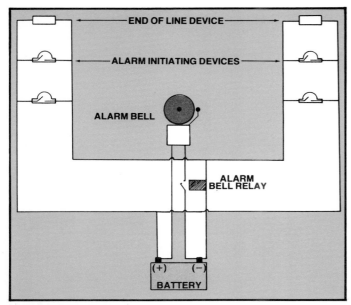

Figure 7.63 In a noncoded system, each alarm-initiating device sends an identical signal.

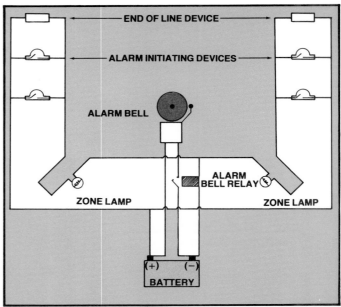

Figure 7.64 A zone noncoded system enables personnel to determine from which zone the alarm has initiated.

The difference between a noncoded system and zone noncoded system is in the alarm circuits. In a zone noncoded system, each circuit serves a zone or area of the building. Each zone is monitored by an indicator lamp placed in the circuit. Under ordinary circumstances, the small flow of current through the supervisory circuit is not strong enough to light the lamp. When one of the alarm initiating devices operates, the current flow increases and lights the lamp. The lamp then trips a relay, which activates the signal device. The same alarm signal sounds for all zones, so it is necessary to check the zone lamp on the control panel to determine which area the alarm is coming from. A schematic of a typical zone noncoded system is shown in Figure 7.64.

Master coded systems are employed in occupancies that use the alarm signals for other purposes. For example, schools use the same bells for sounding class changes as they do for sounding fire alarms. Master coded systems solve this problem by generating a code through the signal devices that gives the fire alarm a distinct and recognizable sound. A schematic of a typical master coded system is shown in Figure 7.65.

A presignal alarm is used in an occupancy in which the potential for panic is high, such as in a hospital. The system responds by first alerting emergency personnel, then alerting the general oc-

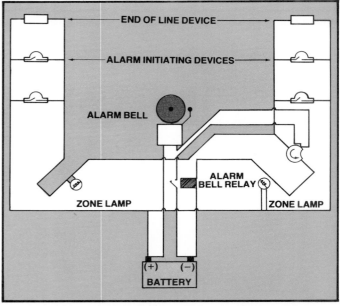

Figure 7.65 Master coded systems are used in schools and other occupancies where bells are in normal usage. The master coder produces a distinct and recognizable alarm signal over the bell system.

cupancy. This presignal gives emergency personnel an opportunity to assist the general occupancy in exiting.

Zone coded systems employ a signal-coding device into the circuit to monitor the zone circuit. The signal device will sound in a specific and unique pattern for each zone. Systems utilizing the zone coded arrangement enable employees on the fire brigade to determine the zone of the emergency simply by listening to the pattern of the signal (Figure 7.66).

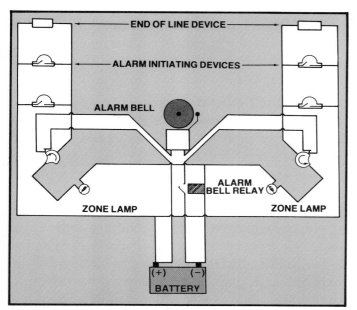

Figure 7.66 A zone coded system can produce a separate audible signal for each zone on the system, enabling personnel to determine the zone of origin.

Proprietary Alarm Systems

Proprietary alarm systems have a central alarm receiving point that is constantly staffed by trained personnel. The building alarm system and the central alarm receiving point are owned by the building occupant or owner. Usually this central point serves a group of buildings. An alarm in one building sounds only in that building and at the central receiving point. The alarm system in the individual building can be zoned or coded similar to a local system. The advantage of this type of system is that it provides 24-hour monitoring with someone responsible for taking specific action. This action may include notifying the fire department or sending a service crew to investigate a trouble signal. For more information concerning proprietary systems, see NFPA 72D, *Installation, Maintenance and Use of Proprietary Protective Signaling Systems.* Figure 7.67 shows a schematic of a typical proprietary system.

Central Station Systems

A central station system ties several separate local alarm systems into a central location. Often, several small industrial plants with one or two buildings each will be protected under one system

since they are not large enough to make staffed alarm centers economically feasible. In these instances, a separate contractor may develop an alarm receiving station and then contract to different plants to provide monitor service. The services provided and the requirements for staffing the station are essentially the same as those for a proprietary system. The difference lies in the fact that a central station receives alarms from different properties not under its ownership. A schematic of a central station layout is shown in Figure 7.68.

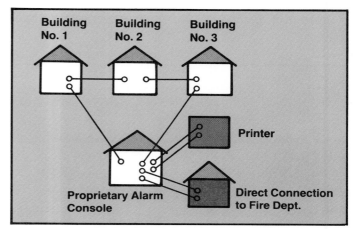

Figure 7.67 A proprietary system features a user-owned control center for monitoring several separate alarm systems on the same property.

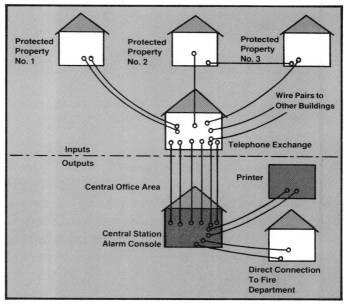

Figure 7.68 A security firm may contract with several occupancies to provide a staffed alarm center.

Remote Station Systems

In areas where the private contracting of central station service has not been established, the community dispatch service, the local fire station, or the 911 center may be equipped to receive alarm signals from local alarm systems. This provides a direct dispatch capability. However, supervision of the systems for maintenance and breakdowns is often a problem. The municipal service is usually unable to provide any maintenance service other than notifying the owner that a trouble signal has been received. In smaller towns or rural counties, the remote station is often located in the local law enforcement dispatcher's office because the fire station or headquarters is not constantly staffed. Figure 7.69 provides a schematic of a remote system.

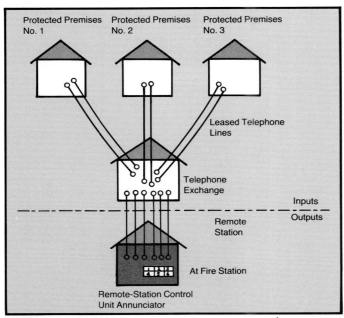

Figure 7.69 A remote system ties into a municipal dispatch facility.

Municipal Fire Alarm Systems

Although the main purpose of municipal fire alarm systems is to allow the public to notify the fire department of an incident, they can be and are used with private fire protection systems. The private alarm system (either a local or proprietary system) can be interconnected to the municipal fire alarm system through a master box. In this arrangement, signals from the private system are transmitted to the fire department alarm office. The private system is then classed as an auxiliary system.

Municipal systems are classified as either Type A (Manual) or Type B (Automatic). The sig-

nal transmitted through a Type A system has to be retransmitted to the fire stations. A Type B system automatically retransmits signals received from system alarm boxes to the fire stations. More information about alarm systems can be found in NFPA 1221, *Installation, Maintenance and Use of Public Fire Service Communication Systems.*

Multiplexing Detection Systems

Multiplexing fire detection and alarm systems are designed to fit an occupancy's present and future protective signaling needs. The term *multiplexing* is defined as the transmission of multiple signals over a single transmission path. An example of this is an FM radio used for the transmission of electrocardiogram (EKG) readings and verbal messages from advanced life support (ALS) ambulances. The radio transmits electrocardiogram signals over one channel and allows voice transmissions to be made over another channel. These two signals are transmitted over a single radio frequency and are separated and reproduced at the receiving end.

In a multiplexing fire alarm system, signals from alarm-initiating devices are transmitted over a single pair of wires. The signals can be reproduced and identified by a single device or group of devices at the fire alarm control panel. The chief advantage of signal multiplexing is that the amount of wire needed to connect alarm-initiating devices to the control panel is minimized, yet alarm devices or zones can still be individually identified. Multiplexing is accomplished through time division or frequency division multiplexing.

In a multiplexing network, the fire alarm control panel contains an array of microprocessors that are programmed by an on-board computer. The computer allows for the programming of detector sensitivity, the sequence that elevators need to be recalled in case of fire, and the number of exhaust fans required to pressurize a single stairwell or series of stairwells. The fire alarm control panel also allows the operator to check the status of any alarm-initiating device.

Multiplexing also allows for system changes due to building expansion or changes in fire or building codes that affect protective signaling systems. The device that permits this versatility is called the *transponder*. A transponder is designed to operate as a single zone control panel. It contains the components necessary to receive and initiate alarm signals in addition to supervising the zone devices and auxiliary functions (elevators, stairwell pressurization). Through the actions and programmed functions of the fire alarm control panel, the transponder converts the alarm-initiating signals into alarm-output signals. However, the transponder will not initiate an alarm unless it receives instructions from the fire alarm control panel.

Within the transponder are two subsystems that control and supervise all fire detection, fire signaling, and auxiliary devices within a particular zone. These two subsystems are called the *digital gathering panel* and the *command receiver*.

The digital gathering panel, or DGP, receives a signal from the alarm-initiating device and transmits it to the fire alarm control panel. This signal (either a diagnostic, trouble, or alarm condition) is evaluated by the fire alarm control panel and retransmitted to the digital gathering panel. If the signal is an alarm condition, the digital gathering panel electronically transfers the change in system status to the command receiver.

The command receiver is the output portion of the transponder. Its function is to convert the signal generated by the fire alarm control panel into an audible and/or visual warning signal that is recognizable by the building occupants. The command receiver, like the digital gathering panel, acts only on the commands of the fire alarm control panel. The command receiver operates any of the auxiliary functions. Figure 7.70 on pg. 156 shows a schematic of a multiplexing fire detection and signaling system.

Alarm-Initiating Devices

All alarm systems, whether they are small local systems or major multiplex high-rise systems, use the same basic alarm-initiating devices. There are three major types of these devices: manually activated devices, products of combustion detectors, and waterflow detectors.

MANUALLY ACTIVATED DEVICES

Manually activated devices are commonly called pull stations (Figure 7.71 on pg. 157). These devices require persons to recognize that a fire emergency exists and to act by using a pull station or, in some cases, a telephone fire alarm device.

PRODUCTS-OF-COMBUSTION DETECTORS

Products-of-combustion detectors function automatically in response to heat, smoke, or flame. Heat, an abundant product of combustion, is detectable by devices that use one or more of the three primary principles of heat physics:

- Heat causes materials to expand.
- Heat causes materials to melt.
- Heated metal has detectable thermoelectric characteristics.

Heat detectors include fixed-temperature devices, bimetallic strips, devices that melt, devices that use expansion of heated solvents, rate-of-rise detectors, combination rate-of-rise/fixed-temperature detectors, visible products-of-combustion detectors, and light detectors. Each is discussed more fully below.

FIXED-TEMPERATURE DEVICES

Fixed temperature fire alarm initiating devices activate at a predetermined temperature. Three basic types of fixed-temperature alarm initiating devices are used:

- A device using a bimetallic strip or disk
- A device using a soft metal alloy or thermoplastic resins that melt
- A device using expansion of heated solvents

Bimetallic strips consist of two metals or metal alloys that are bonded; these metals have different expansion ratios when heated. When the strip is heated, the side with the higher expansion ratio arches toward the side made of the metal with the lower expansion ratio. This action causes an electrical contact to be opened or closed, thus initiating an alarm (Figure 7.72 on pg. 157).

Devices that melt employ soft metal alloys or thermoplastic resins to hold together a two-piece link or latching mechanism. A frequently used metal alloy is solder. When the solder melts, the link separates or the latch is released, causing an alarm signal to sound.

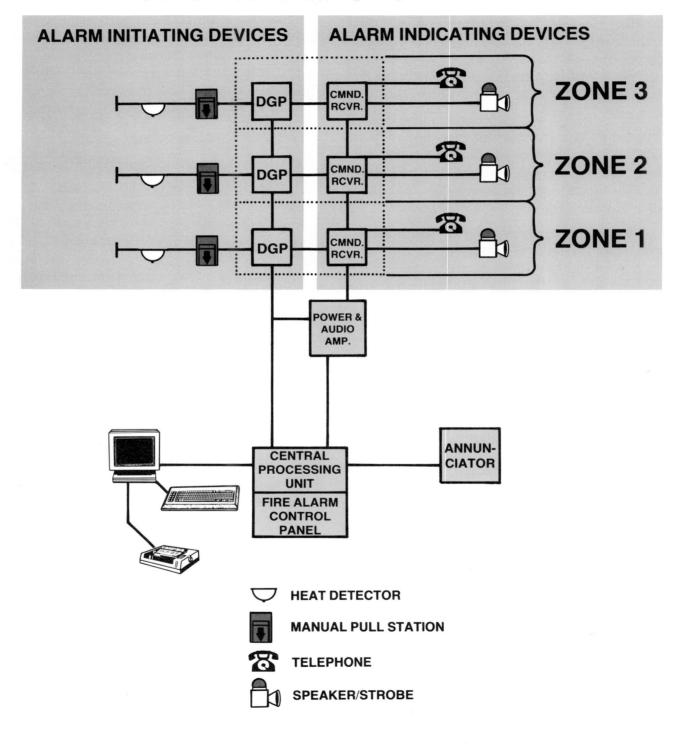

MULTIPLEXING FIRE DETECTION AND ALARM SYSTEM

ALARM INITIATING DEVICES

ALARM INDICATING DEVICES

DGP — CMND. RCVR. — ZONE 3

DGP — CMND. RCVR. — ZONE 2

DGP — CMND. RCVR. — ZONE 1

POWER & AUDIO AMP.

CENTRAL PROCESSING UNIT

FIRE ALARM CONTROL PANEL

ANNUN-CIATOR

HEAT DETECTOR

MANUAL PULL STATION

TELEPHONE

SPEAKER/STROBE

Figure 7.70 A schematic of a multiplexing system.

Figure 7.71 Manual pull stations should be visible and clearly marked.

Another type of device that functions through melting uses thermoplastic covers to insulate wires; these wires are wound around each other to form a tension cable. When the predetermined melting temperature of the thermoplastic is reached, the tension of the wires causes them to twist through the plastic. They short together and send an alarm. A third type of alarm-initiating device that functions as the result of melting sends an alarm when a chemical insulation melts. The insulation does not conduct electricity as a solid, but does act as a conductor when it is a liquid.

Another type of alarm activation device actuates through the expansion of heated solvents. In this type of device, small glass bulbs contain solvents; the bulbs are manufactured to break at predetermined pressures. When the solvent is heated, it vaporizes. The resulting vapor pressure breaks the glass bulb that has been holding two electrical contacts apart. The contacts close, causing an alarm signal to be sent.

Rate-of-rise detectors respond to quick changes in temperature rather than activating at a fixed temperature. Most rate-of-rise detectors have a small chamber filled with air. A small vent from the chamber allows for slow changes in temperatures. If rapid heating occurs, however, the air inside the chamber expands. The small vent cannot relieve the pressure, which forces a diaphragm out. The movement of the diaphragm either opens or closes a set of electrical contacts that initiate an alarm signal (Figure 7.73). A second type uses

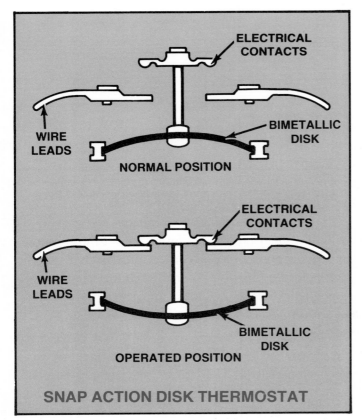

SNAP ACTION DISK THERMOSTAT

Figure 7.72 A bimetallic strip housed in a bowed position can be designed to invert when heated. This principle is used to manufacture restorable fixed-temperature detectors.

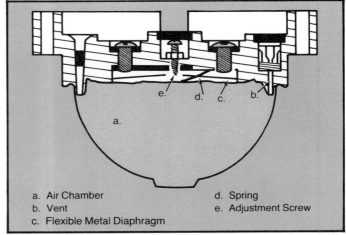

a. Air Chamber
b. Vent
c. Flexible Metal Diaphragm
d. Spring
e. Adjustment Screw

Figure 7.73 A pneumatic rate-of-rise detector utilizes a pocket of expanding air to detect fire. The pocket is vented to allow the detector to adjust to moderate temperature and barometric changes.

bimetallic strips (Figure 7.74) and a third type uses thermoelectric sensors to detect rapid changes in temperature (Figure 7.75). Rate-of-rise detectors tend to react more quickly than fixed-temperature detectors, but they are not quite as reliable.

Combination rate-of-rise/fixed-temperature detectors feature the quicker reaction rate of the rate-of-rise detector coupled with the higher dependability of the fixed-temperature unit (Figure 7.76).

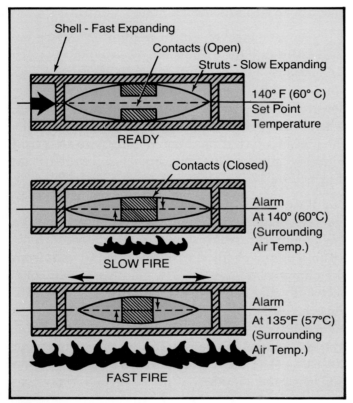

Figure 7.74 A heat-compensated detector uses the expansion characteristics of two different metals to produce one of the most reliable rate-of-rise detectors available.

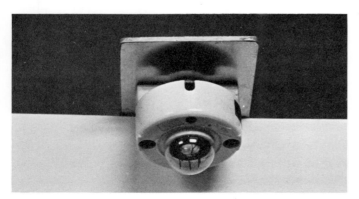

Figure 7.75 A thermoelectric rate-of-rise detector uses the electricity produced when dissimilar metals that have been twisted together are heated.

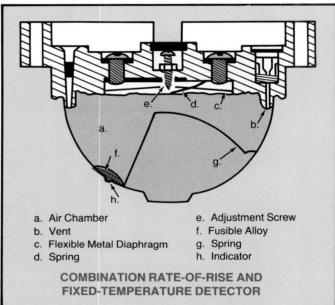

a. Air Chamber   e. Adjustment Screw
b. Vent   f. Fusible Alloy
c. Flexible Metal Diaphragm   g. Spring
d. Spring   h. Indicator

**COMBINATION RATE-OF-RISE AND
FIXED-TEMPERATURE DETECTOR**

Figure 7.76 A tension strip is soldered into a standard rate-of-rise detector to produce a combination rate-of-rise fixed-temperature detector.

VISIBLE PRODUCTS-OF-COMBUSTION DETECTORS

Visible products-of-combustion smoke detectors use a photoelectric cell coupled with a specific light source. The photoelectric cell functions in two ways to detect smoke: beam application and refractory application.

The beam application uses a beam of light focused across the area being monitored onto a photoelectric cell. The cell constantly converts the beam into current, which keeps a switch open. When smoke blocks the path of the light beam, the current is no longer produced and the switch closes. An alarm signal is then initiated (Figure 7.77).

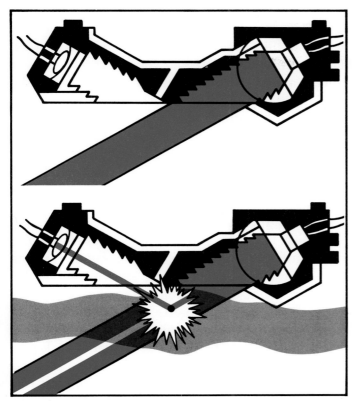

Figure 7.77 A beam application smoke detector is capable of monitoring large open areas. The atmosphere must be kept relatively clean to avoid false alarms.

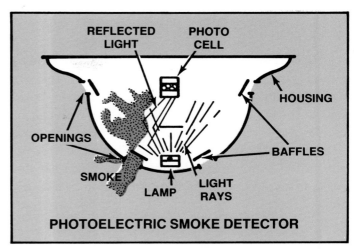

PHOTOELECTRIC SMOKE DETECTOR

Figure 7.78 A refractory-type photocell detector uses the light-scattering qualities of smoke instead of a direct beam to monitor for fire.

The refractory photocell uses a light beam that passes through a small chamber at a point away from the light source. Normally, the light does not strike the photocell and no current is produced. When a current does not flow, a switch in the current remains open. When smoke enters the chamber, it causes the light beam to be refracted (scattered) in random directions. A portion of the scattered light strikes the photocell, causing current to flow. This current closes the switch and activates the alarm signal (Figure 7.78).

INVISIBLE PRODUCTS-OF-COMBUSTION DETECTORS

During a fire, molecules ionize as they undergo combustion. The molecules have an electron imbalance and tend to steal electrons from other molecules. The operation of the invisible products-of-combustion detector utilizes this phenomenon, and for this reason they are commonly called ionization detectors.

The detector has a sensing chamber that samples the air in a room. A small amount of radioactive material adjacent to the opening of the chamber ionizes the air particles as they enter. Inside the chamber are two electrical plates: one is positively charged and one is negatively charged. The ionized particles free electrons, which travel to the positive plate. Thus, a minute current normally flows between the two plates. When ionized products of combustion enter the chamber, they pick up the electrons freed by the radioactive ionization. The current between the plates ceases and an alarm signal is initiated (Figure 7.79).

LIGHT DETECTORS

Light detectors are commonly called flame detectors. There are two basic types: those that detect light in the ultraviolet wave spectrum (UV detec-

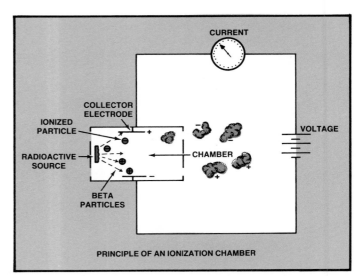

PRINCIPLE OF AN IONIZATION CHAMBER

Figure 7.79 Ionization detectors activate when ionized smoke particles moving into the detection chamber interrupt an electric current, thus initiating an alarm signal.

tors) and those that detect light in the infrared wave spectrum (IR detectors).

Ultraviolet detectors can give false alarms when they are in contact with sunlight and arc welding. Therefore, their use is limited to areas where these and other sources of ultraviolet light can be eliminated (Figure 7.80). Infrared detectors are effective in monitoring large areas. To prevent activation from infrared light sources that are not fires, infrared detectors require the flickering action of flame before they will activate to send an alarm (Figure 7.81).

Figure 7.80 Ultraviolet detectors detect light waves in the spectrum below 4,000 angstroms. With the development of electronic light-discriminating circuits, their use is growing. *Courtesy of Detector Electronics Corp.*

Figure 7.81 Infared detectors detect light above 7,700 angstroms in wavelength. They are good large-area coverage detectors. *Courtesy of Detector Electronics Corp.*

WATERFLOW DETECTOR

The waterflow detector detects movement of water within the sprinkler or standpipe system. Once movement is detected, the detector then gives a local alarm and/or may transmit the alarm. The intent of a waterflow alarm is to transmit an alarm whenever a sprinkler opens or a standpipe valve is opened. This device is usually attached to the main sprinkler riser. Waterflow detectors are susceptible to false alarms caused by pressure fluctuations within the water supply system.

Voice Communication Systems

Most building and fire codes require a voice communication and firefighter telephone system in high-rise buildings. These systems serve two functions: they enable firefighters to give instructions to building occupants and to communicate with the incident commander or other suppression personnel within the building. This network permits clear communications without the problems inherent in using portable radios in high-rise buildings. The voice communication system is usually a part of the overall fire detection and signaling system.

The voice communication system consists of a series of highly reliable, fire-resistive audio speakers, handset stations or wall jacks, and a communication console that usually contains spare handsets (Figure 7.82). The speakers and telephones are located at strategic points throughout the building such as in elevator cars and stairwells. The fire department communication console is usually located in the main entrance lobby or an area designated as the building's command center. Some systems have fire communications stations located on each floor, or zone, to which fire department personnel can go to assume command of a particular floor. Since complete evacuation of a high-rise building is not always feasible, the occupants can be instructed to move to areas of safe refuge. Communications with the building occupants can be maintained to allow for further relocation if necessary.

Though some systems use permanently installed wall-mounted telephones, several manufacturers now offer a plug-in style telephone. The handset plugs into the special wall jacks located in de-

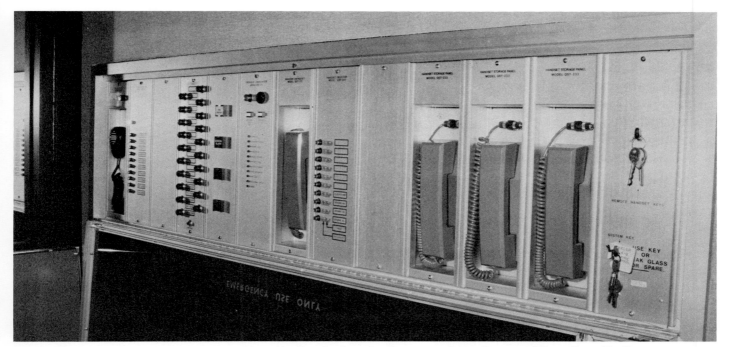

Figure 7.82 Voice communications systems provide an ideal location for an interior command post. *Courtesy of Plano, Texas Fire Department.*

signated building areas. This arrangement gives firefighters greater mobility and eliminates the need for installing permanent telephones in areas that are subject to vandalism or misuse.

Automatic Telephone Dialers

An automatic telephone dialer connects a building fire detection or alarm system to a standard residential or business telephone. This arrangement eliminates the cost of leased lines necessary for other systems that transmit alarms to locations away from the premises. When the detection or alarm system activates, the automatic telephone dialer dials a preprogrammed telephone number, usually the number of the fire department or central fire alarm facility (Figure 7.83). A recorded message is replayed three or four times to ensure the dispatcher receives the entire message, especially the address. At the present time, many cities and even some states have outlawed the use of automatic telephone dialers because they tie up emergency telephone circuits leading into the fire alarm center.

SUPERVISION OF ALARM CIRCUITS

When a high degree of reliability is required, alarm circuits may be electronically supervised. These types of circuits are designed and equipped

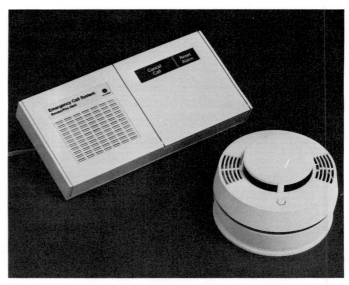

Figure 7.83 Automatic dialers are commonly used to protect private residences and small businesses. *Courtesy of American Telephone and Telegraph.*

to automatically detect faults in the system circuitry and transmit a trouble signal. Such circuits are said to be "supervisory circuits."

Since the fire department is not responsible for fire alarm system maintenance, except for municipal fire alarm systems, supervisory circuits usually terminate at a central supervisory station where operators initiate action to have circuits repaired when the trouble signal is received.

Inspecting and Testing Alarm and Detection Devices

Fire inspectors must have some basic knowledge of fire alarm and detection systems to be able to inspect them and to observe if tests are being performed properly. It is not the responsibility of the inspector to perform maintenance; however, properly trained personnel are required for this task. These individuals may be employed full time by management or may work under a service contract with a company specializing in this service.

Fire inspectors should ensure that qualified personnel routinely inspect all circuit wiring for proper support, wear, punctures, cracks, insects, or any other defects that may render the system ineffective. In places where circuits are enclosed in conduit, the conduit should be checked for solid connections and proper support.

Inspectors should check to determine that all sounding devices, such as bells, gongs, and buzzers are free from paint and dirt, which may interfere with their operation. All equipment should be kept free of dust, dirt, and similar foreign materials. **NOTE:** Inspectors should encourage those cleaning this equipment to use a vacuum cleaner, rather than wiping the dust and having it settle on electrical contacts or relays.

Control panels, annunciator boxes, recording instruments, and other devices should not have such objects as extra relays, rectifiers, light bulbs, and testing equipment stored on or in them. These items can foul moving parts or cause electrical shorts that can cause the system to fail. Many units with lockable doors have ample room inside to store these items.

Where batteries are used as an emergency power source, they should be checked for clean contacts. The float balls should indicate that the battery is well charged. Batteries and rectifiers should be kept free from dust and other materials.

Detectors are a vital part of the alarm system. The reliability of the detector is, in fact, the main factor in the reliability of the alarm system. For this reason, detectors require testing when they are first installed, at specified intervals, and after fires. Procedures should be in accordance with NFPA 72H, *Testing Procedures for Local, Auxiliary, Remote Station and Proprietary Protective Signaling Systems.*

The following information pertains to testing detectors in general:

- Detectors should not show any sign of damage.

- Detectors should not have been painted unless the testing laboratory has found them to be unaffected by painting. No attempt should be made to remove paint.

- Detectors on systems that are being restored to service after a period of use should be checked by a testing laboratory.

- Detectors should not be perceptibly corroded.

- Detectors on circuits that have been subjected to overvoltage surges or lightning damage should be sent to a recognized testing laboratory for testing.

- Detectors subjected to other conditions that may permanently affect their operation, such as grease or corrosive atmospheres, should also be sent to a laboratory for testing.

The procedures for testing specific types of detectors are as follows:

Nonrestorable Fixed-Temperature Detectors. These detectors cannot be tested periodically because the application of heat, which is required for testing, will destroy the unit. For this reason, tests are not required until 15 years after installation. After 15 years, usually 2 percent of the detectors are removed and replaced with new detectors. The detectors that are removed are shipped to a nationally recognized testing laboratory for testing. If any detector fails, then additional detectors are removed and sent for testing. Cable-type line detectors must have the loop resistance tested semiannually.

Restorable Detectors. Restorable detectors can be tested with a portable heat source such as a hairdryer or heat lamp with a temperature shield. One detector on each signal circuit should be tested semiannually. A different detector should be tested each time.

Fusible-Link Detectors. Fusible-link detectors with REPLACEABLE links should be tested semiannually by having the link removed and observing whether or not the contacts close. After the test, the fusible link is reinstalled. It is recommended that links be replaced every five years.

Pneumatic-Type Detectors. These may be tested with a heating device or with an approved pressure pump. **NOTE:** If a pressure pump is used, the manufacturer's instructions must be followed. Testing should be conducted on a semiannual basis.

Smoke Detectors. Smoke detectors should be tested seminannually according to the manufacturer's instructions. Instruments for testing performance and sensitivity are generally available from the manufacturer. Blowing cigarette smoke into the detector is NOT considered an acceptable test.

Flame Detectors. These devices are usually intricate and expensive. Instructions from the manufacturer for test procedures, training, and testing equipment, if obtainable, should be required by the bid specifications for the flame detectors. Inspectors should determine if testing devices are available and, if so, determine if procedures are being followed. Testing may be included in a service contract.

A permanent record of all detector tests should be maintained for a minimum of five years. The record should include the date of the test, the detector type and location, type of test, results, and person performing the test.

INSPECTING CONTROL EQUIPMENT

Inspecting alarm systems includes checking the control equipment. This includes the local annunciator panels and signal switching or transmitting devices, printers, annunciator consoles, sounding devices, and power sources at proprietary, remote, and central station locations. When necessary, inspectors should witness the signal switching to transmitting device test, which will also give them the opportunity to test recording mechanisms. They should check to see that initiating devices are clean and do not stick or bind. When alarms are tested, they must transmit the correct number of signals. At this time, the inspector can

also evaluate auxiliary devices such as local evacuation alarms or air-handling equipment. All devices must be restored to proper operation after testing. The inspector can check the local annunciator panel during the restorable detector tests, as previously described, or by bridging contacts.

Receiving devices should also be checked. The proper signal and/or number of signals should be received or recorded. Signal impulses should be definite, clear, and evenly spaced so that each coded signal can be identified. There should be no sticking, binding, or other irregularities. At least one complete round of printed signals should be clearly visible and unobstructed by the receiver at the end of the test. The time stamp should clearly indicate the time of a signal and should not interfere in any way with the recording device. Inspection personnel should make sure that all alarm indicating devices, which may include gongs, electronic sirens, chimes, or pulsating high intensity lights (strobe), are clean and operable. They should check that the supervisory power source is operating properly by viewing the pilot light and/or gauge. If an emergency power supply is installed, they can test it by interrupting the power source to ensure that the unit will convert to battery power. After all tests are complete, the devices should be returned to their normal standby condition.

WATER SUPPLY SYSTEMS AND WATER SUPPLY ANALYSIS

Water is one of the most important tools firefighters use to control and extinguish fire. The mere physical properties of water make it an excellent extinguishing agent. Water, at ordinary temperatures, is a heavy and stable liquid. When water is converted from liquid to vapor (steam), its volume increases about 1,700 times. This huge volume of water (saturated steam) displaces an equal volume of air surrounding a fire. The reduced volume of air helps smother the fire and in most cases, the surface of the burning material is cooled below the temperature at which it gives off sufficient vapors to ignite. Because of water's importance in fire fighting, inspectors must understand the theory, application, and devices used to transport water.

Inspectors must be familiar with the types of water distribution systems in their local communities in order to ensure that the systems are adequate to handle emergency situations. Inspectors can usually obtain information on virtually any aspect of the local water supply network from the water department. The local water department is usually a separate city utility whose main function is to provide safe water for human use. As with all other city organizations, it is important that the fire department maintain a good working relationship with the water department.

Water can be obtained through surface supplies (rivers or lakes) or group supplies (wells or water-producing springs). There are four components of an effective water distribution system:

- Water supply source
- Processing or treatment facilities
- Means or methods of moving the water
- Delivery system, including storage (Figure 7.84)

Water Distribution Systems

There are three types of distribution systems: gravity, direct pumping, and combination.

A true *gravity system* delivers water from the source to the distribution system without pumping equipment (Figure 7.85). The natural pressure created by a difference in elevation provides the pressure within the distribution system. When elevation pressure cannot provide sufficient pressure for the community, a pump is placed relatively close to the water source to create the pressure within the distribution system. This is called a *direct pumping system* (Figure 7.86). Most communities use *combination systems* — both gravity and pumping — to provide adequate pressure (Figure 7.87 on pg. 166). Water is pumped into the distribution system and into elevated storage tanks (gravitational pressure).

When the consumption demand is greater than the rate at which the water is pumped, the water flows from the storage tanks into the distribution system. Conversely, when demand is less, the water flows back into the storage tanks. Elevated storage reservoirs are usually constructed of steel or concrete. They vary in height and can hold as much as 2,000,000 gallons (7 570 000 L), depending on pressure desired.

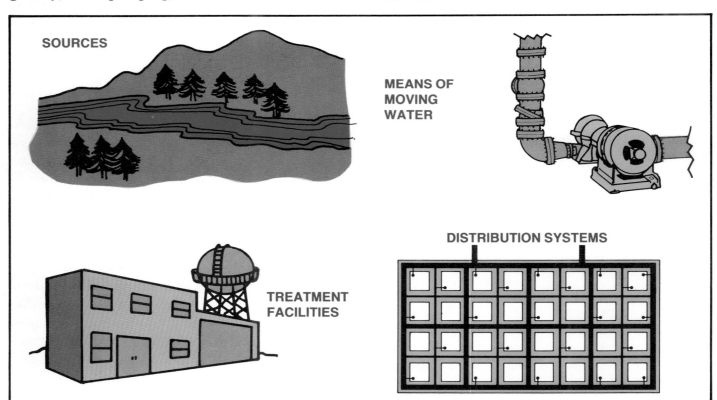

Figure 7.84 There are four main components in a water distribution system.

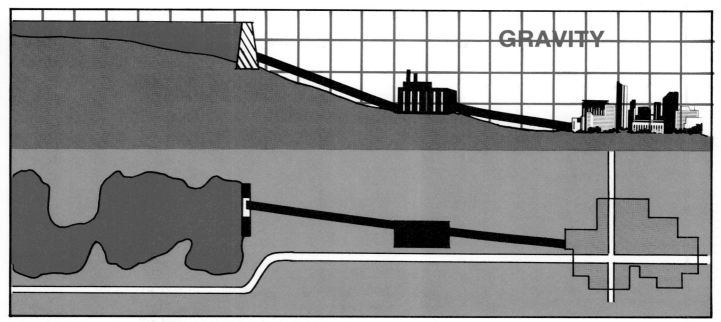

Figure 7.85 The gravity system is used where the water source is elevated.

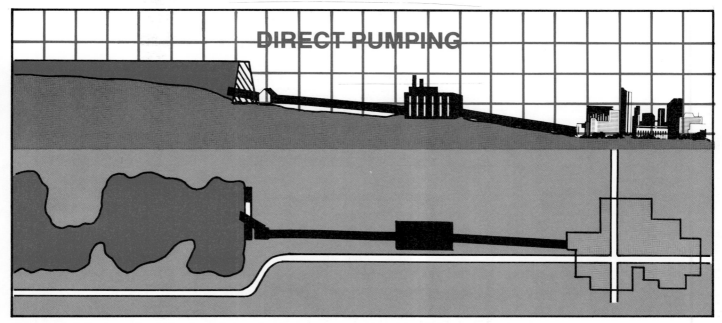

Figure 7.86 A direct pumping system is used when elevation pressure cannot provide sufficient water pressure.

The distribution system receives the water from the pumping station and delivers it throughout the area to be served. Fire hydrants, gate valves, elevated storage, and reservoirs are supplementary parts of the distribution system. The term "grid" is sometimes used to describe the network of water mains that make up a water distribution system (Figure 7.88 on pg. 166).

When water flows through pipes, a pressure loss occurs due to the movement of the water against the inside of the pipe (friction loss). A fire hydrant that receives water from one side (called a dead-end hydrant) has less available water than a fire hydrant supplied from two or more directions (called a circulating or looped feed hydrant). The distribution system consists of three main feeders:

- Primary feeders — Largest pipes (mains) with relatively widespread spacing. These feeders convey large quantities of water to

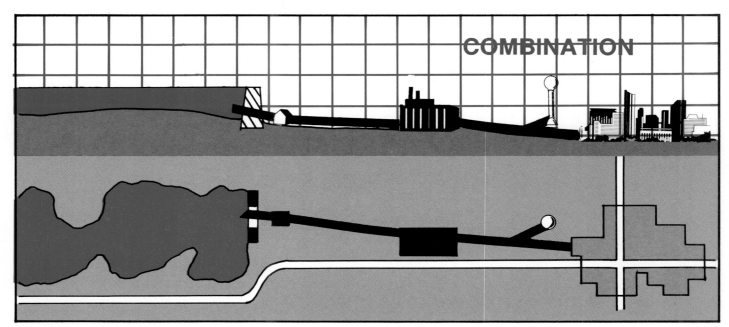

Figure 7.87 A combination of direct pumping and gravity allows water to be stored during periods of low demand and to be pumped into the distribution system when needed.

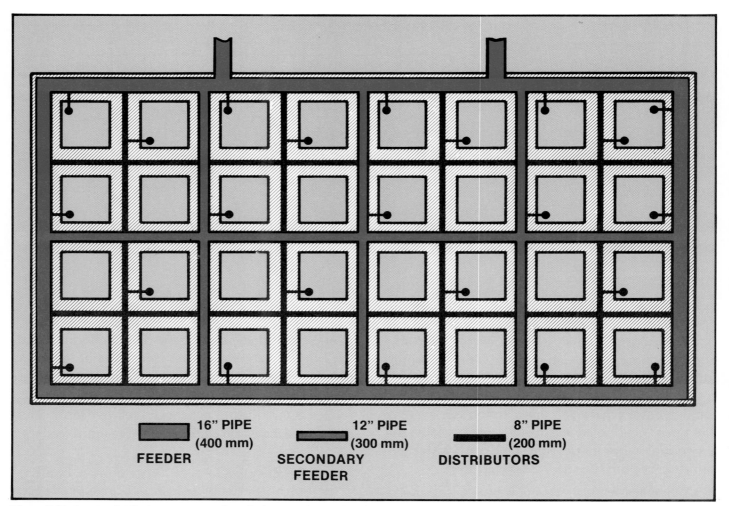

16" PIPE
(400 mm)
FEEDER

12" PIPE
(300 mm)
**SECONDARY
FEEDER**

8" PIPE
(200 mm)
DISTRIBUTORS

Figure 7.88 A water distribution system consists of primary and secondary feeders and distribution mains.

various points of the system for local distribution to the smaller mains.

- Secondary feeders — Intermediate pipes within the primary feeder network that reinforce the grid with a concentrated supply of water.
- Distributors — Smallest of the mains that serve the individual fire hydrants and blocks of consumers.

The ability to deliver adequate water depends upon the capacity of the system's network of pipes. Today, 8-inch (200 mm) pipe is becoming the *minimum* size used due to its increased flow capability over 4- and 6-inch (100 mm and 150 mm) pipes. The following minimum pipe dimensions (inside diameter) are recommended for good water supply distribution:

- RESIDENTIAL DISTRICTS should use 8-inch (200 mm) pipe unless a 6-inch (150 mm) pipe will complete a good grid and is cross-connected at intervals not exceeding 600 feet (183 m).

- SHOPPING CENTERS AND INDUSTRIAL AREAS should use 12-inch (305 mm) pipe, or 8-inch (200 mm) pipe if it will complete a good grid. Larger sized mains may be needed depending upon the size, layout, and occupancy of the structure.

- MULTIPLE HOUSING DEVELOPMENTS should use at least 8-inch (200 mm) pipe; in many cases, however, industrial area standards are needed.

Access to the underground water distribution network is made through the hydrant. The two main types of modern fire hydrants are dry-barrel hydrants and wet-barrel hydrants (Figure 7.89). Regardless of the design or type, the hydrant outlets are considered to be standard if there is at least

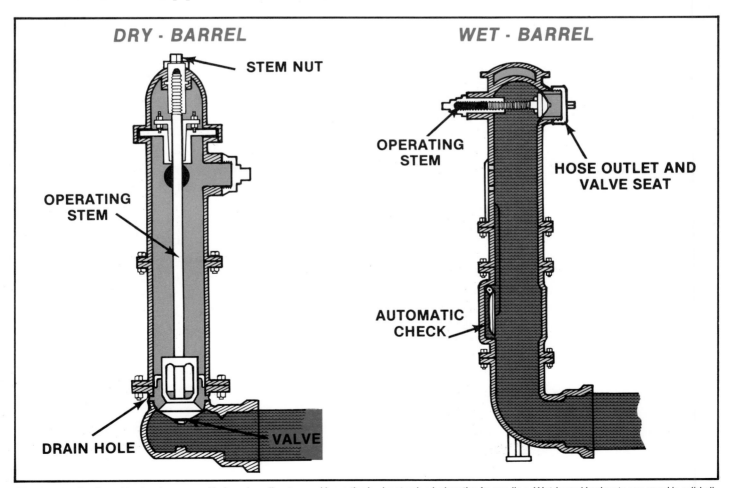

Figure 7.89 Dry-barrel hydrants are used in freezing climates and have the hydrant valve below the freeze line. Wet-barrel hydrants are used in mild climates and have the valves at the outlets.

one large outlet (4 or 4½ inches [100 mm or 115 mm]) for pumper supply and two outlets for 2½-inch (65 mm) hose. Hydrant specifications require a 5-inch (125 mm) valve opening for standard 3-way hydrants and a 6-inch (150 mm) connetion to the water main. The threads on all hydrant outlets must conform to those used by the local fire department. The principal items covered by the standard are the number of threads per inch and the outside diameter of the male thread. For exact details, refer to NFPA 1963, *Screw Threads and Gaskets for Fire Hose Connections*. In general, hydrant bonnets, barrels, and foot pieces are made of cast iron. The important working parts are usually made of bronze, but valve facings may be made of rubber, leather, or a composition material.

Hydrant location is usually determined by the type, size, occupancy location, and type of occupancy. There should not be more than 800 feet (245 m) between hydrants. In closely built areas, 500 feet (150 m) is a more acceptable standard. In general, hydrants are normally installed approximately 50 feet (15.2 m) from the structure to be protected. Where hydrants are located on a private water system and hoselines are to be used directly from the hydrants, hoselines should be kept relatively short, preferably not over 250 feet (75 m).

Dry-barrel hydrants are used in areas that have freezing temperatures at least part of the year under normal circumstances. The dry-barrel hydrant has a base valve located below the frost line; the stem nut to open and close the base valve is located on the top of the hydrant. It is classified as a compression, gate, or knuckle-joint type valve that either opens with or against the water pressure. Any water remaining in a closed dry-barrel hydrant will drain through a small valve that opens at the bottom of the hydrant when the main valve approaches a closed position.

The wet-barrel hydrant usually has a compression valve at each outlet, but may have another valve in the bonnet that controls the water flow to all outlets. This type of hydrant features the valve at the hose outlet and is used in mild climates where typical weather conditions are above freezing.

One of the most important periodic maintenance considerations is to check for leaks in the following areas:

- The main valve when it is closed
- The drip valve when the main valve is open but the outlets are capped
- The water mains near the hydrant

The American Water Works Association has adopted specifications for a national standard hydrant for ordinary waterworks service. These specifications are designed to produce a hydrant that is free from difficulties, such as trouble in opening and closing, interior mechanical parts that can work loose, leakage, excessive friction loss, failure to drain properly, and loose nipples. These specifications may be obtained from: The Amercian Water Works Association, 6666 West Quincy Avenue, Denver, Colorado 80235.

The actual flow of water from a hydrant may vary due to such conditions as feeder main location, incrustations, and deposits. Firefighters can make better tactical decisions if they know at least the relative available water flow of different hydrants in the vicinity. To address this problem, NFPA standardized a color code system (NFPA 291, *Fire Flow Testing and Marking of Hydrants*) to mark hydrants according to capacity (Figure 7.90). The colors may vary due to geographic location, but the main intent of any color scheme is simplicity.

The main function of water main valves is to control water flow through the mains as circumstances dictate. Valves should be operated at least once a year to ensure they are in good working condition. Valve spacing should be planned so a minimum length of the water distribution system will be out of service if a cutoff procedure is initiated.

The maximum lengths for valve spacing should be 500 feet (150 m) in high-value districts and 800 feet (245 m) in other areas, as recommended by Commercial Risk Services, Inc. (formerly Insurance Services Office) engineers.

Valves for water systems are broadly divided into indicating and nonindicating types (Figure 7.91). An indicating valve shows whether the gate

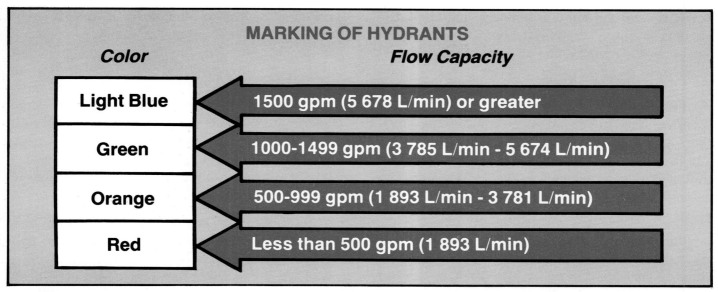

MARKING OF HYDRANTS

Color	Flow Capacity
Light Blue	1500 gpm (5 678 L/min) or greater
Green	1000-1499 gpm (3 785 L/min - 5 674 L/min)
Orange	500-999 gpm (1 893 L/min - 3 781 L/min)
Red	Less than 500 gpm (1 893 L/min)

Figure 7.90 Color coding hydrants allows responding personnel to easily determine how much water will be available at a particular location.

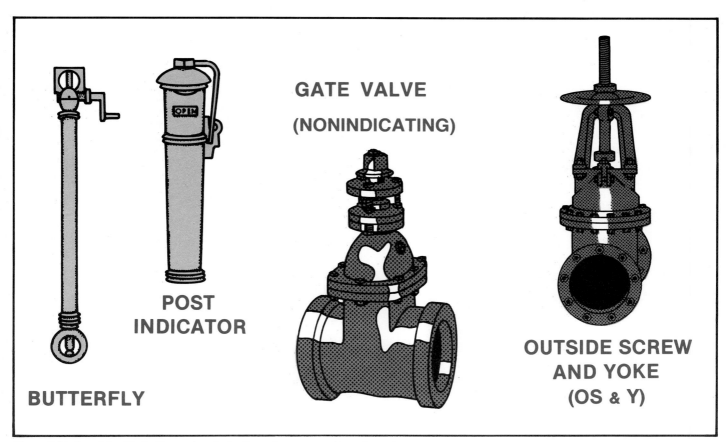

GATE VALVE (NONINDICATING)

POST INDICATOR

OUTSIDE SCREW AND YOKE (OS & Y)

BUTTERFLY

Figure 7.91 A variety of indicating and nonindicating valves are used in water mains.

valve seat is open, closed, or partially closed. Valves in private fire protection systems are usually of the indicating type. Valves in public water systems are usually of the nonindicating type, except for some valves in treatment plants and pump sta-

tions. Two common indicator valves are the post indicator valve (PIV) and the outside screw and yoke valve (OS&Y).

Nonindicating valves in a water distribution system are usually buried or installed in manholes. If a buried valve is properly installed, the valve is operated through a valve box by a special socket wrench on the end of a reach rod. Control valves may be gate valves or butterfly valves. Gate valves are usually the nonrising stem type, which requires a valve key for operation. As the valve nut is turned, the gate either rises or lowers to control the water flow. Butterfly valves are tight closing and they usually have rubber or a rubber composition seat that is bonded to the valve body. The valve disk rotates 90 degrees to open or shut the valve.

Inspectors should be aware of the consequences of stuck or partially closed valves. If a valve is partially closed, it would not be noticed during normal use of water; however, the high friction loss will prevent the fire department from obtaining sufficient water to combat a fire. This situation can be prevented by accurate and routine inspections.

Obstruction of Water Mains

After several years, fire flow tests may show a progressively inadequate flow through water mains. Investigation may reveal that the mains are obstructed. Obstructions may be a result of one or more of the following:

- *Incrustation*
  — Caused by tubercular corrosion or rust
  — Caused by chemical constituents of water
  — Caused by biological or organism growth
  — Caused by biodegradation of water agents and pipe materials.

- *Sedimentation Deposits*
  — Sedimentary decay (mud, clay, leaves)
  — Foreign matter other than sediment

- *Valves*
  — Closed
  — Partially Closed

- *Foreign Matter*
  — Stones, tools, wood, or lead

Although incrustation and sedimentation both result in deposits on the inside walls of water mains, there are differences between the two processes. Incrustations are usually caused by a pro-

gressive growth of rust deposits on the iron pipes or the accumulation of various salts due to oxidation. This process may also result from certain biological reactions produced by organisms present in most water supplies. Sediment deposits are mud, clay, dead organisms, and decayed vegetation. These deposits are formed mechanically by a process of precipitation. Sedimentation deposits are found chiefly in the bottom of water mains; incrustations can form around the entire inside wall of the water main.

Foreign matter other than deposits may also obstruct water mains. Chunks of lead (from ball and spigot joints), boards, crowbars, pickhandles, stones, and various other materials have been found in water mains. It is also possible that a combination of incrustation, sedimentation, partially closed valves, and foreign matter can increase friction loss and decrease water flow.

There are methods of cleaning water mains; however, municipal water departments and fire departments do not usually have the tools necessary to perform this task. Cleaning water mains is generally contracted to private companies that specialize in this type of work.

Conducting Waterflow Analysis
PHYSICAL INSPECTION OF FIRE HYDRANTS

In most cities, repair and maintenance of fire hydrants are the responsibility of the water department. The utility companies are better equipped to do this work than any other agency. Although the water department may technically be responsible for maintenance, it is the fire departments that depend on hydrants in top operating condition during a fire. This requires the fire department to perform periodic testing of the hydrants and water supply system.

The materials needed to complete inspection of fire hydrants include the following:

- Notebook

- Gauging device for checking threads

- Can of lubrication oil

- Pot with a mixture of light lubrication oil and graphite

- Small, flat brush

- Gate valve key

- Pressure gauge and tapped hydrant cap

- Pitot gauge for pressures up to 200 psi (1 380 kPa)

- 12-quart (11.4 L) pail

- Hydrant wrench

When hydrant inspections are made, inspectors should observe the following conditions:

- Check for any obstructions near the hydrant such as sign posts, utility poles, shrubbery, or fences (Figure 7.92).

- Check the direction of the hydrant outlet(s) to ensure they face the proper direction and that there is clearance between the outlet and the surrounding ground. The clearance between the bottom of the butt and the grade should be at least 15 inches (380 mm) Figure 7.93).

- Check for mechanical damage to the hydrant such as dented outlets or rounded (stripped) stem nuts (Figure 7.94).

- Check the condition of the paint for rust or corrosion.

- Check water flow by having the hydrant fully opened and then checking its ability to drain.

Figure 7.93 Insufficient ground clearance makes this hydrant almost useless.

Figure 7.94 Damaged hydrants should be repaired immediately so they are usable during an emergency.

Figure 7.92 Obstructions can make a hydrant hard to locate or difficult to attach to.

Inspectors should notify the water utility company when hydrant inspections will be made (at least twice a year), and they should be kept informed concerning the route that is to be taken. A record of the hydrant inspection and maintenance performed should be filed for future reference (Figure 7.95). When accurate testing, observation, and records have been compiled, inspectors can evaluate the operational readiness of a water supply system.

HYDRANT RECORD

LOCATION _____ HYDRANT NO. _____

POSITION _____ MAKE _____

INSTALLED _____ TYPE _____ TURNS TO OPEN _____ R. _____ L. _____

SIZE OF LEAD _____ SIZE OF MAIN _____

VALVE IN LEAD _____ FT. _____ TURNS TO OPEN _____ R. _____ L. _____

BENCH MARK _____ ELEV. _____

PRESSURE TESTS

DATE	STATIC PRESSURE	FLOW PRESSURE	GPM	DATE	STATIC PRESSURE	FLOW PRESSURE	GPM

REMARKS

_____

_____

_____

RECORD OF MAINTENANCE

WORK PERFORMED _____ DATE _____

Flowed

Lubricated

Cap Gasket Replaced

Bonnet Gasket Replaced

Valve Leather Replaced

Drain Valve Replaced

Cap Replaced

Lead Valve Operated

Painted

Raised

Moved

Figure 7.95 A sample inspection record form.

DETERMINING AVAILABLE WATER SUPPLY

It is important for every fire officer to know the overall capacity of the water system and its flow in given areas. Fire fighting defenses cannot be planned intelligently without this information. In order for fire service personnel to determine the quantity of water available for fire protection, it is necessary to conduct fire flow tests on the water distribution system. These tests include the actual measurement of static (normal operating) and residual pressures, equipment needed, and the formulas and calculations used to determine available water.

Fire flow tests are made to determine the rate of water flow available for fire fighting at various locations within the distribution system. By measuring the flow from hydrants and recording the pressures corresponding to this flow, the number of gallons available at any pressure or the pressure available at any flow can be determined through calculations or graphical analysis.

Before conducting a flow test, a responsible water department official should be notified since the opening of hydrants may upset the normal operating conditions in a water supply system. Notification is also important because water service personnel may be performing maintenance work in the immediate vicinity; therefore, the results of the flow test would not be typical for normal conditions. This practice of proper notification will also promote a better working relationship between the water department and fire service personnel.

Knowing the capacity of a water system is just as important as knowing the capacities of pumpers and water tanks. This knowledge is also essential when making pre-incident plans. The results of fire flow tests can be used to an advantage by both the fire and water departments of a municipality. Fire officers who are familiar with fire flow test results are better qualified to locate pumpers at strong locations on a distribution system and avoid weak locations. Since test results indicate weak points in a water distribution system, they can be used by water works personnel when improvements in an existing system are planned and when extensions to newly developed areas are designed. Tests that

are repeated at the same locations year after year may reveal a loss in the carrying capacity of water mains and a need for strengthening certain arterial mains. Flow tests should be run after any extensive water main improvements, after extensions have been made, or at least every five years if there have been no changes.

USING THE PITOT TUBE AND GAUGE

Using a pitot tube and gauge to take a flow reading is not difficult, but it must be used properly to obtain accurate readings (Figure 7.96). A good method of holding a pitot tube and gauge in relation to a hydrant outlet or nozzle is illustrated in Figure 7.97. Note that the pitot tube is grasped just behind the blade with the first two fingers and

Figure 7.96 Pitot tubes can be used to accurately determine the velocity pressure from a discharge. The velocity pressure can then be used to determine the flow from the discharge.

Figure 7.97 When holding the pitot tube in this manner, the little finger is used to steady the instrument.

thumb of the left hand while the right hand holds the air chamber. The little finger of the left hand rests upon the hydrant outlet or nozzle tip to steady the instrument.

Unless some effort is made to steady the pitot tube, the movement of the water will make it difficult to get an accurate reading. Another method of holding the pitot tube is illustrated in Figure 7.98. The left hand fingers are split around the gauge outlet and the left side of the fist is placed on the edge of the hydrant orifice or outlet. The blade can then be sliced into the stream in a counterclockwise direction. The right hand once again steadies the air chamber. The procedure for using a pitot tube and gauge is as follows:

Step 1: Open the petcock on the pitot tube and make certain the air chamber is drained. Then close the petcock.

Step 2: Edge the blade into the stream, with the small opening or point centered in the stream and held away from the butt or nozzle approximately one-half the diameter of the opening (Figure 7.99). For a 2½-inch (65 mm) hydrant butt, this distance would be 1¼ inch (32 mm). The pitot tube blade should now be parallel to the outlet opening with the air chamber kept above the horizontal plane passing through the center of the stream. This will increase the efficiency of the air chamber and help avoid needle fluctuations.

Step 3: Take and record the velocity pressure reading from the gauge. If the needle is fluctuating, read and record the value located in the center between the high and low extremes.

Step 4: After the test is completed, open the petcock and be certain all water is drained from the assembly before storing.

COMPUTING HYDRANT FLOW

The easiest way to determine how much water is flowing from the hydrant outlet(s) is to refer to prepared tables for nozzle discharge. These tables have been computed by using a formula for gallons per minutes (L/min) flow when the flow pressure is

Figure 7.98 Holding the pitot tube in this fashion gives the operator maximum control over the device.

Figure 7.99 Once the inspector has a good grip on the pitot tube, it may be inserted into the stream.

known. The formula may be stated in this manner: Flow rate is equal to a constant multiplied by the coefficient of discharge, multiplied by the diameter of the orifice squared, multiplied by the square root of the pressure.

The formula is written as follows:

$$GPM = (29.83) \times C \times d^2 \times \sqrt{P}$$
$$L/min = (0.0667766) \times C \times d^2 \times \sqrt{P}$$

Where: C = The coefficient of discharge
d = The actual diameter of the hydrant or nozzle orifice in inches (mm)
P = The pressure in psi (kPa) as read at the orifice.

NOTE: 29.83 (0.0667766) is a constant derived from the physical laws relating water velocity, pressure, and conversion factors that conveniently leave the answer in gallons per minute (Liters per minute).

This formula was derived by assigning a coefficient of 1.0 for an ideal frictionless discharge orifice. An actual hydrant orifice or nozzle will have a lower coefficient of discharge, reflecting friction factors that will slow the velocity of flow. The coefficient will vary with the type of hydrant outlet or nozzle used. When using a hydrant orifice, the operator will have to feel the inside contour of the hydrant to determine which one of the three types of hydrant outlets is being used (Figure 7.100). When a nozzle is used, the coefficient of dis-

charge will depend upon the type of nozzle. Consult the manufacturer's recommendations for determining the coefficient of discharge for a specific nozzle.

The flow formula also depends upon the actual internal diameter of the outlet or nozzle opening being used. A ruler with a scale that measures to at least sixteenths of an inch (mm's) should be used to measure the diameter of the outlet or nozzle opening (Figure 7.101).

Assuming a 2½-inch (65 mm) hydrant outlet is used that has an actual diameter of 2 7/16 inches (2.44 inches [62 mm] with a C factor of 0.80, and a flow pressure of 10 psi (69 kPa) read from the pitot gauge, the waterflow equation would read:

$$GPM = 29.83 \times C \times d^2 \times \sqrt{P}$$
$$GPM = 29.83 \times 0.80 \times (2.44)^2 \times \sqrt{10}$$
$$GPM = 449.28 \text{ or} \approx 450$$

$$L/min = 0.0667766 \times C \times d^2 \times \sqrt{P}$$
$$L/min = 0.0667766 \times 0.80 \times (62)^2 \times \sqrt{69}$$
$$L/min = 1705.78 \text{ or} \approx 1700$$

Generally, 2½-inch (65 mm) outlets should be used to conduct hydrant flow tests. This is because the stream from a large hydrant outlet (4 to 4½ inches [100 mm to 115 mm]) contains voids, that is, the entire stream of water is not solid. For this reason, the above formula alone will not give accurate results for flows using large outlets. If it is necessary to use the large outlets, a correction factor

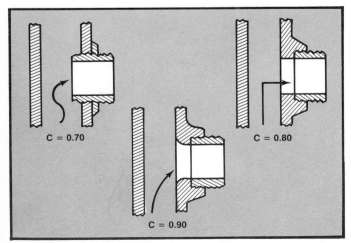

Figure 7.100 Different hydrant outlets have different coefficients of discharge used in water flow formulas. The bottom example features a smooth and rounded outlet.

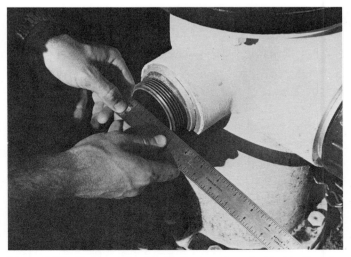

Figure 7.101 The inspector should measure the actual inside diameter of the outlet in order to be able to calculate water flows accurately.

can be used to give more accurate results. The flow (as determined by gpm = 29.83 x C x d^2 x $\sqrt{P}$ or L/min - 0.0667766 x C x d^2 x $\sqrt{P}$) should be multiplied by one of the following factors, corresponding to the velocity pressure measured by the pitot tube and gauge:

TABLE 7.5
VELOCITY PRESSURE FACTORS

VELOCITY PRESSURE	FACTOR
2 psi (13.8 kPa)	0.97
3 psi (20.7 kPa)	0.92
4 psi (27.6 kPa)	0.89
5 psi (34.5 kPa)	0.86
6 psi (41.4 kPa)	0.84
7 psi (48.3 kPa) or over	0.83

From Table 7.5, a flow of 6 psi (41.4 kPa) through a 4-inch (100 mm) outlet is indicated as 1,050 gpm (3 974 L/min). However, tests have shown that only 84 percent of this quantity is actually flowing due to voids in the water stream. Accordingly, actual flow is 1,050 x 0.84 = 883 GPM (3 974 x 0.84 = 3 338 L/min).

These formulas allow the computation of total flow from the flowing hydrants when performing an area fire flow test. They also would indicate the flow from the hydrant at the time of the test.

REQUIRED RESIDUAL PRESSURE

As a result of experience and water system analysis, fire protection engineers have established 20 psi (138 kPa) as the minimum required residual pressure when computing the available water for area flow test results. This residual pressure is considered enough to overcome friction loss in a short 6-inch (150 mm) branch, in the hydrant itself, and in the intake hose, as well as allowing a safety factor to compensate for gauge error. Many state health departments require this 20 psi (138 kPa) minimum to prevent the possibility of external water being drawn into the system at main connections. Pressure differentials can result in water main collapse or create cavitation, which is the implosion of air pockets drawn into pumps. A more

common occurrence is that pumpers working at these low system pressures may be pumping near the water main's capacity. If a valve on the pumper is shut down too quickly, a water hammer is created. This sudden surge in pressure may be transferred to the water main, resulting in damaged or broken mains or connections.

FLOW TEST PROCEDURES

When testing the available water supply, the number of hydrants to be opened will depend upon an estimate of the flow available in the area; a very strong probable flow requires several hydrants to be opened for a more accurate test. Enough hydrants should be opened to drop the static pressure by at least 10 psi (69 kPa). If more accurate results are required, the pressure drop should bring the residual pressure as close as possible to 20 psi (138 kPa). The flow available at 20 psi (138 kPa) can be determined by dropping the residual pressure to exactly 20 psi (138 kPa), by making a graphical analysis at any residual pressure, or by formula calculations.

Another problem that may be encountered is that water mains may contain such low pressures that no flow pressure will register on the pitot gauge. If this occurs, straight stream nozzles with smaller than 2½-inch (65 mm) orifices must be placed on the hydrant outlet to increase the flow velocity to a point where the velocity pressure is measurable (Figure 7.102). It should be noted that these straight stream nozzles will require an ad-

Figure 7.102 In cases where a poor stream of water is discharging from the hydrant, a nozzle may be added to the discharge to boost velocity pressure.

justment in the waterflow calculation that must include the smaller diameter and the respective coefficient of friction.

Flow tests are sometimes conducted in areas very close to the base of an elevated water storage tank or standpipe. This can result in flows that are quite large in gallons per minute (L/min). It should be realized that such large flows can only be sustained as long as there is sufficient water in the elevated tank or standpipe. It is advisable to conduct an additional flow test with the storage tank shut off to determine the quantity of water available when the storage has been depleted.

During a flow test, the static pressure and the residual pressure should be taken from a fire hydrant as close as possible to the location requiring the test results. This hydrant is commonly called the "test" hydrant. The "flow" hydrants are those hydrants where pitot readings are taken to find their individual flows. These readings are then added to find the total flow during the test.

In general, when flow testing a single hydrant, the test hydrant should be between the flow hydrant and the water supply source. In other words, the flow hydrant should be downstream from the test hydrant. When flowing multiple hydrants, the test hydrant should be centrally located relative to the flow hydrants. (NOTE: Water is actually never discharged from a test hydrant, rather, a capped gauge is placed on a discharge and the hydrant is opened fully.)

The procedure for conducting an available water test is as follows:

Step 1: Locate personnel at the test hydrant and at all flow hydrants to be used.

Step 2: Remove a hydrant cap from the test hydrant and attach the pressure gauge cap with the petcock in the open position. After checking the other caps for tightness, slowly open the hydrant several turns. Once the air has escaped and a steady stream of water is flowing, the petcock should be closed and the hydrant opened fully (Figure 7.103).

Step 3: Read and record the static pressure as seen on the pressure gauge.

Step 4: The individual at the flow hydrant(s) should remove the cap(s) from the outlet(s) to be flowed. When using a hydrant outlet, the hydrant coefficient and the actual inside diameter of the orifice should be checked and recorded. If a nozzle is placed on the outlet, its coefficient and diameter should be checked and recorded.

Step 5: Open flow hydrants as necessary and take and record the pitot reading of the velocity pressures (Figure 7.104). The individual at the test hydrant should simultaneously

Figure 7.103 After air has escaped from the hydrant, have the petcock closed and the hydrant opened fully.

Figure 7.104 Once the flow hydrant has been fully opened, record the pitot reading of the velocity pressure.

read and record the residual pressure (Figure 7.105). (NOTE: the residual pressure should not drop below 20 psi (138 kPa) during the test. If this happens, the number of flow hydrants must be reduced.)

Figure 7.106 Once the test is complete and the flow hydrant is turned off, check it for proper drainage.

Figure 7.105 While water is flowing from the flow hydrant, take a reading of the residual pressure from the test hydrant.

Step 6: Slowly close the flow hydrant to prevent water hammer in the mains. After checking for proper drainage, replace and secure all hydrant caps. Report any hydrant defects.

Step 7: Check the test hydrant for a return to normal operating pressure, then close the hydrant. The petcock valve should be opened to prevent a vacuum on the pressure gauge. Remove the pressure gauge. After checking for proper drainage, replace and secure the hydrant cap. Report any hydrant defects (Figure 7.106).

Flow Test Precautions

Certain precautions must be observed before, during, and after conducting flow tests in order to avoid injuries to those participating in the test or passersby. Efforts must also be made to minimize damage to property from the flowing stream. Both pedestrian and automobile traffic must be controlled during all phases of the testing. This may

require assistance from the local law enforcement agency. Other safety measures include tightening caps on hydrant outlets not being used, not standing in front of closed caps, and not leaning over the top of the hydrant when operating it. Property damage control measures include opening and closing hydrants slowly to avoid water hammer, not flowing hydrants where drainage is inadequate, and always remembering to check downstream to see where the water will flow. Since flowing water across a busy street could cause an accident, take proper measures beforehand to slow or stop traffic. Water should not be flowed during freezing weather unless proper drainage minimizes icing problems. A good rule to follow is this: when in doubt, do not flow! If problems exist in conducting a flow test, give thought to their solution so the test can be completed without disruptions or property destruction.

COMPUTING AVAILABLE WATERFLOW TEST RESULTS

There are three ways to compute waterflow test results: graphical analysis, mathematical formula, and nomograph. Each method of computation is discussed below.

Determining Available Water by Graphical Analysis

The waterflow chart in Figure 7.107 is a logarithmic scale that has been developed to simplify the process of determining available water in an area. The chart is accurate to a reasonable degree if one uses a fine-point pencil or pen when plotting results. The figures on the vertical and/or hori-

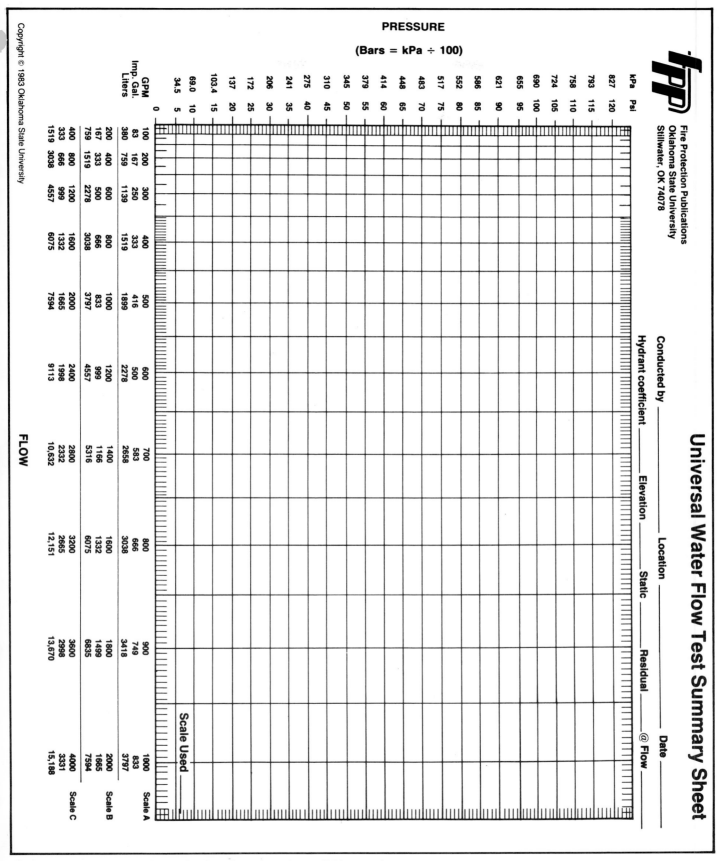

Figure 7.107 The waterflow chart makes it easier to determine available water for an area.

zontal scales may be multiplied or divided by a constant, as may be necessary to fit any problem.

The procedure for graphical analysis is as follows:

Step 1: Determine which gpm (L/min) scale should be used.

Step 2: Locate and plot the static pressure on the vertical scale at 0 gpm (0 L/min).

Step 3: Locate the total water flow measured during the test on the chart.

Step 4: Locate the residual pressure noted during the test on the chart.

Step 5: Plot the residual pressure above the total water flow measured.

Step 6: Draw a straight line from the static pressure point through the residual pressure point on the waterflow scale.

Step 7: Read the gpm available at 20 psi (138 kPa) and record the figure. This reading represents the total available water that can be relied upon.

Examples of graphical analysis are given below for waterflow tests using one and two outlets.

Example 1 (U.S.): One Outlet
Test Hydrant = 50 psi static and 25 psi residual
Flow Hydrant #1 = Using one 2½-inch outlet, with C = 0.9, pitot reading = 7 psi, and actual discharge diameter = 2.56 inches.
$$(29.83)\,(0.9)\,(2.56)^2\,(\sqrt{7}) = 466 \text{ gpm}$$

Flow Hydrant #2 = Using one 2½-inch outlet, with C = 0.8 pitot reading = 9 psi, and actual discharge diameter = 2.44 inches.
$$(29.83)\,(0.8)\,(2.44)^2\,(\sqrt{9}) = 426 \text{ gpm}$$

Total Water Flow = 466 + 426 = 892 gpm

Example 1 (Metric): One Outlet
Test Hydrant = 345 kPa static and 173 kPa residual
Flow Hydrant #1 = Using one 65 mm outlet, with C = 0.9 pitot reading = 48 kPa, and actual discharge diameter = 66.5 mm.
$$(0.0667766)\,(0.9)\,(66.5)^2\,(\sqrt{48} = 1\,841 \text{ L/min}$$

Flow Hydrant #2 = Using one 65 mm outlet, with C = 0.8 pitot reading = 62 kPa, and actual discharge diameter = 63.5
$$(0.0667766)\,(0.8)\,(63.5)^2\,(\sqrt{62}) = 1\,691 \text{ L/min}$$

Total Flow = 1 841 + 1 691 = 3 532 L/min

Figures 7.108 and 7.109 on pg. 182 show that the test results are plotted for graphical analysis of the water supply. The static pressure of 50 psi (345 kPa) is plotted at 0 gpm (0 L/min). The residual pressure of 25 psi (173 kPa) is above the total measured flow of 892 gpm (3 532 L/min), Scale A. (**NOTE:** It is important to understand that pitot pressures are never plotted on the graph; only the flow that corresponds to the pitot pressures is used). A line drawn through the static and residual pressure points now represents the water supply at the test location. It is easy to note that approximately 978 gpm (4 000 L/min) would be available at 20 psi (138 kPa). This figure represents the minimum desired intake pressure.

Example 2 (U.S.): Two Outlets (Figure 7.110 on pg. 183)
Test Hydrant = 90 psi static and 50 psi residual
Flow Hydrant = Using two 2½-inch outlets, with each C = 0.9, pitot reading for each is 17 psi, and an actual diameter of 2.56 inches.
$$(29.83)\,(0.9)\,(2.56)^2\,(\sqrt{17}) = 725 \text{ gpm x two}$$
outlets = 1 450 gpm
Total Water Flow = 1 450 gpm

Example 2: Two Outlets (Metric) (Figure 7.111 on pg. 184)
Test Hydrant = 621 kPa static and 345 kPa residual
Flow Hydrant = Using two 65 mm outlets, both with C = 0.9, pitot reading for each = 117 kPa, and the actual diameter = 66.5 mm.
$$(0.0667766)\,(0.9)\,(66.5)^2\,(\sqrt{117}) = 2\,875 \text{ L/min x}$$
two outlets = 5 750 L/min
Total Water Flow = 5 750 L/min

This example shows that the waterflow scale must be changed so a line can be drawn down to the 20 psi (138 kPa) level. The available water rate at 20 psi (138 kPa) in this case would be approximately 1,970 gpm (7 450 L/min).

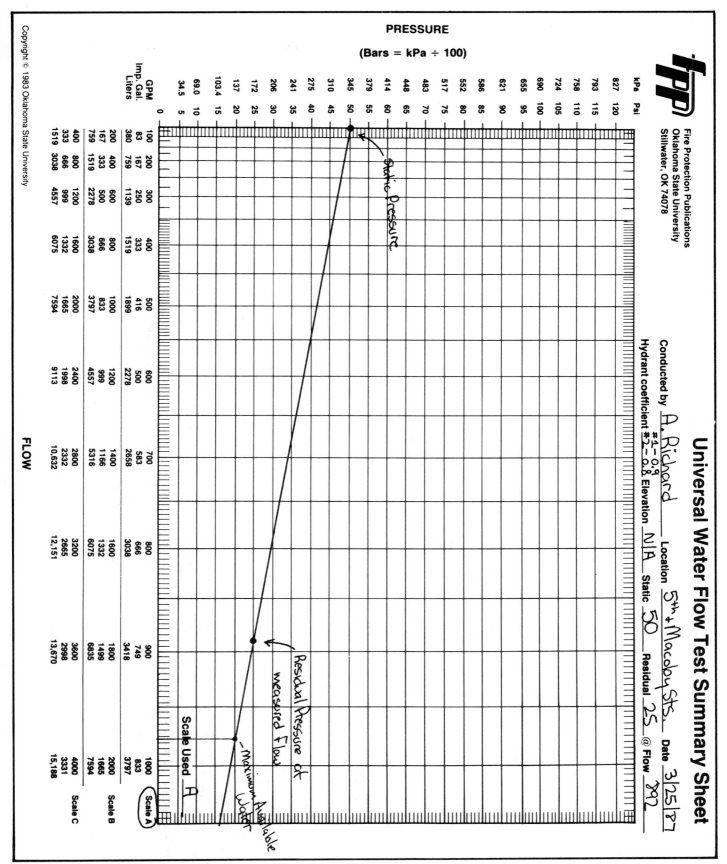

Figure 7.108 A graphical analysis of Example 1 (U.S.).

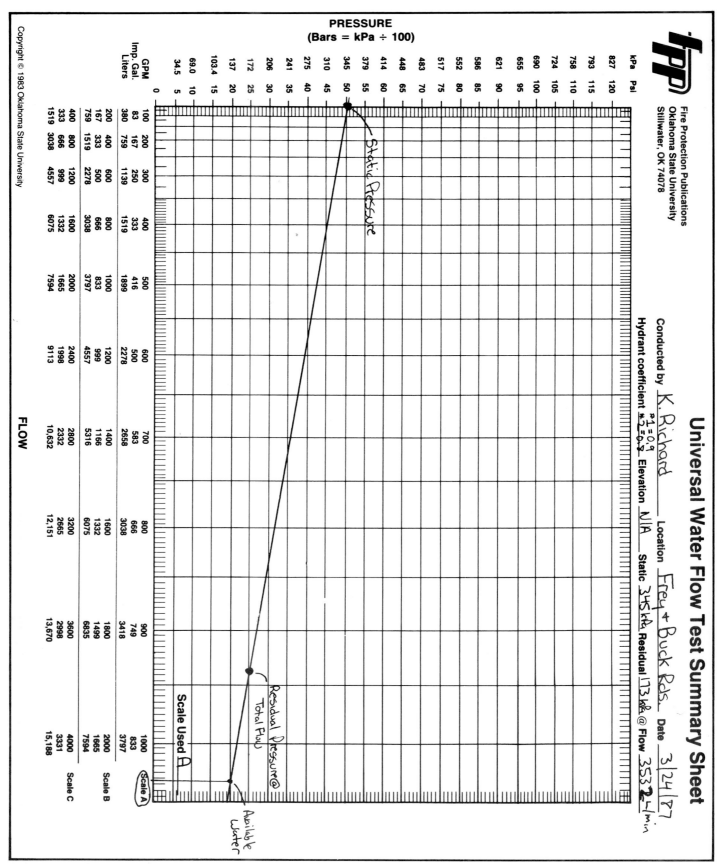

Figure 7.109 A graphical analysis of Example 1 (metric).

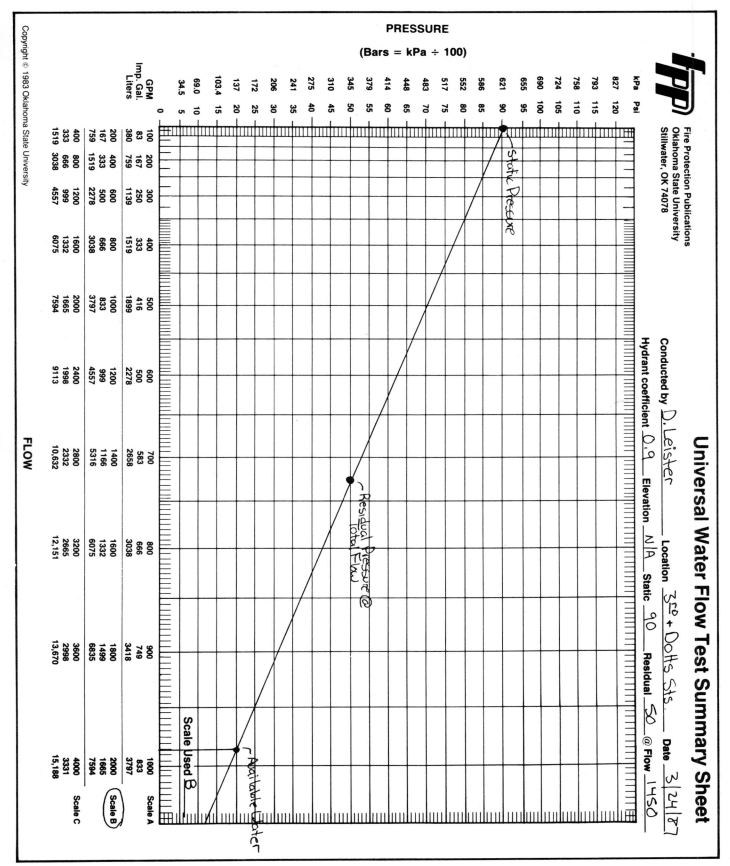

Figure 7.110 A graphical analysis of Example 2 (U.S.).

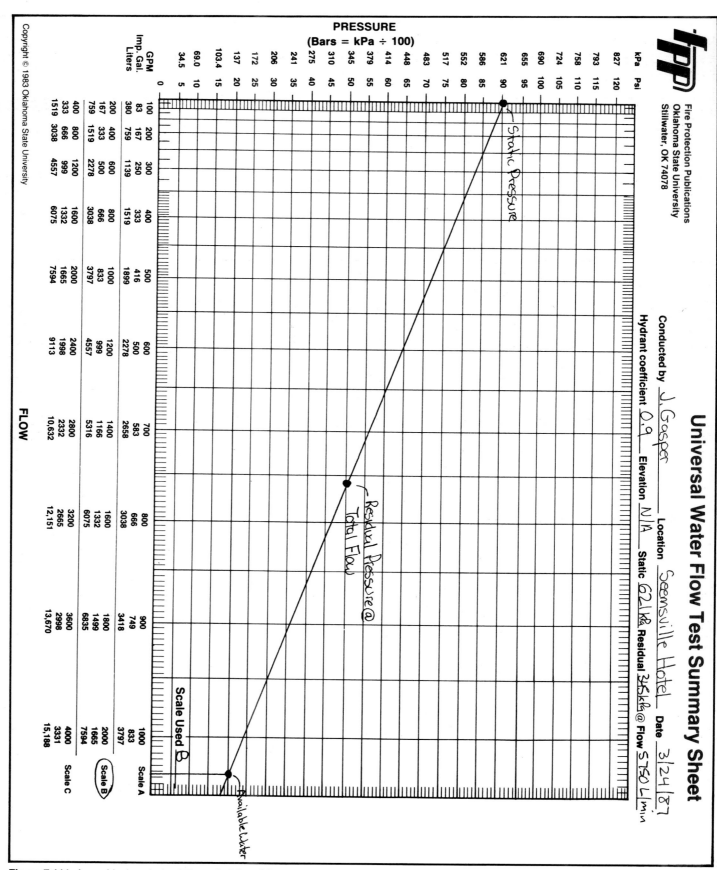

Figure 7.111 A graphical analysis of Example 2 (metric).

Determining Available Water by Mathematical Method

A variation of the Hazen-Williams formula for determining available water is written as follows:

$$Q_r = \frac{Q_f \times h_r^{0.54}}{h_f^{0.54}}$$

Where Q_r = Flow available at desired residual pressure

Q_f = Flow during test

h_r = Pressure drop to residual pressure (normal operating pressure minus required residual pressure)

h_f = Pressure drop during test (normal operating pressure minus residual pressure during flow test)

The values for h_r or h_f to the 0.54 power are listed in Table 7.6.

Using the values from the first example, in addition to a normal operating pressure of 55 psi:

Q_f = 892 gpm

h_r = 55 psi — 20 psi = 35 psi

h_f = 55 psi — 25 psi = 30 psi

Under h at 35, $h^{0.54}$ = 6.82. Under h at 30, $h^{0.54}$ = 6.28.

So, Q_r = 892 $\frac{\times\ 6.82}{6.28}$

Q_r = 967 gpm

TABLE 7.6
VALUES FOR COMPUTING FIRE FLOW TEST RESULTS

h	$h^{0.54}$	h	$h^{0.54}$	h	$h^{0.54}$	h	$h^{0.54}$	h	$h^{0.54}$	h	$h^{0.54}$	h	$h^{0.54}$
1	1.00	26	5.81	51	8.36	76	10.37	101	12.09	126	13.62	151	15.02
2	1.45	27	5.93	52	8.44	77	10.44	102	12.15	127	13.68	152	15.07
3	1.81	28	6.05	53	8.53	78	10.51	103	12.22	128	13.74	153	15.13
4	2.11	29	6.16	54	8.62	79	10.59	104	12.28	129	13.80	154	15.18
5	2.39	30	6.28	55	8.71	80	10.66	105	12.34	130	13.85	155	15.23
6	2.63	31	6.39	56	8.79	81	10.73	106	12.41	131	13.91	156	15.29
7	2.86	32	6.50	57	8.88	82	10.80	107	12.47	132	13.97	157	15.34
8	3.07	33	6.61	58	8.96	83	10.87	108	12.53	133	14.02	158	15.39
9	3.28	34	6.71	59	9.04	84	10.94	109	12.60	134	14.08	159	15.44
10	3.47	35	6.82	60	9.12	85	11.01	110	12.66	135	14.14	160	15.50
11	3.65	36	6.93	61	9.21	86	11.08	111	12.72	136	14.19	161	15.55
12	3.83	37	7.03	62	9.29	87	11.15	112	12.78	137	14.25	162	15.60
13	4.00	38	7.13	63	9.37	88	11.22	113	12.84	138	14.31	163	15.65
14	4.16	39	7.23	64	9.45	89	11.29	114	12.90	139	14.36	164	15.70
15	4.32	40	7.33	65	9.53	90	11.36	115	12.96	140	14.42	165	15.76
16	4.47	41	7.43	66	9.61	91	11.43	116	13.03	141	14.47	166	15.81
17	4.62	42	7.53	67	9.69	92	11.49	117	13.09	142	14.53	167	15.86
18	4.76	43	7.62	68	9.76	93	11.56	118	13.15	143	14.58	168	15.91
19	4.90	44	7.72	69	9.84	94	11.63	119	13.21	144	14.64	169	15.96
20	5.04	45	7.81	70	9.92	95	11.69	120	13.27	145	14.69	170	16.01
21	5.18	46	7.91	71	9.99	96	11.76	121	13.33	146	14.75	171	16.06
22	5.31	47	8.00	72	10.07	97	11.83	122	13.39	147	14.80	172	16.11
23	5.44	48	8.09	73	10.14	98	11.89	123	13.44	148	14.86	173	16.16
24	5.56	49	8.18	74	10.22	99	11.96	124	13.50	149	14.91	174	16.21
25	5.69	50	8.27	75	10.29	100	12.02	125	13.56	150	14.97	175	16.26

Metric Example:

$Q_r = 3\ 532$ L/min

$H_r = 380$ kPa $- 138$ kPa $= 242$

$H_f = 380$ kPa $- 173$ kPa $= 207$

NOTE: When doing these problems in metrics, quite often the figures obtained will be higher than those provided in Table 7.6. It will be necessary to use a calculator to determine $h^{0.54}$.

$$Q_r = 3\ 532 \times \frac{19.38}{17.81}$$

$$Q_r = 3\ 843 \text{ L/min}$$

Determining Available Water by Nomograph

The flow available at any desired residual pressure can also be determined by nomograph (Figure 7.112). The nomograph shown here is based on the Hazen-Williams derivation used in the mathematical method of determining available water.

The procedure for determining available water by nomographic analysis is as follows:

Step 1: Using a straightedge, connect points h_r and h_f. Using a sharp point pencil or pen, mark the intersection on line S.

Step 2: Rotate the straightedge about the intersection point on S until it is in line with the observed flow on the Q_f scale.

Step 3: Make the intersection of this line on the Q_r scale. This Q_r point is the available flow at the assumed pressure drop.

Again using the data from the mathematical example, the following results are obtained: Lining up h_r and h_f, locate the intersection point on line S. Rotate the straightedge about this point until it lines up with the observed flow (891.7 gpm [3 375.4 L/min]). The available flow is determined by extending the line to the Q_r scale (970 gpm [3 660 l/min]).

Any of the three methods described above will enable individuals to determine available water flow resulting from the test as well as available flow at any system pressure. With this information, several questions can be answered:

- Does the water system meet the needed fire flow for the occupancy or area?

- Is the available system flow adequate to provide sprinkler system demands in the area?

- Where will water system improvements be needed?

Each type of determination method has its advantages. Graphical analysis provides a visual graph that indicates system capability at any pressure. The nomograph is often useful for field measurements and provides close approximations. Mathematical determination provides the most exact answer for any given situation and may be desirable for use in computer applications where large amounts of flow test data must be analyzed. Fire departments should choose their analysis method based on individual needs. (**NOTE:** Each of the available waterflow determination methods gives nearly identical answers. Discrepancies found in some of the metric examples are due to the rounded off factors used currently as well as those in use during the printing of the Universal WaterFlow Test Summary Sheet.)

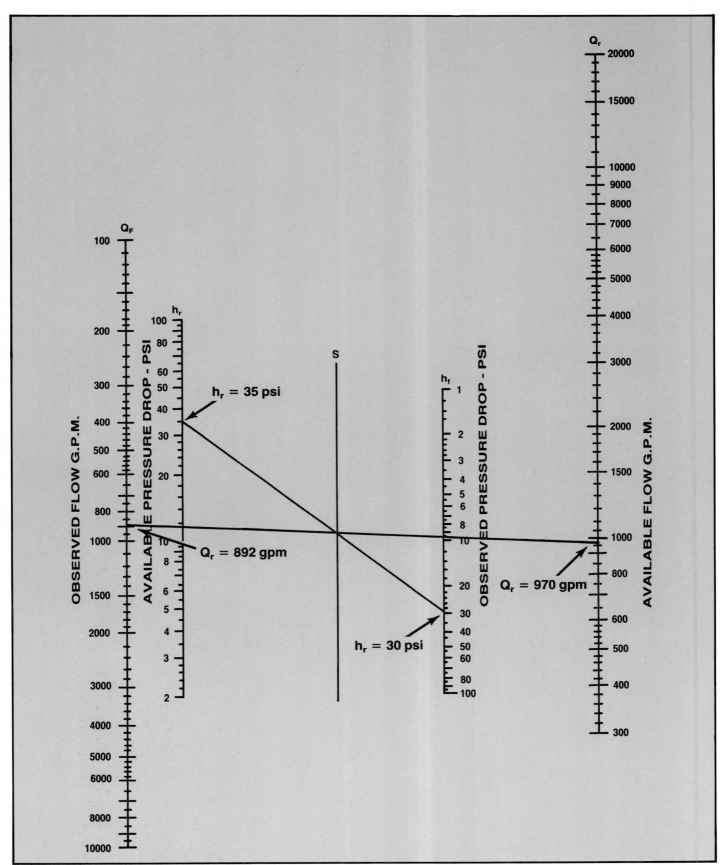

Figure 7.112a A nomograph can also be used to determine the flow available at a desired residual pressure.

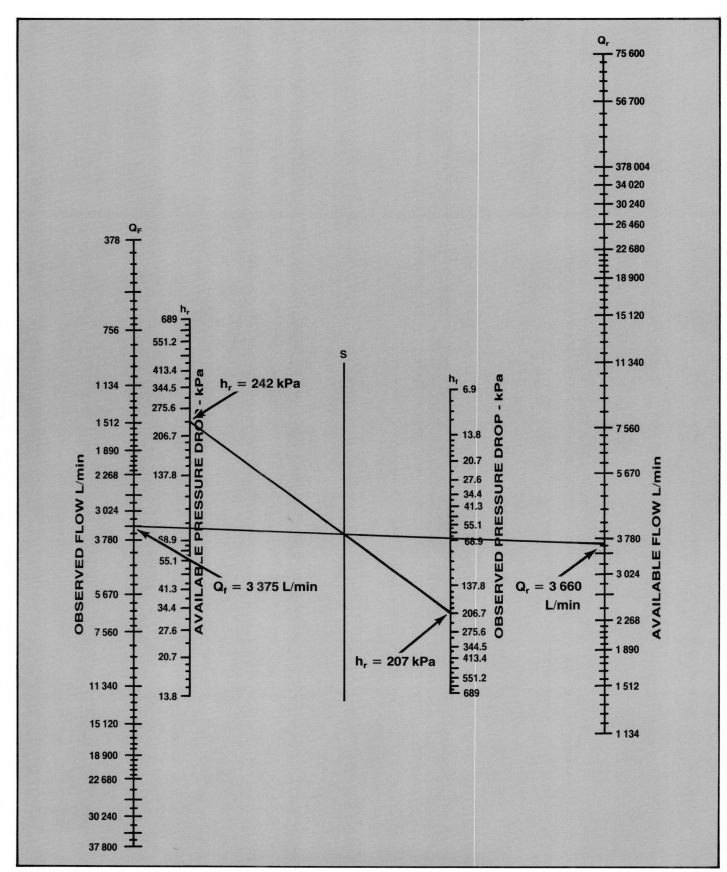

Figure 7.112b A nomograph using metric figures.

8

Plans Review

10-90

This chapter provides information that addresses performance objectives described in NFPA 1031, *Standard for Professional Qualifications for Fire Inspector* (1987), particularly those referenced in the following sections:

Fire Inspector II

4-9.2 Plans and Specifications.

4-9.2.1

4-9.2.2

Chapter 8
Plans Review

Many fire agencies do not have plans review authority; therefore, code enforcement often is a reactive process, beginning after a building is constructed and ready for occupancy. Many of the problems encountered under this system can be avoided if code enforcement is begun *before* construction — at the plans review stage. To begin a proactive system of code enforcement, the fire department and the building inspection agency that has plans review authority must establish a cooperative relationship. With this type of system, the fire service plans review process can be incorporated into the process of applying for a building permit.

The obvious advantage of establishing a plans review process is that it enables the fire service reviewer to point out discrepancies before construction begins. Correcting these problems prior to the start of construction improves the efficiency and cost effectiveness of the project. This will also improve the fire department's image with developers, which will help in forming harmonious relations between the two in future contacts.

When no fire protection engineer is employed for plans review, the task is given to fire inspectors, who must be well prepared before attempting to examine any plans. This preparation requires that inspectors have a thorough knowledge of the applicable codes and knowledge of how to use them. Just as hoselines and a water supply are basic tools of suppression forces, building codes and their application are the basic tools of the plans reviewer.

In addition to having a thorough knowledge of codes, fire inspectors must develop a good working relationship with all individuals and agencies involved in constructing the building. These persons and agencies may include government and regulatory agencies, architects, contractors, and the individuals for whom the building is being constructed.

Fire inspectors may also want to acquaint fire suppression forces with the proposed building project. By looking at the plans, firefighters can provide feedback about difficulties that may be encountered during fire suppression activities (Figure 8.1).

REVIEWING ARCHITECTURAL DRAWINGS

The plans and specifications of buildings are shown in working drawings made by engineers, architects, draftsmen, or inspectors for the purpose of

Figure 8.1 One of the inspector's responsibilities is to ensure that the fire department is informed of potential hazards associated with a particular occupancy. *Courtesy of Springfield, Illinois Fire Department.*

communicating construction of various building types (Figure 8.2). There are four main views of working drawings:

- The Plan View
- The Elevation View
- The Sectional View
- The Detailed View

Each drawing has a title block that contains specific information about the drawing and the project (Figure 8.3). The format, or location, of the information is determined by the designers, although the information is the same. The title block always contains the following information:

1. The title of the drawing, such as ELEC-TRICAL FLOOR PLANS — 3RD AND 4TH FLOORS, the site plan, foundation plan, or floor plan.

2. The description of the project, such as CORDELL NORTH RENOVATION.

Figure 8.2 Blueprints communicate specific construction details.

3. The scale of the drawing. For example, ⅛" = 1'0" (⅛-inch drawn on the paper is equal to 1 foot of actual construction). Scaling the drawing provides a view in exact proportion to the actual size of the building. Sometimes the scale is placed on the drawing rather than in the title block.

4. The revisions block dates the drawing and any revisions (if applicable) for distribution to various people involved on the project.

5. The name of the firm producing the drawings, such as ARCHITECTURAL SERVICES.

6. The name (usually initials) of the person performing the drafting, and the person responsible for checking the final drawings.

7. The sheet number in the set. The first number is the drawing number, and the second number denotes the total number of sheets in the set, for example E8 OF 9 SH.

8. If the plans were drawn by a licensed architect, the title block should have the architect's official stamp as issued by a state architectural registration board. The registered architect in this title block is David Scheirman, 4715.

Large construction projects are usually divided into four or more basic areas, such as architectural, structural, mechanical, and electrical. Sheets containing a specific type of information are marked accordingly. For example, architectural information would be marked A1, A2, A3, structural sheets S1, S2, S3, mechanical sheets M1, M2, M3, electrical sheets E1, E2, E3, and so on.

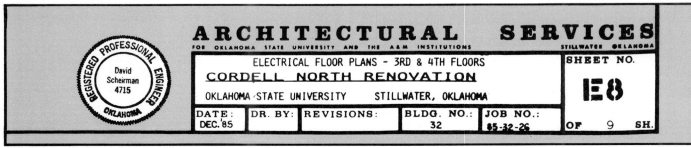

Figure 8.3 A title block provides information about the drawing.

The Plan View

The plan view is a two-dimensional view of the site or the building as seen from directly above the area. A plan view is extremely useful in giving information concerning the overall layout of the site or building. The fire inspector will commonly review three basic plan views: the site plan, the basement plan, and the floor plan.

The site, or plot, plan is usually the first sheet of a set of drawings (Figure 8.4). It identifies conditions currently existing on the site and relates information needed to locate the building. This information includes the north direction, lot dimensions, utility lines (gas, water, sewer, power, and telephone), grade level, contour lines, structures to

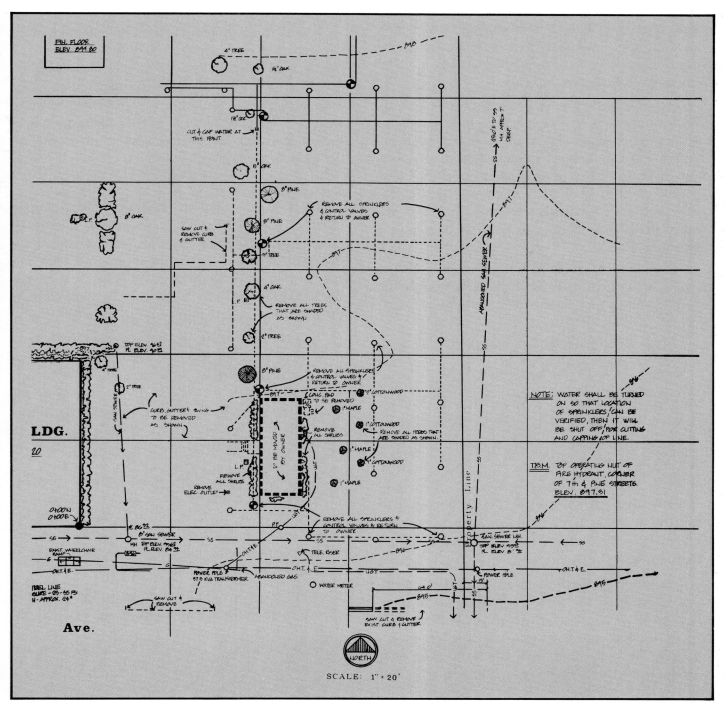

Figure 8.4 The site plan is a view of the proposed construction site as seen from above.

be removed, concrete walks and driveways, and areas to be landscaped.

The north direction symbol usually points toward the top of the sheet. This conforms to universal standards used in the drawings of all maps.

The lot dimension is often shown as a broken line with the dimensions expressed in feet and decimal fractions of a foot (meters and millimeters). The property lines are legal instruments that define the land boundaries for a particular site. Most building codes base fire separation on the location of property line versus the location of the building in relation to that line.

The placing of the building on a site plan is called "dimensioning." The objective of dimensioning is to *clearly* place the building on the site. This is usually accomplished by placing the dimensions some distance away from the shapes on the drawing through the use of extension lines. (**NOTE:** this can be accomplished without extension lines.) Dimensions are expressed in feet and inches (meters and millimeters).

The location of service lines (water, sewer, power, gas, and so on) is usually shown by a broken line marked at intervals with a letter. For example:

Water lines ---w---w---w---w---w---w---w---w

Sewer lines --ss--ss--ss--ss--ss--ss--ss--ss--ss

Power lines ---p---p---p---p---p---p---p---p

Gas lines ---g---g---g---g---g---g---g---g

In addition to the character in the broken line, a legend on the map also identifies the type of service line.

Structures or objects that must be removed from the area are also included on the site plan. These items are drawn with a broken line. Brief identification is also permitted.

Contour lines display the existing grade elevations and grade lines show the planned elevations after grading is completed. The numbers are given in feet and decimal fractions of a foot (or meters and decimal fractions of a meter) and are related to a benchmark. A benchmark is the point from which a surveyor begins measuring. It is usually a stationary object, such as a fire hydrant or man-

hole, and will be noted on the site plan. The grade level usually appears in a grid pattern that provides a total picture of the finished grading on the site.

The location of sidewalks, driveways, and graveled areas are not usually shown, except when they are included in a builder's contract. When they are included, their sizes and locations are relative to the lot lines or building dimensions.

Area landscaping is normally started when construction is complete or nearly so. Areas to be landscaped are often shown on site plans to illustrate the total concept created by the architect. The location of trees and other potential obstructions should be checked to ensure that landscaping does not block fire department connections or aerial ladder placement.

THE BASEMENT PLAN

The basement plan shows the belowground view of the building. The thickness and external dimensions of the basement walls are given, as are floor joist locations, strip footings, and other attached foundations. Since footings are hidden under the slab or the ground, they are displayed by broken lines (Figure 8.5).

If the construction project is a small-scale commercial project, the location of doors and windows, if any, drains, furnace, hot water tank, chimney, taps, and so forth are shown in the basement plan. These items would be shown separately in large construction projects.

THE FLOOR PLAN

The floor plan, like the basement plan, is viewed from the top. The floor plan provides information for constructing external walls, internal partitioning, doors and windows, ceiling and roof joists or trusses, cabinets, closets, shelving, electrical outlets, and fixtures (Figure 8.6 on pg. 196).

The usual symbol used for exterior walls is a pair of parallel lines. The materials from which the wall will be constructed are sometimes listed on a floor plan; usually, however, construction materials are listed in a section view of the wall.

The interior walls (partitions) are also drawn with parallel lines. The function of each room is listed on the sheet where the room is located. Examples of these are office numbers and proposed uses. When examining plans for code room size requirements, an allowance must be made for the thickness of an interior wall. Codes typically specify the rooms that are to be measured from the surface of a finished wall.

The symbol for a standard hinged door is a single line drawn from the hinge side of the doorway. An arc indicates the swing of the door. Codes also require doors to be installed in certain locations and to swing in a certain direction.

The windows and doors on the drawings are coded by letters enclosed in a circle. Window sizes

are found on some floor plans, but are usually found in elevation views.

Elevation View

The elevation view is a two-dimensional view of the building as seen from the exterior (Figure 8.7 on pg. 197). There are four exterior elevation views on any rectangular building and they can be labeled in one of two ways:

Method One: If a wall is facing north, it is labeled the north elevation. The other views are then labeled according to the direction they face.

Method Two: The front elevation faces the front of the street and the rear elevation faces the

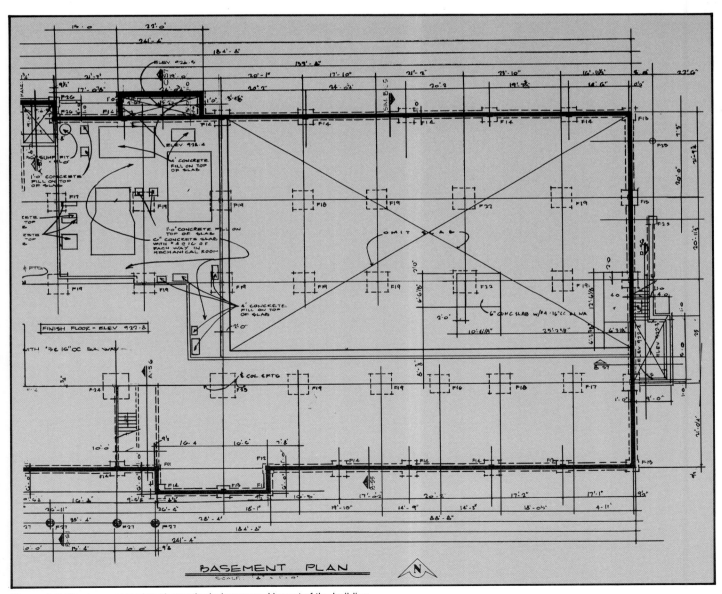

Figure 8.5 The basement plan shows the belowground layout of the building.

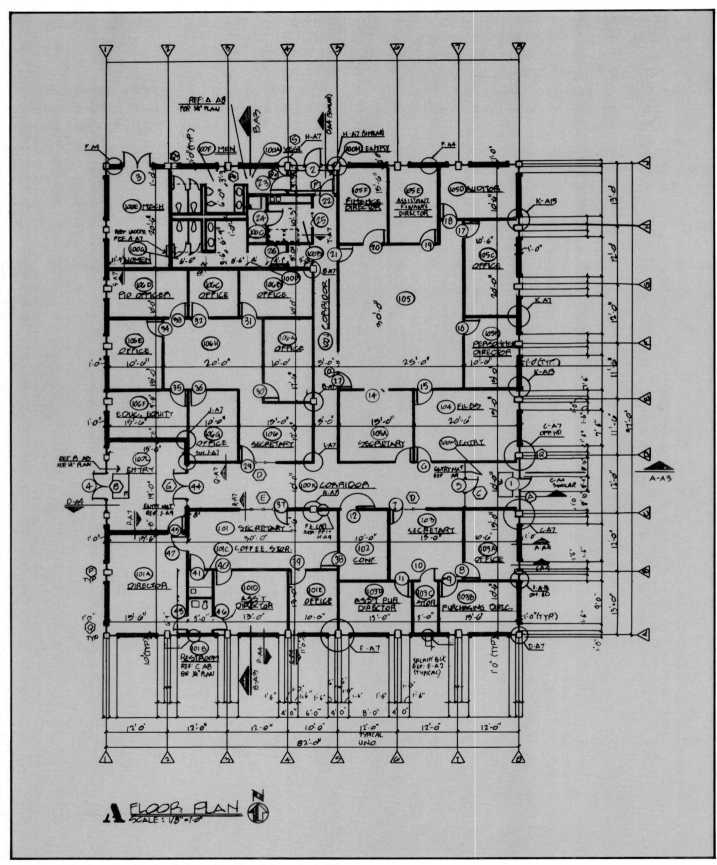

Figure 8.6 The floor plan is a detailed top view of each floor of a building.

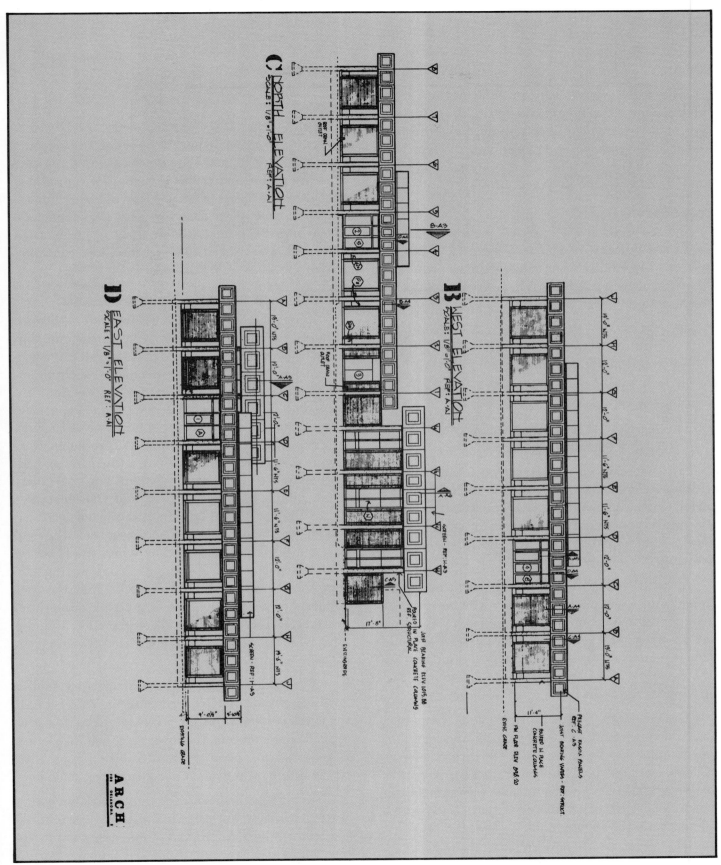

Figure 8.7 The elevation view is a two-dimensional view of the entire building.

back. When viewed from the front of the building, the left elevation is on the left side of the building and the right elevation is on the right side of the building.

Elevation views contain information about the exterior components of the building such as type of exterior finish or combinations of finishes, and locations, sizes, and types of doors and windows. As in the site plan, an elevation view of the grade level is shown as a dark line.

Sectional View

A sectional view is a vertical view of a building as if it were cut into two parts (Figure 8.8). The purpose of a sectional view is to show the internal construction of each assembly. There are three types of cross-section views:

- *Horizontal* — Cross section of the complete building from exterior wall to exterior wall and from foundation to roof.

- *Typical* — Cross section of a common construction feature, such as a wall.

- *Detail* — Cross section of some construction feature to show exact detail.

The exterior wall construction detail is found in the sectional view. The details include the exterior finish, sheathing, studs, plates, insulation (thermal), vapor barrier, interior finish, and base detail. The materials used are often listed on the drawing.

Interior walls, including interior finish, insulation (sound), studs, and plates can be determined from the cross section and are generally listed on the drawing. The process of listing the materials used in constructing a wall is referred to as "calling up" the wall. This same procedure is used to "call up" floor systems and ceiling/roof assemblies.

Construction of the finished floor, subfloor, size and type of floor joists, bridging and location, and floor support beams are also shown on the typical cross-section view. Roof pitch, roofing, eave projection, sheathing, eave troughing, soffit and fascia detail, roof structure (truss or rafters), and insulation may also be shown.

The foundation wall height and thickness, width and thickness of footings (pad and strip), lo-cation and size of weeping tile, location and depth of granular fill, thickness of floor slabs, location of expansion joints, and protective coatings should all be included on the cross section.

Detailed View

Sometimes it becomes necessary to show a feature in a larger scale because enough information cannot be crowded into the space of the small-scale drawings. This additional information is provided with detailed views (Figure 8.9 on pg. 200). For example, a floor plan drawing may show that one corner of a room has built-in shelving. A detailed view will be required to show the exact construction of the shelves themselves.

DEVELOPING A PLANS REVIEW SYSTEM

In developing a plans review system, fire inspectors should first identify typical problems that could be prevented if a plans review system were in place. Examples of these problems include inadequate water supply for fire protection, insufficient exits, restricted access for emergency vehicles, and use of undesirable building materials. After they have clearly defined the problems, they should formulate and write down the goals and objectives of the plans review system. The goals state what should be accomplished and the objectives define the operational guidelines needed to attain the goals. A well-written objective states who is to perform what function and by when the function should be performed. Standards for evaluating performance should also be indicated. The standard should indicate the average plan turnaround time, procedures for correcting discrepancies, and the amount of time that will be dedicated to reevaluating corrected plans.

After initial planning, fire inspectors should develop methods for coordinating the efforts of all the agencies that take part in the plans review process. To have an effective program, fire inspectors must maintain good working relationships with the building, planning, and zoning departments.

Fire inspectors should develop a written policy stating which agency has the authority to enforce code regulations during the plans review stage. As previously pointed out, some fire agencies do not have enforcement authority and may have to con-

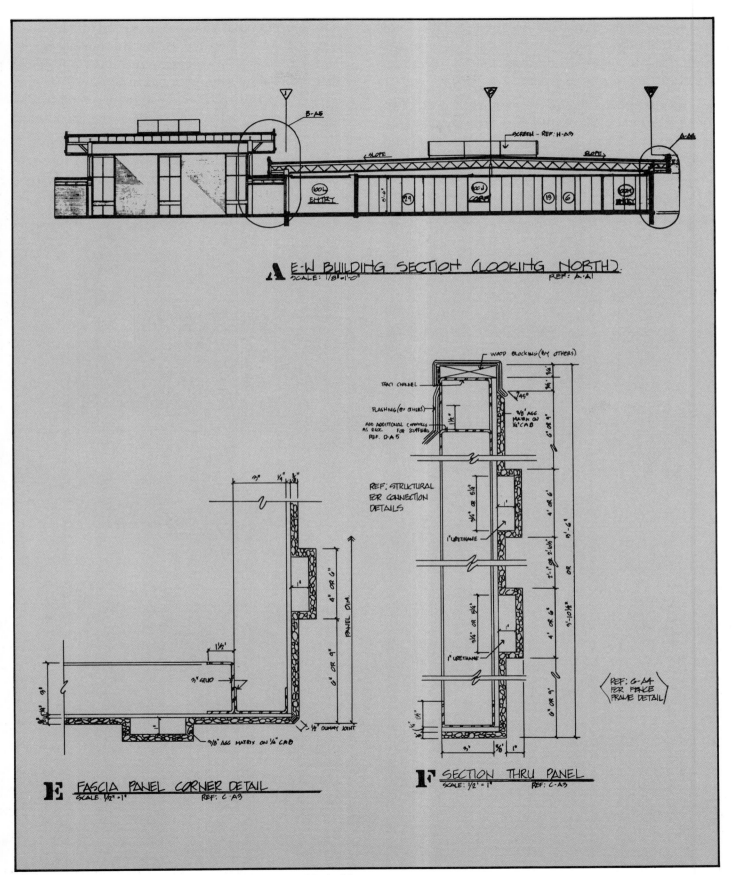

Figure 8.8 The sectional view displays the internal structure of the building.

duct plans reviews on an advisory basis. This is not an effective way to ensure that desired changes are made. It is important that a legal method of enforcing requirements be in place.

Fire inspectors should establish a procedure that spells out the minimum requirements for submitting plans for review. The requirements should clearly define the information to be provided before the plans will be reviewed. The agencies involved in the plans review process should decide how many sets of plans are to be submitted and how the plans will be distributed to the appropriate agencies for review. If possible, requirements for submitting the plans should be put in booklet form and distributed to all architects, engineers, contractors in the area, and other contractors who are named to area building projects.

Good record-keeping procedures are an essential part of any plans review program. Both formal and informal communications should be

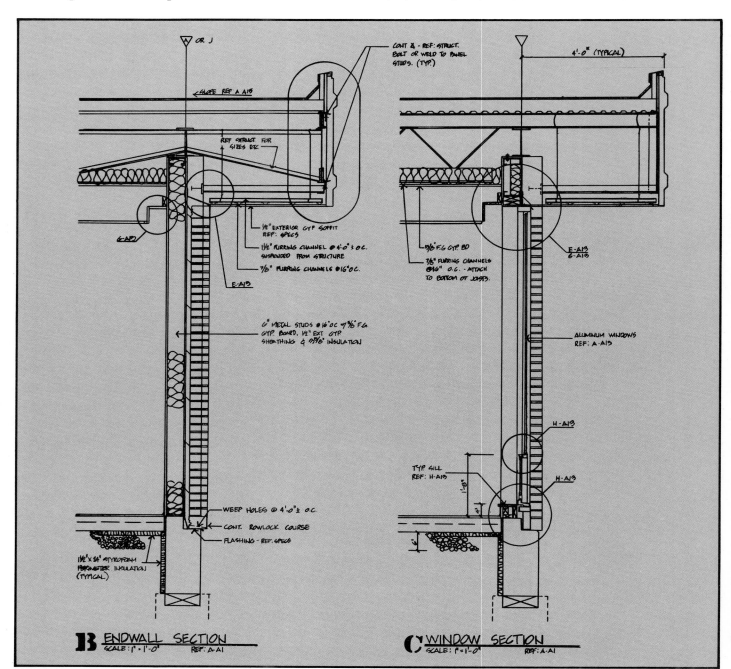

Figure 8.9 A detailed view displays exact construction details of a particular area of a structure.

documented because fire inspectors may need these records for a performance evaluation, legal action, or future reference. Informal communications include telephone conversations and discussions at meetings while formal communications include plans review letters, drawings, and specifications. All correspondence should be reviewed and properly filed. In addition to organizing correspondence, the record-keeping system should provide a means for fire inspectors to quickly obtain and refer to drawings and specifications.

While planning the record-keeping system, fire inspectors should develop the format for a plans review letter. The plans review letter should be clear, concise and should contain the following information:

- Reviewer's name
- Name of the project
- Date review is started and completed
- Scope of work to be performed
- Description of the building and fire protection features to be provided
- Applicable codes that must be followed
- Comments as a result of the plans review
- Specific code references for each comment

Those who carry out plans review need equipment and reference materials. These include an architect's scale, a calculator having scientific functions, a well-lit drawing table, the applicable code books, and design manuals. Code books and design manuals can be obtained from the model code organization, trade association, manufacturers, and independent testing laboratories.

Developing a plans review system requires advance planning and cooperation between agencies. Fire inspectors need flexibility and creativity to develop effective systems that meet the needs of each jurisdiction. After fire inspectors establish a system, they should initiate a public information program targeted at those who are most affected — architects, engineers, contractors, and developers. They should explain the system to them in detail and encourage comments. These groups are more apt to cooperate if they are familiar with the sys-

tem, have input into the system's design, and know what their responsibilities are.

Reviewing building plans for code compliance requires an examination of several areas: site plans, structural drawings, mechanical systems, electrical systems, and fire protection features. The following sections highlight some of the design and safety elements that fire inspectors need to check during the review process.

REVIEWING SITE PLANS

Site plans provide a view of the proposed construction in relation to existing topographical conditions. Fire inspectors should pay close attention to accessibility, water supply, and general building parameters.

Accessibility refers to the ability of fire department apparatus to get close enough to the building during an emergency. The following is a list of questions that fire inspectors should answer concerning building accessibility:

- Are the fire department access roadways wide enough, and do they have an adequate road base to permit apparatus to operate and to pass?
- Are there any physical, topographical, or architectural obstructions that may restrict the access of emergency vehicles?
- Are the intersections looped to permit traffic flow and is an adequate turning radius provided?
- Are there enough access roads and are they within the minimum building access distance requirements?
- Are surrounding grades too steep?

The site plans review also involves an assessment of the proposed water supply system. Fire inspectors should determine if water is available in sufficient quantities and pressures to satisfy potential emergency demands. After inspectors have determined the minimum fire flows, they can verify the adequacy of the water supply by examining the type and location of hydrants (both public and private), pump systems, cross-connected water supplies, and the location and size of public water mains. Records of waterflow testing performed in the area in question will be invaluable, assuming they are recent and accurate. It is often necessary to use hydraulic calculations to verify the adequacy of water supplies.

Last, fire inspectors should review the general building parameters: location, occupancy, fire and land-use zones, property line clearances, and special hazards. It is important to determine the use of the building in order to assign an occupancy classification. The occupancy classification often dictates the total allowable area, maximum height, exposure protection, occupant load, necessity for fixed fire protection, and other factors unique to each building. Most cities have fire and land-use zones; therefore, fire inspectors must check to see that the building is located in the proper zones. During the exterior wall examination, the reviewer should determine if there is adequate property clearance to prevent fire spread. Finally, fire inspectors must identify any special hazards, such as tank installation or high-piled stock, and check for any corrections that are required.

REVIEWING STRUCTURAL DRAWINGS

Structural drawings indicate the way in which the building is to be built, what materials are to be used, and how the building is intended to be used (Figure 8.10). Fire inspectors reviewing structural drawings must determine if the building can be built and used as proposed. Building size, occupancy load and class, construction classification, exit systems, compartmentation, and other assorted considerations are all factors in this determination. Since the inspector's task is to verify the information that the architect has provided, the construction documents should contain all of the needed facts.

Reviewing the overall size of the building in terms of height and area should be the first step. This will provide information that will be essential in carrying out later steps.

The code requirements concerning building height limits were initially developed to ensure that fire hose streams could reach upper-story fires, that occupants would be able to exit the build-

Figure 8.10 Architectural drawings should be reviewed to determine the type of construction, intended use, and size of the building. *Courtesy of Edward Prendergast.*

ing swiftly, and that the building not pose a hazardous exposure to surrounding buildings.

Today, building height restrictions are based on the occupancy classification and construction of the building. When determining the height of a building, fire inspectors must understand how the jurisdictional code treats the following three factors: lowest point to consider, highest point to consider, and automatic sprinkler protection. Various codes handle sloping grades, basements, parapet walls, and penthouses differently when determining the lowest or highest points. In addition, many codes permit an additional story if complete sprinkler protection is provided.

Similar to height limitations, area limitations are intended to ensure effective fire suppression. Such factors as the capabilities of the fire department water supply, the community's layout, fire risk, and climatic conditions should be considered during an evaluation of area limitations. Each of the model codes define allowable area differently; therefore, fire inspectors should use the definition from the appropriate code. The net floor area within the building will also be used in determining the maximum occupant load allowed in the structure or a certain part of the structure.

Allowances for additional areas also vary among codes. Area increases or decreases are based on automatic sprinkler protection, accessible building perimeter, and the distance between buildings. These increases and decreases are usually expressed in percentages.

The architect should determine an occupancy classification for the proposed building and identify the intended use of each room or section of the building. Fire inspectors should evaluate the fire and life hazards throughout the building to verify that the occupancy classification is correct. It is extremely important for the reviewer to be familiar with the applicable codes in order to make accurate judgments about mixed occupancies and special use requirements.

The various classifications for building construction consider only the factors necessary to define the building types. As with occupancy classification, the intended construction should be identified on the drawings. To verify the construc-

tion classification, fire inspectors should refer to the appropriate fire-resistance rating lists. (**NOTE:** Refer to Chapter 5, Building Construction for Fire and Life Safety for more information about construction classifications.)

After verifying that the project can be built, inspectors should determine the occupant load before evaluating the life safety and fire protection features of the building. Fire inspectors can use the occupant density or occupant load factors given in the codes to determine the occupant load. Often, the architect is required to provide the occupant load calculations for fire inspectors to verify.

After fire inspectors have determined the occupant load, they should evaluate the means of egress. It is the reviewer's responsibility to identify the various components of the exit system and to verify that each means of egress provides a continuous path of travel to a place of safe refuge. Typically, codes require exits to be separated from all other parts of the building by construction having a specified fire-resistance rating. The architect should provide details of the separation walls as well as door assembly information. Fire inspectors should verify that there are no unnecessary openings, all penetrations are sealed, and the fire-resistance ratings of the doors are correct. Separation from the exit discharges and exit accesses must also be verified. For more information about both occupant load and means of egress, see Chapter 6 Occupancy Classifications and Means of Egress.

Fire inspectors should evaluate other related code requirements applying to means of egress such as those concerned with interior finish, headroom, elevation changes, and obstructions. Means of egress should be highlighted on the architectural, mechanical, and electrical drawings so inspectors can detect penetrations or other code violations easily.

Once the means of egress have been identified and the acceptability of each component has been verified, the reviewer should evaluate the capacity of the means of egress. Fire inspectors must evaluate the exit capacity according to the specifications of the appropriate codes. The architect may be required to provide the exit capacity calculations so that the inspector can verify that the exit capacity is sufficient. When determining the adequacy of exit signs, the inspector should examine the following components: location, spacing, color, illumination, design, and size. Individual codes will outline the requirements for exit signs in terms of those components.

Another aspect of fire protection to be examined during the architectural drawing review is building compartmentation. Building compartmentation includes fire barriers, smoke barriers, the protection of vertical openings and concealed spaces, and protection from hazards. The theory behind building compartmentation is to limit fire and resulting products of combustion to one area of the building should they occur.

The reviewer should verify the fire-resistance rating of the construction separating the vertical openings, such as stairways and elevator shafts, from the remaining parts of the building. It may be helpful to highlight the vertical openings with a different color than the one used to highlight the exiting system. Atriums are vertical openings that are permitted if they are provided with adequate smoke control, automatic sprinkler protection, and fire barriers (Figure 8.11). Escalators may be permitted, although they may not serve as part of a means of egress. The escalator openings must be protected by rolling shutters, automatic sprinklers, partial enclosures, or a combination of sprinklers and mechanical ventilation (Figure 8.12). It may be necessary to refer to the mechanical and electrical plans to verify that the level of protection is adequate. For more information regarding escalators, consult NFPA 101, *Life Safety Code,* Chapter 6.

Fire inspectors should locate and check fire and smoke barriers for continuity, fire-resistance rating, and opening protection. It is often difficult to verify fire stopping in concealed spaces and penetrations. The architect should provide detailed plans that clearly show the fire stopping. When evaluating hazardous areas, fire inspectors should verify that adequate separation or automatic sprinkler protection has been provided.

Additional aspects that fire inspectors must consider are special hazards, interior finishes, fur-

Figure 8.11 Atriums are vertical openings which, if not properly protected, provide an avenue for smoke and fire movement. *Courtesy of Edward Prendergast.*

nishings and decorations, insulation materials, and materials that produce smoke and toxic gases.

Interior finish requirements usually apply only to walls and ceilings because floor coverings are tested using a different standard. As reviewers, fire inspectors must be aware of the way in which the applicable code treats the specific interior finish ratings, cellular or foam plastic materials, incidental trim, and fire-retardant coatings. In addition, codes often permit concessions involving interior finish when automatic sprinklers are provided. If furnishings and decorations are included in the construction documents, fire inspectors should review the specifications and evaluate exit access.

The increased use of insulation materials to conserve energy has created some fire problems. Insulation can accelerate the spread of fire by im-

Figure 8.12 An escalator enclosure protected by automatic sprinklers. *Courtesy of Edward Prendergast.*

pending heat dissipation; furthermore, some insulation materials are highly flammable (Figure 8.13). Fire inspectors should be aware of these factors and be familiar with any code requirements concerning insulation.

Fire inspectors also must be aware that many new materials release harmful quantities of smoke and toxic gases when they burn. Therefore, there are regulations excluding the use of certain materials with high smoke production or smoke densities. It is the responsibility of fire inspectors to check for these materials during the review of the structural drawings.

The inspector should also verify that portable fire extinguishers, fire detection and signaling systems, or automatic extinguishing systems, if required, are located and installed according to the appropriate standards.

REVIEWING MECHANICAL DRAWINGS

Mechanical systems are numerous and sometimes complex. It is therefore necessary that when fire inspectors review them, they gather as much information as possible to aid in evaluating the life safety hazards of the system.

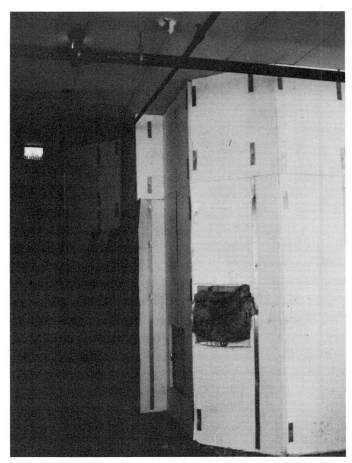

Figure 8.13 Insulation materials can accelerate the growth and spread of fire. *Courtesy of Edward Prendergast.*

It is important to understand the operation of the building's mechanical system and the factors that lead to the spread of smoke and fire throughout a building. While mechanical systems cannot prevent all smoke from filtering throughout a building, the quantity of smoke that does escape should be within human tolerances. The type of fuel to be used determines the type of heating, ventilating, and air conditioning (HVAC) equipment needed. The type of fuel also determines the type of venting needed, required clearances from combustibles, and quantities of air needed for proper combustion of the fuel. Most HVAC code requirements are based on the following NFPA standards:

- NFPA 211 *Standard for Chimneys, Fireplaces, Vents, and Solid Fuel Burning Appliances*

- NFPA 54 *National Fuel and Gas Code*

- NFPA 31 *Standard for the Installation of Oil Burning Equipment*

- NFPA 89M *Manual for Clearances for Heat Producing Appliances*

- NFPA 97M *Standard Glossary of Terms Relating to Chimneys, Vents, and Heat Producing Appliances*

Fire inspectors can use these standards as a guide during their review of the mechanical plans.

During the review of the mechanical system, the inspector should verify the style and design of the heating, ventilating, and air conditioning (HVAC) system (Figure 8.14). The refrigerant used in the system should be checked for possible health or fire hazards. Fire inspectors should be familiar with code requirements concerning the automatic shutdown of heating systems to control the spread of fire and smoke. This shutdown can be accomplished with heat-sensing devices or smoke detectors inside the duct work (Figure 8.15 on pg. 208). Inspectors must also determine if the system can be operated in a total exhaust mode. The total exhaust mode could be crucial in removing heat and smoke from a building during fire conditions. Fire and smoke dampers should be placed where the duct work penetrates fire-resistive walls or floors. The location of these features can be determined from the architectural drawings. The exit areas should not be penetrated by the duct work, nor should they be used for supply or return air. Additional references for damper placement and installation of duct work are NFPA 90A, *Standard for the Installation of Air Conditioning and Ventilating Systems* and the local mechanical code.

Usually, smoke control systems are used in buildings involving either a large number of people or that have a large quantity of combustibles such as shopping malls and buildings with open atriums. Most codes require that the smoke control system in these buildings provide a minimum of six air changes per hour. Smoke control systems must be engineered and tested and involve not only the mechanical systems, but the doors, partitions, windows, shafts, ducts, fan dampers, wires, controls, and pipes. For additional information about smoke control systems, consult NFPA 90A, *Standard for Installation of Air Conditioning and Ventilating Systems*, Appendix C, "Smoke Control." Additional information can be obtained from the American Society of Heating, Refrigeration, and Air Conditioning Engineers (ASHRAE) manual.

Exit stairwell pressurization is another method of smoke control. The intent of stairwell pressurization is to maintain a smoke-free atmosphere for occupants to exit during a fire. Stairwell pressurization systems must also be engineered and tested to ensure that the applied pressure does not make it difficult, if not impossible, for the average person to open stairwell doors.

REVIEWING ELECTRICAL DRAWINGS

General construction plans should include as much information about the electrical systems as possible. During the review of the electrical systems, the inspector must verify that illumination of exit ways is adequate. It is extremely important that fire inspectors take a detailed and systematic approach to the review of these systems. Exit illumination must be continuous during the time that the building is occupied. The stairs, aisles, corridors, ramps, escalators, and passageways leading to an exit must be lit (Figure 8.16 on pg. 208). Fire inspectors should evaluate such requirements

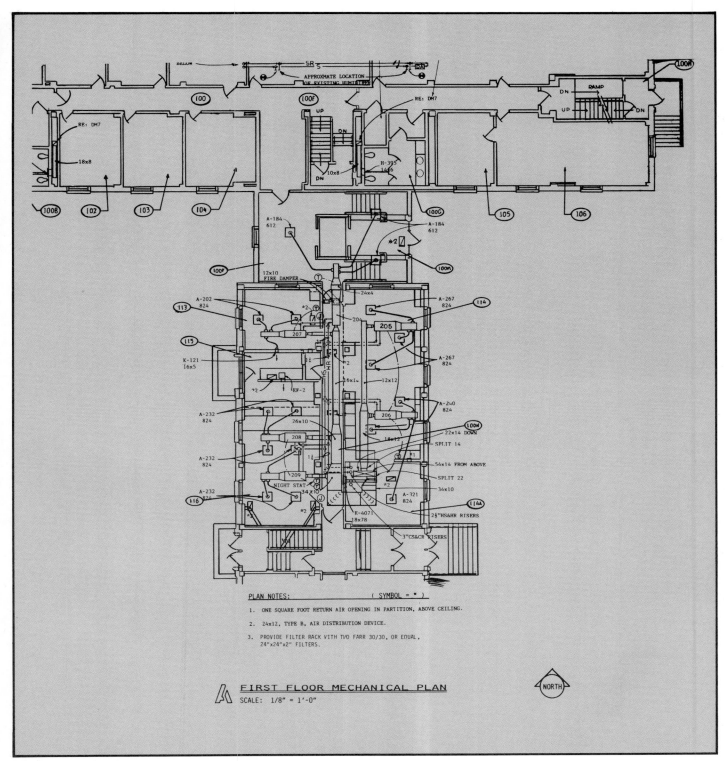

Figure 8.14 Mechanical plans display heating, ventilation, and air conditioning (HVAC) specifications.

as the level of illumination, power sources, emergency power, and special considerations for theaters or concert halls according to the appropriate code. General fire alarm requirements and emergency control systems must also be assessed.

Detailed information about fire alarm plans and communications systems is not usually provided on the general construction plans. Generally, separate plans must be submitted and approved before these systems can be installed. Plans

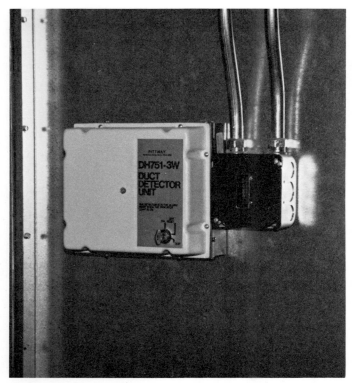

Figure 8.15 Some heating systems must be able to shut down automatically in response to heat or smoke.

for alarm systems should contain enough information for fire inspectors to evaluate the following areas:

- Signal initiation
- Signal notification
- Supervision of alarm systems
- Power supply
- Elevator control
- Automatic door closers
- Stair pressurization
- Smoke control
- Damper control
- Initiation of automatic extinguishing equipment
- Fire pumps
- Doors that unlock automatically when the alarm activates
- Lightning protection

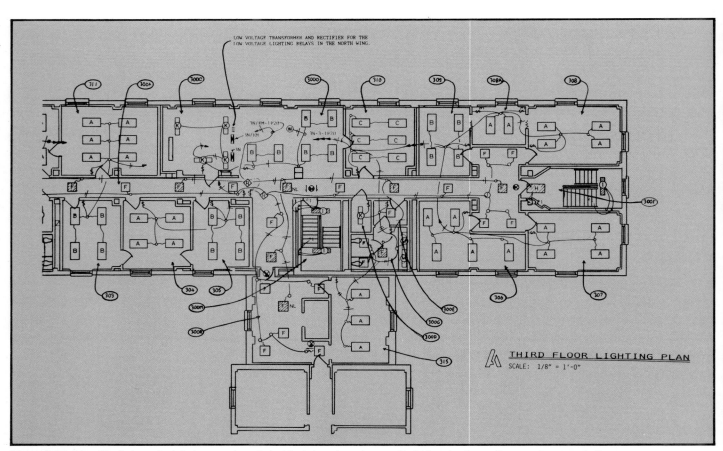

Figure 8.16 It is critically important during a review of electrical plans to make sure that all parts of an exit access be properly lit.

Each of these areas relates to the building's electrical system and must be examined in detail. It is helpful to use a checklist to ensure that nothing is overlooked.

REVIEWING FIRE PROTECTION FEATURES
Sprinkler Systems

A thorough, accurate review of sprinkler systems plans is necessary to ensure code compliance. Inspectors should be familiar with different types of sprinkler systems, their operation, and the specifications of each type of system. In addition to knowing national code requirements, fire inspectors should be aware of local ordinances, amendments, and other requirements that may affect the installation of the sprinkler system. They should develop a checklist that lists the specific requirements for sprinkler systems and the code sections that relate to the requirements to use as a guide during sprinkler plans review. See Appendix H for an example of a checklist.

Before the sprinkler review begins, the inspector should determine which standard applies and what the basic requirements are. This involves classification of the hazard by occupancy (light, ordinary, or extra hazard) and by storage commodity (shelf, high-piled, rack, or warehouse storage), or by both. The inspector should evaluate mixed occupancies very carefully to verify that the correct guidelines have been followed. Next, the inspector should determine whether the system has been designed according to the pipe schedule or has been hydraulically calculated. Drawings for hydraulically calculated systems should be accompanied by the calculations proving that the design complies with the code. The system designer should provide all the necessary information, including calculations and applicable data, on the drawings or in the specifications. All calculations, including computer printouts, must be reviewed carefully for accuracy.

Fire inspectors should check several system design factors such as the extent of coverage, type of system, size of system, information about sprinklers and piping, water supply connections and valves, and the fire department connection. Sprinklers should be installed throughout the building or area to be covered. This includes closets, stairwells, storage areas, walk-in freezers, and concealed spaces such as areas above suspended ceilings. Partial protection is permitted by most codes, but fire inspectors must consult the authority having jurisdiction in each case and review the code for any special requirements.

The size of each sprinkler system is based on the total square feet (square meters) of floor area controlled by a single riser and control valve. Regulations concerning the size of the system determine the number of individual systems and the number of system control valves needed. Fire inspectors should examine the specifications to verify the temperature rating, type, orifice size, area of coverage, use in special areas, spacing, and location of sprinklers. The drawings should clearly indicate all connections, piping, valves, drains, and gauges throughout the system (Figure 8.17 on pg. 210). Often, the inspector's test connection, drainage provisions, and the fire department connection are left off the drawings; therefore, fire inspectors should be sure to verify their presence and correct location.

Water supplies for sprinklers must be of sufficient capacity and pressure to satisfy the tabular requirements for pipe schedule systems or the calculated requirements for hydraulically designed systems. The sprinkler system designer should include a graph that shows the water demand compared to the available water supply. Flow tests are used to measure the pressure and flow available at the city main. Information indicating when, where, and by whom the last flow test was conducted should be provided. In addition to checking the pressure and the flow, fire inspectors should check the total capacity of the water supply. It must be adequate for the expected duration of fire fighting operations listed in the codes.

Special Extinguishing Systems

Special extinguishing systems are often required in areas where special hazards exist or where the use of water is undesirable. These situations call for the use of Halon, foam, carbon dioxide, or dry or wet chemical extinguishing systems. Fire inspectors should be familiar with each type of system and the applicable requirements.

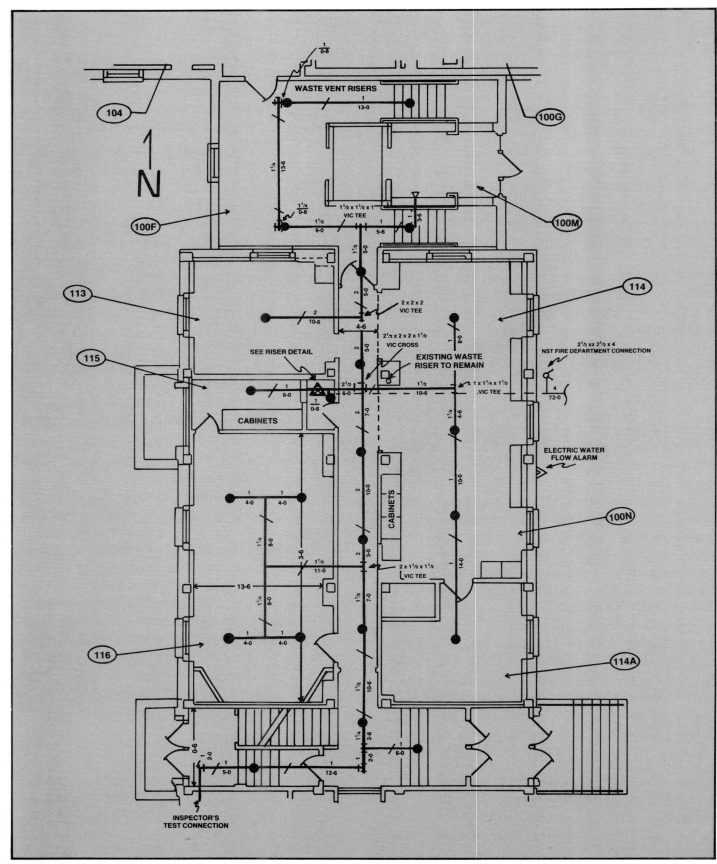

Figure 8.17 Automatic sprinkler system plans must be inspected to make sure they meet the appropriate specifications.

The architect's specifications should be very detailed and include the following information:

- Definition of the area or equipment to be protected
- Type of system (local application or total flooding)
- Type of extinguishing agent being used
- Amount of agent required
- Concentration of extinguishing agent to be developed
- Storage container size
- Type of expellent gas
- Rate of discharge
- Duration of flow
- Location and type of piping included, whether engineered or pre-engineered
- Location and type of discharge nozzles
- Method of actuation and auxiliary alarm functions, such as shutting off ventilation
- Type of presignaling devices used, if required
- Area and volume of the protected space

The architect should also include the calculation sheets or computer data sheets used to design the system.

Fire Alarm and Detection Systems

There are numerous types of fire alarm and detection systems that can be employed for the protection of a building, its contents, and the people who are in the building. From the buildings occupants' standpoint, the most important alarm system is the local alarm system. Local alarm systems are used to alert the occupants of a building that it is necessary for them to leave the building. The three most common methods of initiating a signal are manual operation of a pull box, automatic operation of heat or smoke detectors, or the actuation of sprinklers. All alarm equipment must be listed by UL or a similar testing agency for use as fire protective signaling devices.

The first method of actuating an alarm is by use of a manual pull box. Manual pull boxes must be highly visible and easily accessible. Instructions for actuating an alarm should be printed directly on the box. Fire inspectors should be aware of specified maximum travel distances to pull boxes and code requirements concerning pull box placement for accessibility to the handicapped.

A signal can also be initiated by smoke or fire detectors that automatically activate the local alarm system. The operation of the detectors can be affected by several factors: ceiling construction, forced ventilation, and placement of the detectors. Therefore, fire inspectors should be familiar with the types of detectors specified and the code requirements for their use.

Local alarm signals can also be initiated by a flow switch in sprinkler water supply lines that operates when a sprinkler opens. When a flow switch operates, the alarm sounded in the building should sound the same as with any other alarm-initiating device.

Once the detection has actuated, several types of devices are used to warn occupants of a building or the persons in charge that a problem exists. These methods include audible signaling devices and visual signaling devices. Signaling devices used for alarm notification and evacuation purposes must be listed by UL or FM as fire protective signaling devices. Fire inspectors must verify the types of devices to be used and their locations. Audible devices may include bells, buzzers, electronic tones, or other recognizable sounds. The distinctiveness of the noise, as well as its level, are important factors in choosing acceptable audible warning devices. Visual warning devices include flashing or strobe lights. They are designed to complement audible devices to provide warning for the hearing impaired.

Trouble signals are nonemergency signals designed to warn the occupants of the building and those responsible for maintenance of the fire alarm system that there is a malfunction within the system. A supervisory circuit within the system monitors itself for problems such as a power outage, problems with backup power supplies, or damaged wiring. Trouble signals should be distinct from alarm signals and should also operate a visual signal. The switches for silencing audible signs must not turn off the visual signs until the problem has been corrected.

Complex occupancies that need clarification as to what type of device is operating or the location of the device may require an annunciator panel. Examples of these types of buildings are institutions, hospitals, and high-rise buildings. The annunciator panel should be placed as close as possible to the entrance of the building most likely to be entered by fire personnel. This will allow fire personnel to quickly see which section of the building is involved upon their arrival. The annunciator panel may indicate other important information, such as the shutdown of HVAC systems, the activation of the sprinkler system, or control of the elevators for use by fire personnel.

To review the alarm and detection system, fire inspectors need information from the specifications, floor plans, equipment list, and symbol list. The specifications should include the type and gauge of wire, protection provided for the wire (conduit or raceways), wiring methods, and methods of supervision. All fire alarm systems must have electrical supervision. Fire inspectors should use the floor plans to verify the location and number of alarm-actuating and signaling devices. By using the symbols list, they determine if approved equipment is being used. It is the responsibility of fire inspectors to require inspection of the alarm system during construction and to observe the final acceptance test to ensure compliance.

Portable Fire Protection Equipment

The inspector should check the plans to make sure all portable fire protection equipment required by the codes will be placed in the building. This includes hand-held fire extinguishers, axes, standpipe hoselines, and any other particular portable equipment that may be required. The inspector should check to see that all maximum travel distances to these pieces of equipment are within the distances set forth by the code. The inspector should also check to see that equipment to be placed in the building is of the proper type to protect the hazards found in the building.

Plans Review Letter

Once the inspector has finished reviewing all the plans, a letter should be prepared with all the findings noted. The plans review letter, like the inspection report, should inform and analyze. The fire inspector's responsibility is to verify code compliance and to point out any discrepancies, not to recommend changes. Comments should be restricted to written provisions of the code in order to avoid liability. The responsibility for code compliance is assigned to the architect or engineer.

Inspectors should send copies of the review letter to all parties interested in or affected by the comments. The record-keeping system should include a schedule for follow-up reviews. A plans review letter is not always necessary; in many cases, the letter can be replaced with a checklist that serves as a guide to ensure that all applicable codes have been followed. Using checklists is becoming more common because of the complex nature of the plans review process.

The plans review letter should note that the plans review process is only an *initial compliance inspection*. The plans review should not be considered as the "final sign-off" for the certificate of occupancy. Final sign-off by the fire inspector can only be given after an official test of the fire protection equipment installed in the finished building is performed.

9

Hazardous Materials

This chapter provides information that addresses performance objectives described in NFPA 1031, *Standard for Professional Qualifications for Fire Inspector* (1987), particularly those referenced in the following sections:

Fire Inspector I

3-12.1 Properties of Flammable and Combustible Liquids.

3-12.1.1

3-12.1.2

3-12.1.3

3-12.1.4

3-12.1.6

3-12.8 Flammable and Combustible Liquids Fire Extinguishment.

3-12.8.1

3-12.8.2

3-12.8.3

3-12.9 Flammable and Combustible Liquids Labeling.

3-12.9.1

3-13.1 Properties of Compressed and Liquefied Gases.

3-13.1.1

3-13.1.2

3-13.1.3

3-13.1.4

10-90

3-13.7 Fire Extinguishment of Compressed and Liquefied Gases.

3-13.7.1

3-14.1 Properties of Explosives.

3-14.2

3-15.1 General.

3-15.1.1

3-15.1.3

3-15.2 Specific Hazardous Materials.

3-15.2.1

3-15.2.2

3-15.2.3

3-15.2.4

3-15.2.5

3-15.2.6

3-15.2.7

3-15.2.8

3-15.3 Combustible Metals.

3-15.3.1

3-15.4 Combustible Dusts.

3-15.4.1

3-15.4.2

10-90

Fire Inspector II

4-3 Compressed and Liquefied Gases.

4-3.1

4-4 Explosives, Including Fireworks.
(See NFPA 495, *Code for the Manufacturing, Transportation, Storage and Use of Explosive Materials,*
and NFPA 1124, *Code for the Manufacture, Transportation, and Storage of Fireworks.***)**

4-4.4

4-5.1 Natural and Synthetic Fibers.

4-5.1.2

10-90

Chapter 9
Hazardous Materials

Hazardous materials range from chemicals in liquid or gas form to radioactive materials to disease-causing agents. Moreover, hazardous materials pose risks to health and safety in many ways. They may be flammable, explosive, toxic, corrosive, radioactive, unstable, or reactive. Because of the dangers posed by hazardous materials, it is the inspector's job to see that they are stored and handled in accordance with the appropriate regulations.

A complete analysis and description of all types of hazardous materials is beyond the scope of this text; however, this chapter describes the properties of major types of hazardous materials. These include flammable and combustible liquids; compressed and liquefied gases; corrosive, reactive, unstable, and toxic materials; oxidizers; radioactive materials; plastics; explosives; combustible metals and dusts; and natural and synthetic fibers. Chapter 10 will concentrate on the storage, handling, and use of these materials.

Additional information concerning regulations, code requirements, and emergency actions may be obtained from the following agencies:

- National Fire Codes — NFPA
- U.S. Department of Transportation (DOT)
- Federal Emergency Management Agency (FEMA)
- Other U.S. Government Agencies such as Federal Bureau of Alcohol, Tobacco, and Firearms (BATF), Environmental Protection Agency, Coast Guard, Occupational Safety and Health Administration (OSHA), National Institute of Occupational Safety and Health (NIOSH)

- CHEMTREC (1-800-424-9300)
- National Response Center
- Institute of Makers of Explosives
- Association of American Railroads
- State Fire Marshal's Office
- State or County Health Department
- Police Agencies

IDENTIFICATION OF HAZARDOUS MATERIALS

The labeling and placarding of hazardous materials may fall under several systems of marking: bulk, nonbulk, and international. Identification systems have been developed by the United Nations, the U.S. Department of Transportation, and the National Fire Protection Association.

The United Nations has developed a system to identify the presence of hazardous materials that are transported internationally. This system identifies the hazard class as well as the specific commodity for bulk quantities. Hazardous materials are divided into nine numbered classes as follows:

Class 1 — Explosives
Class 2 — Gases (compressed, liquefied, or dissolved under pressure)
Class 3 — Flammable combustible liquids
Class 4 — Flammable solids or substances
Class 5 — Oxidizing substances
        Division 5.1 — Oxidizing substances or agents
        Division 5.2 — Organic peroxides
Class 6 — Poisons and infectious substances
Class 7 — Radioactive substances
Class 8 — Corrosives

Class 9 — Miscellaneous dangerous substances

For bulk shipments, the numbering system identifies the specific hazardous material. Each material that is considered hazardous has been assigned a four-digit number. By using the class number and the four-digit specific material number, people can immediately identify the hazardous material from reference lists. Examples of specific materials and their class and identification numbers are shown in Table 9.1.

TABLE 9.1
EXAMPLES OF HAZARDOUS MATERIALS
BY CLASS AND IDENTIFICATION NUMBER

Product	Hazard Class	Identification Number
Gasoline	3	1203
Uranium Hexafluoride	7	9173
Vinyl Chloride	2	1086
Hydrochloric Acid	8	1789
Carbolic Acid (Phenol)	6	1671
Phosphorus	4	1381
Ammonium Nitrate	5.1	1942
Oleum	8	1831
Black Powder	1	
Styrene	3	2055
Lauroyl Peroxide	5.2	2124
Carbon Tetrachloride	9	1846

In the United States, the transportation of hazardous materials is regulated by the U.S. Department of Transportation (DOT). These regulations include such elements as packaging, labeling, vehicle marking, and billing. The present DOT system for identifying materials uses the United Nations' system of class and commodity identification numbers coupled with its own North American numbers for local products. The DOT system uses both placards and labels for identification purposes.

Placards are diamond-shaped signs that are affixed to each side and to each end of any motor vehicle or rail carrier transporting hazardous materials (Figure 9.1). The placards must be used according to the DOT regulations for transportation of hazardous materials by motor vehicle or rail carrier. The placard indicates the primary class of the hazard of the cargo and, in some cases, the specific hazardous material being transported. The DOT transportation shipping labels are part of or are affixed to cartons, crates, or other packing materials. The labels are similar in shape and color to placards.

To complete the identification system, the Department of Transportation has issued the *DOT Hazardous Materials Emergency Response Guide (ERG)* (Figure 9.2). The ERG lists hazardous materials by both names and four-digit numbers. It also gives guidelines for the initial actions fire inspectors can take in handling an incident involving hazardous materials.

Figure 9.1 Hazardous materials placards should be highly visible so emergency personnel can readily identify the product.

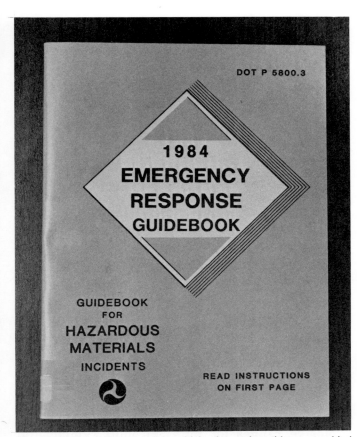

Figure 9.2 Hazardous materials guidebooks, such as this one provided by the United States Department of Transportation, provide emergency personnel with vital information about handling hazardous materials incidents.

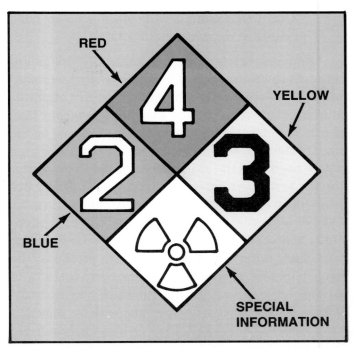

Figure 9.3 The NFPA 704 system provides information on the hazards to health, flammability, and reactivity of materials, as well as special information such as reactivity with water or radioactivity. *Reprinted with permission from NFPA 704-1985, Standard for the Identification of Fire Hazards of Materials, Copyright 1985, National Fire Protection Association, Quincy, MA 02269. This warning system is intended to be interpreted and applied only by properly trained individuals to identify fire, health, and reactivity hazards of chemicals. The user is referred to a certain limited number of chemicals with recommended classifications in NFPA 49 and NFPA 325.*

The National Fire Protection Association has developed a nationally recognized labeling system for fixed facilities such as buildings or tanks. This system is known as NFPA 704, *System for the Identification of the Fire Hazards of Materials*. The 704 label is diamond-shaped, like a placard, but is divided into four different colored sections, which denote health, flammability, reactivity, and special hazards. Blue is used for the health hazard, red for flammability, and yellow for reactivity. These three sections are further classified with a number from 0 to 4; 4 is the most hazardous and 0 is the least hazardous. The bottom section contains special information such as radiation hazard, reactivity with water, or polymerization. An example of an NFPA 704 label is shown in (Figure 9.3).

FLAMMABLE AND COMBUSTIBLE LIQUIDS

Flammable liquids are any liquids having a flash point below 100°F (38°C) and having a vapor pressure not exceeding 40 psi (256 kPa). Flamma-ble liquids are divided into Class IA, IB, and IC liquids:

- Class IA flammable liquids have flash points below 73°F (23°C) and boiling points below 100°F (38°C).

- Class IB flammable liquids have flash points below 73°F (23°C) and boiling points at or above 100°F (38°C).

- Class IC flammable liquids have flash points at or above 73°F (23°C) and below 100°F (38°C).

Combustible liquids have flash points at or above 100°F (38°C). Combustible liquids are further divided into Class II, IIIA, and IIIB liquids:

- Class II combustible liquids have flash points at or above 100°F (38°C) and below 140°F (60°C).

- Class IIIA combustible liquids have flash points at or above 140°F (60°C) and below 200°F (93°C).

● Class IIIB combustible liquids have flash points at or above 200°F (93°C).

Remember from Chapter 3 that the *flash point* refers to the minimum temperature at which a liquid fuel gives off sufficient vapors to form an ignitable mixture with the air near the surface. Flash points depend on the atmospheric pressure at the time of the test, the oxygen concentration, the type of test used, and the skill of the individual performing the test. Therefore, if the atmospheric pressure is actually lower than the pressure was during the test, it is possible that flammable vapors will be emitted at a much lower temperature. Listed flash points are determined by one of two methods: the open cup test or the closed cup test. Results of studies have indicated that the flash points determined by the open cup method will be approximately 10 to 15° F (-12°C to -9°C) higher than those determined by the closed cup method for the same material.

For liquids, the *fire point* refers to the temperature at which a liquid fuel will continue to burn. For all other fuels, the ignition temperature is the temperature at which the fuel will continue to burn.

Boiling Point

The boiling point of a liquid is that temperature at which the vapor pressure of the liquid is equal to the external pressure applied to it. In many cases, this external pressure will simply be the atmospheric pressure at the given location. When this state of equilibrium exists, boiling occurs. The bubbles produced during boiling are bubbles of vapor that have formed within the liquid and are rising to the surface, where they can escape into the atmosphere.

Specific Gravity

The specific gravity of a flammable or combustible liquid refers to the ratio of the weight of the liquid to an equal volume of water. Water is assigned a value of one. Liquids with specific gravities of less than one are therefore lighter than water and will float on water. Liquids with a specific gravity greater than one will cause water to float on their surface.

Flammable and Explosive Limits

Before any flammable or combustible liquid ignites, it must first emit a certain amount of vapors. These vapors then mix with an oxidizing agent to form a flammable vapor concentration. The flammable and explosive limits of flammable or combustible liquids are the upper and lower concentrations of the vapor that will produce a flame at a given pressure and temperature. These limits are expressed as a percent of the vapors in the mixture with an oxidizer. For example, 16 to 26 percent are the flammable limits of anhydrous ammonia. The lowest concentration of vapors in an oxidant that will ignite is known as the lower explosive limit, or LEL. Mixtures below the LEL are said to be too lean to burn. The highest concentration of vapors in an oxidant that will ignite is known as the upper explosive limit, or UEL. Vapor concentrations above this point are said to be too rich to burn. Variations in temperature will affect explosive limits: when the temperature increases, the range will increase, and when the temperature decreases, the range will decrease. Examples of the explosive ranges for butyl alcohol, xylene, and gasoline are shown in Table 9.2.

TABLE 9.2 EXAMPLES OF EXPLOSIVE RANGES		
	Lower Explosive Limit	Upper Explosive Limit
Butyl alcohol	1.4	11.2
Xylene	1.1	7.0
Gasoline	1.4	7.6

Temperature/Pressure Effect

When the temperature of a flammable or combustible liquid is raised, more vapors are emitted from the liquid. If the liquid is in a closed container, the pressure inside the container will increase. If the temperature rise is significant, the resulting pressure rise inside the container may be sufficient to cause container failure, often with explosive force. Such explosions are known as Boiling Liquid Expanding Vapor Explosions (BLEVEs).

Extinguishing Fires in Flammable and Combustible Liquids

Fire extinguishing equipment for use on flammable or combustible liquids includes hand-operated portable extinguishers, wheeled extinguishers, local application fixed systems, and total flooding fixed extinguishing systems (Figure 9.4a,b). The suppression agents used with this equipment include various dry chemicals, carbon dioxide (CO_2), foam, and halogenated compounds. Sprinkler protection is sometimes provided. Sprinklers are not necessarily recommended for extinguishing fires involving flammable liquids because water may not be a suitable extinguishing agent. Application of water may also cause the flammable liquid to overflow its container, thereby spreading the fire. Sprinklers are highly recommended, however, for containing fires and for protecting surrounding equipment and structural components.

Extinguishing methods for flammable or combustible liquids include removing the fuel supply, excluding the air or oxidizing means, and cooling the liquid to stop the flammable vapors from being produced. Areas utilizing total flooding fixed extinguishing systems should have a predischarge warning alarm so that personnel can leave the area rapidly.

Fire inspectors must check that all types of extinguishing equipment are properly maintained and installed and that there are no obstructions that would render the equipment inoperative or ineffective. NFPA 10, *Portable Fire Extinguishers,* contains additional information about fire extinguishers.

COMPRESSED AND LIQUEFIED GASES

Compressed gases are those materials which, at a normal temperature, exist as gases when pressurized within a container. In contrast, liquefied gases are ones which, at normal temperatures, exist in both liquid and gaseous state when pressurized in a container. Furthermore, this condition will continue as long as some amount of the liquid remains in the container.

Gases are often classified by their principal use. Fuel gases (which are, of course, flammable) are generally used to produce heat, light, or power.

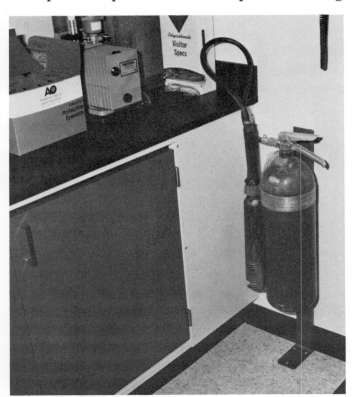

Figure 9.4a Fire extinguishers should be placed as close to the hazards they are protecting as possible. *Courtesy of Plano, Texas Fire Department.*

Figure 9.4b Larger fire extinguishers are often placed on wheeled carts to make them more maneuverable. *Courtesy of Edward Prendergast.*

10-90

Examples of fuel gases are liquefied natural gas and propane. Industrial gases are used in heat treating, chemical processing, refrigeration, water treatment, welding, and cutting. Examples of industrial gases are MAPP gas, acetylene, freon, and chlorine. Medical gases are used in doctors' offices and hospitals for providing basic life support and anesthesia. Medical gases include oxygen, cyclopropane, and nitrous oxide.

Vapor Density

Vapor density is used to evaluate the relative weights of gases in much the same way as specific gravity is used to evaluate liquids. Vapor density is the weight of a gas as compared to the weight of air. Air is assigned a vapor density of one. Gases with vapor densities less than one will rise and can be expected to concentrate near the ceiling or to form vapor clouds before they dissipate.

Temperature/Pressure Relationships

According to Charles' Law, if the volume of a gas is kept constant and the temperature is raised, the pressure will increase in direct proportion to the increase in temperature. This relationship is given by the formula:

$$P^1 T^1 = P^2 T^2$$

where P^1 = original pressure
T^1 = original temperature
P^2 = final pressure
T^2 = final temperature

Thus, if a gas in a cylinder or in any other type of unfired pressure vessel is heated, the pressure within the vessel will also increase, causing pressure relief devices to activate. If heating continues and pressure is not relieved, the container will rupture.

A second important relationship is known as Boyle's Law. This law states that the volume of gas will vary inversely with the applied pressure, if the gas is confined. This relationship can be shown by the formula:

$$P^1 V^1 = P^2 V^2$$

where P^1 = original pressure
V^1 = original volume
P^2 = final pressure
V^2 = final volume

Therefore, doubling the applied pressure reduces the volume by one-half.

Cryogenic Liquids (Refrigerated Liquids)

Gases can be converted to liquid form through refrigeration. This process is known as cryogenics. Cryogenic liquids are produced as noted at extremely low temperatures: -130°F (-90°C) and colder. (**NOTE:** Cryogenic liquids are also known as refrigerated liquids, especially when in transit.) One advantage of using cryogenic liquids is the ability to modify a material's liquid-to-gas volume ratio. For example, 862 volumes of gaseous oxygen can be liquefied to a single volume by reducing the gas temperature below its boiling point. A cryogenic cylinder of liquid oxygen can hold 12 times more gas than a pressurized cylinder of oxygen can hold.

Besides the savings in storage space and weight they offer, cryogenic liquids are valuable simply because they are so cold. For example, liquid nitrogen is used to freeze liquids for emergency pipeline repairs; to solidify gum-like materials, such as plastics and cosmetics, before they are ground; and for medical purposes.

HAZARDS OF CRYOGENIC LIQUIDS

Basically, the hazards of cryogenic liquids can be reduced to three categories:

- The inherent hazard of the particular gas, which may be intensified when it is in liquid form. Liquid hydrogen and methane are still flammable. Liquid oxygen will support combustion with great intensity. Liquid fluorine is both reactive and toxic.

- The tremendous liquid-to-vapor ratio.

- The extremely low temperature.

All of the cryogenic liquids, except oxygen, are either asphyxiants or are toxic. There is also the possibility that a gas with a lower boiling point than oxygen will extract the oxygen from the air exposed to it, reducing concentrations to levels that will not support life.

Because of their extremely low temperatures, all cryogenic liquids can inflict severe freeze burns (similar to severe frostbite) to exposed flesh. When small amounts are spilled on skin, they tend to move across the flesh quickly. In large amounts, they cling to skin because of their low surface tension. The gases produced by cryogenic liquids are also extremely cold. Inhaling these gases can severely damage the respiratory tract. Vapors can damage the eyes by causing the water in the eyes to freeze.

Cryogenic liquids will refrigerate the moisture of the air and create a visible fog. The fog normally extends over the entire area containing cryogenic vapors.

STORAGE AND HANDLING OF CRYOGENIC LIQUIDS

Because of the temperature and vapor ratio, cryogenic liquids must be stored in special containers. These containers are typically double-walled and have a vacuum in the area between the walls. An insulating material, such as perlite or a multi-layered, aluminized Mylar film stored under vacuum pressure, fills the space between the inner shell and the outer vessel jacket. The inner tank is built of corrosion-resistant stainless steel, aluminum, or various copper alloys. Carbon steel is not used because it becomes too brittle at these temperatures.

There are several types of cryogenic containers: spherical tanks, cylindrical tanks (either vertical or horizontal), and dewars. Heat from the atmosphere will penetrate the most well-insulated tank. The temperature differential between ambient air and the product can be several hundred degrees. Some of the cryogenic liquid will absorb heat and evaporate, but most of the fluid remains at the same temperature. Through this process of self-refrigeration, the temperature and vapor pressure inside a cryogenic container remain uniform. If the container remains unused for a long time, a slow rise in temperature can be expected.

Several pressure relief devices are used to control pressure in cryogenic containers. The most common types used are pressure relief valves, frangible disks, and a safety vent in the insulation space. A pressure relief valve is a device that opens well below any pressure that would threaten the integrity of the tank. A frangible disk ruptures if the pressure rise is excessive or if the pressure re-lief valve malfunctions. The insulation space also guards against leakage from the inner vessel.

If cryogenic liquids are trapped anywhere and expand, they can cause a violent pressure explosion. Their liquid-to-vapor ratios can create havoc when the product is heated in a confined space. This is particularly true in piping. A pressure relief device should be installed on every length of pipe between two shutoff valves. All pipes should be sloped up from the container to avoid the possibility of trapped fluids. Pipe and its associated fittings should be manufactured from stainless steel, aluminum, copper, or Monel.

Extinguishing Fires Involving Gases

The primary goal in controlling fire emergencies involving flammable gases is to stop the flow of the gas. Gas that continues to escape after the fire is extinguished may rapidly produce an explosive atmosphere. Conventional agents, such as dry chemicals, carbon dioxide, and halogenated agents, are usually effective on gas fires. Water may be applied as a cooling medium through hose streams, monitor nozzles, fixed spray systems, and sprinkler systems (Figure 9.5). Automatic sprinkler protection has proven to be very effective in cooling exposed tanks and avoiding pressure build-up.

Figure 9.5 Water may be applied from the side to cool a flammable gas storage tank.

OTHER HAZARDOUS CHEMICALS

Corrosives

Corrosives are those chemicals that cause visible destruction or irreversible harm to skin tissue. Liquid corrosives that are leaking will also have a severe corrosion rate on steel. Corrosives include inorganic acids (such as hydrochloric acid, nitric acid, sulfuric acid, hydrofluoric acid, and perchloric acid), halogens (such as bromine, iodine, fluorine, and chlorine), and several strong bases (such as sodium hydroxide, potassium hydroxide, calcium oxide, calcium hypochlorite, hydraxine, and hydrogen peroxide). Corrosives may also be strong oxidizers; however, they are classified as corrosives because of their hazardous effect when someone inhales, ingests, or touches them.

The principal hazard associated with inorganic acids involves the possibility of leakage. If this occurs, the acids may come in contact with other materials, resulting in a fire or explosion. The halogens are noncombustibles, but may support combustion or cause certain materials to spontaneously ignite. Fluorine and iodine may become explosive under certain conditions. Strong bases have oxidizing capabilities that may accelerate the burning of various materials or cause them to burn with explosive force.

Reactive Materials

Reactive refers to the material's ability to release energy as a result of a chemical reaction. Two types of chemical reactions must be considered: reactivity and self-reactivity. Reactivity refers to the reaction of two or more chemicals releasing energy and the ease with which the reaction takes place. Self-reactivity refers to the internal activity of a material and the ease with which a reaction may be initiated. Either type of reaction may be intensified by increased heat or pressure, possibly to explosive levels. Currently, there are no formal hazard levels for these conditions, with the exception of radioactive materials and explosives.

Reactive materials that are of particular concern to the inspector are those materials that react with air or water. Reactive materials include certain alkalies (such as sodium hydroxide and potassium hydroxide), aluminum trialkyls (such as triethylaluminum), anhydrides (such as acetic anhydride and propionic anhydride), carbides (such as sodium hydride and lithium hydride), oxides (such as calcium oxide, sodium hydrosulphite, metallic sodium, and red and white phosphorus). Some of these materials, known as pyrophorics, will spontaneously ignite when exposed to air. Others may ignite and explode when exposed to small quantities of water. During such a reaction, the mass quantities of heat that are produced may lead to self-ignition or to the ignition of nearby combustibles.

Reactive materials are categorized by degrees based on their ability to undergo a chemical reaction with another substance and to release some type of energy during this process. Classifications include such categories as detonates readily with an initiating source, normally stable but will undergo a chemical change without detonating, normally stable but may react with air or water, or normally stable and will not react with air or water. Fire inspectors must carefully consider such classifications when evaluating housekeeping practices, transportation and handling procedures, and fire control efforts.

Unstable Materials

Unstable materials are those materials that are capable of undergoing chemical changes or decomposition in the presence of a catalyst or without a catalyst. These materials may spontaneously decompose, explode, polymerize, or react, often with violent results. Some of the more common unstable materials include acetaldehyde, ethylene oxide, hydrogen cyanide, benzoyl peroxide, nitromethane, ethyl acrylate, and iospropyl ether. As with reactive materials, fire inspectors must carefully evaluate storage and handling procedures, housekeeping practices, and fire control efforts for unstable materials.

Toxic Materials

Toxic materials are commonplace in industrial and residential occupancies. They include such common items as disinfectants, weed killer, rodent killers, insecticides, and various cleaning solvents. These materials are further classified as poisons, irritants, or anesthetics. They include such chemicals as morphine, strychnine, atropine, aniline, arsenic acid, benzene, toluene, xylene, creosote, sodium fluoride, hydrogen cyanide, and cyanogen chloride.

Materials classified as poisons include those substances that are capable of producing serious illness or death once they enter the bloodstream. Irritants produce some type of local inflammation or irritate the skin, respiratory membranes, or eyes. Anesthetics reduce the muscular powers of vital body functions and result in loss of consciousness, suppressed breathing, or even death.

Toxic materials may enter the body through absorption, ingestion, injection, or inhalation. It is imperative that fire inspectors be aware of the possible toxic effects of a material, and make sure that sufficient personal protective equipment is available. They must also consider the threshold limit value (TLV). This value is the concentration of a given toxic material that generally may be tolerated without ill effects. Threshold limit values (TLV's) are published by the American Conference of Governmental Industrial Hygienists and by NIOSH. These values may serve as guides for fire inspectors in determining the maximum allowable concentration of a substance to which an individual may be repeatedly exposed without harmful effects. The inspector must check for proper ventilation in these areas, along with the adequacy of exit requirements. The action of these products under fire conditions and the products of combustion will also be noteworthy.

The toxicity of a material refers to the ability of that substance to do bodily harm by causing some form of chemical action within the body. Toxic liquids commonly enter the body through the skin, eyes, lungs, or digestive tract, eventually entering the bloodstream. Because of the many avenues of entry, the body organs affected by the substance may be remote from the point of contact.

From an exposure standpoint, toxicity may be divided into acute or chronic forms. Acute toxicity refers to the effects caused by a single dose or exposure. Depending upon the inherent concentrations and rates of exposure, the effects may be as minor as a simple headache or as serious as death. Acute

toxicity is expressed in terms of LD_{50} and LC_{50} for purposes of hazard evaluation. LD_{50} refers to the ingested dose of a given substance that was lethal to 50 percent or more of the animals tested when they swallowed or ate the substance. LC_{50} refers to the concentration in the air of a given substance that killed 50 percent or more of the test animals when they inhaled or absorbed the vapors, fumes, or mists of the substance.

Chronic toxicity refers to the effects of the substances upon repeated exposure over a long period of time. Chronic exposure is most often found in industrial situations where toxic substances are part of the daily or regular process. Fire inspectors should determine the maximum allowable concentration to which an individual may repeatedly be exposed without harmful effects.

Oxidizers

Oxidizers are chemicals that provide oxygen for combustion, either as a result of exposure to heat or through natural decomposition. Oxidizers are not necessarily combustible in themselves; however, by releasing oxygen, they support combustion and increase the intensity of burning. When in contact with materials such as oils, greases, hydrocarbons, and various cleaning agents, oxidizers react chemically to ignite the material. This reaction can result in an explosion. When exposed to heat, oxidizers decompose, releasing oxygen, which accelerates the fire. The accelerated fire produces more heat, causing the oxidizer to further decompose, again adding to the intensity of the fire. Some oxidizers decompose spontaneously and may do so with violent or explosive results. Oxidizers that inspectors are most likely to come in contact with include sodium nitrate, potassium nitrate, ammonium nitrate, cellulose nitrate, nitric acid, ammonium nitrite, sodium peroxide, potassium peroxide, strontium peroxide, hydrogen peroxide, barium peroxide, benzoyl peroxide, potassium chlorate, potassium perchlorate, sodium chlorite, calcium hypochlorite, ammonium perchlorate, ammonium dischromate, sodium perchlorate, magnesium perchlorate, potassium permanganate, potassium persulfate, and perchloric acid.

Radioactive Materials

Since radioactive materials are becoming increasingly common, fire inspectors must have a working knowledge of the hazards associated with these materials. Inspectors can expect to find radioactive materials in areas of electrical power generation, nondestructive testing, plastics, manufacturing, monitoring of chemical processes, treatment of disease, laboratories performing age determination, and food preservation (Figure 9.6). Since there is a life hazard presented by the release of radioactive materials, it is important that fire inspectors use all practical techniques to prevent a fire in or around processes using these materials. Radioactive materials that may be commonly found in the workplace include radium, plutonium, cobalt, uranium, cesium, and indium.

Figure 9.6 Signs should be placed at the entrance to any area containing radioactive materials. *Courtesy of Edward Prendergast.*

How hazardous a radioactive material is depends greatly upon whether the material is in a solid, liquid, or gaseous state, as well as how the material is being contained and handled. The type of radiation present is another major consideration. There are three types of radioactive particles: alpha particles, beta particles, and gamma particles. Alpha radiation can be stopped by a sheet of paper and therefore has the least amount of penetrability. Outside the body, this type of radiation is relatively harmless. If radioactive dusts containing alpha particles are inhaled or ingested, however, they may cause severe sickness or death. Beta radiation is more penetrating; beta particles can usually be stopped with a thin piece of sheet metal. Gamma particles have much greater penetration properties than alpha or beta particles and generally must be stopped by some type of lead shield.

Alpha, beta, and gamma rays are measured in "roentgens," which represent the amount of radiation absorbed at any one time in a specific situation. Radiation doses are measured in "rems," or the unit doses that will produce a specific effect in an exposed person. Measuring devices include personal film badges, pocket dosimeters, Geiger-Mueller survey meters, scintillation meters, and thermoluminescent dosimeters, or TLD's.

Plastics

Plastics, with the exception of cellulose nitrates, are classified as ordinary combustibles. The term *plastic* encompasses a variety of materials including acetals, acrylics, alkyds, alkyls, cellulosics, cellulose nitrates, epoxies, fluroplastics, furane, ionomers, melamines, methylpenetene, nylon, olefins, phenolics, polyallomers, polybutylene, polycarbonates, polyester, polyethylene, polypropylene, polystyrene, polyurethane, silicones, ureas, and vinyls. Plastics are found in many forms, including solid sections, pellets, powders, sheets, foams, molded forms, and synthetic fibers. Plastics have many uses: building construction, thermal or acoustical insulation, decorative trims, roofing materials, ceiling and wall panels, imitation wood, light fixtures, resilient flooring, wiring insulation, electrical conduits, piping, heating ducts, housings and parts of equipment and appliances, textiles, furniture, in automobiles, packaging materials, sports equipment, toys, photographic film, and many other applications.

Plastic materials are characterized as either thermoplastic or thermoset materials. Thermoplastic materials have a tendency to melt and flow during a fire situation, thereby presenting an additional way for flames to spread. Thermoset materials tend to retain their shape when exposed to a fire. Some plastics can be made less hazardous if fire-resistant additives are introduced into them. Various epoxies, polyester, polyurethanes, polyvinyl chloride, polyethylene, and cellulose acetate are all materials to which fire-retardant chemicals may be added.

Plastics may behave in a variety of ways during a fire, depending upon the physical form of the material, its chemical makeup, and the type of fire exposure. While many plastics have a higher ignition temperature than wood or other cellulosic products, they can often be ignited with a small flame. Once ignited, plastics may burn with a flame spread rate as much as ten times as great as that for common wood surfaces. Burning plastics sometimes generate large amounts of sooty, dense, black smoke. Depending upon the material, this smoke may include such toxic gases as hydrogen cyanide, hydrogen chloride, and phosgene.

Fire inspectors must exercise caution if catalysts, particularly peroxygen compounds, are located nearby plastics. These materials are very commonly used by plastics manufacturers and may decompose with explosive force when exposed to a fire situation. Dust explosions are also a possibility, especially in operations involving pulverized material or finishing processes that involve sanding and grinding.

Model building codes often require that foamed plastic used for interior insulation be protected by sheets of ½-inch (13 mm) gypsum board or another thermal barrier. These additional materials are designed to lessen the probability that the foamed plastic will ignite. Except for noncombustible or fire-resistive occupancies, model building codes accept steel or aluminum-sheathed building panels that are insulated with foamed plastic. Acceptance is given provided that the flame spread classification of the foam core is 25 or less and that the space is protected by automatic sprinklers.

Extinguishing methods for plastic materials are the same as for ordinary Class A combustibles and include automatic sprinkler protection, portable fire extinguishers, and hose and standpipe systems.

EXPLOSIVES

An explosive is any material or mixture that will undergo an extremely fast self-propagation reaction when subjected to some type of energy. This reaction produces heat and creates a sudden buildup of gas. The release of this gas pressure produces the explosive force. Explosive materials include explosives, water gels, and detonators.

There are three groups of explosives: primary or initiating high explosives, secondary high explosives, and low explosives. Primary or initiating high explosives are easily detonated by applying small amounts of heat, mechanical shock, or pressure. Their chief function is to initiate detonations in secondary high explosives. The major ingredients in primary explosives include but are not limited to lead azide, lead styphnate, and mercury fulminate. Electric blasting caps and detonating cord delay connectors are both examples of primary high explosives (Figure 9.7 on pg. 228).

Secondary high explosives are much less sensitive to heat and mechanical shock, but are considerably more powerful than primary explosives. Secondary explosives are usually detonated by use of primary explosives. Products such as TNT, nitroglycerin, and dynamite are considered to be secondary high explosives and can be found in commercial blasting and military operations (Figure 9.8 on pg. 228).

Low explosives or propellants usually operate by deflagration (rapid burning) rather than by detonation. (Detonation is defined as the introduction of a combustion zone into an unreacted explosive medium.) Major propellants include such materials as solid rocket fuels, black powder, and smokeless powder. Under the proper conditions, some low explosives may detonate.

When explosive materials are exposed to fire situations, all will react with varying degrees of sensitivity. Of course, there is always the possibility that they *will* explode, rendering any direct fire suppression efforts too dangerous for firefighters. The only effective way for fire inspectors to prevent disaster is to eliminate all sources of ignition.

Fire inspectors also need to know about small arms ammunition, small arms primers, and smokeless propellants. Small arms ammunition

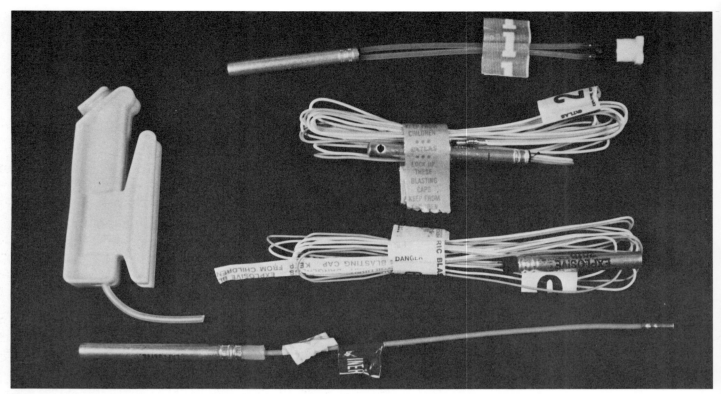

Figure 9.7 Explosives can come in many shapes and sizes. *Courtesy of The Institute of Makers of Explosives.*

Figure 9.8 TNT, nitroglycerin, and dynamite are used in commercial blasting and military operations. *Courtesy of Atlas Powder Co., Tamaqua, Pennsylvania.*

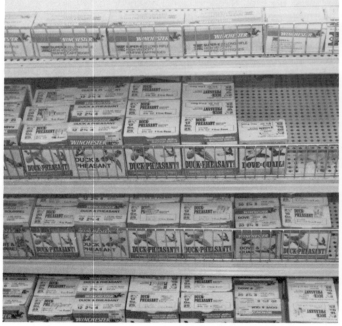

Figure 9.9 Many retail stores have at least a small quantity of boxed ammunition on hand. *Courtesy of Mount Prospect, Illinois Fire Department.*

includes shotgun, pistol, rifle, or revolver cartridges as well as cartridges for power devices and industrial guns (Figure 9.9). Small arms ammunition primers are basically small, percussion-sensitive charges that are encased in a cap. Ammunition primers are used to ignite the propellant power of the respective device. Smokeless propellants are solid materials used in various types of small arms ammunition, rockets, and so on. Military ammunition containing bursting charges, tracer, incendiary, or pyrotechnic projectiles are usually covered under specific regulations based on the inherent hazards associated with these items.

Some additional sources of technical data on explosives are

- Institute of Makers of Explosives (I.M.E.), Washington, D.C.
- Federal Bureau of Investigation, Bomb Data Center
- International Association of Bomb Technicians and Investigators
- E.I. DuPont de Nemours, Wilmington, Delaware
- Motorola Telecommunications
- U.S. Army-Explosives Ordinance Disposal (E.O.D.)
- Bureau of Mines, U.S. Department of Interior
- Bureau of Alcohol, Tobacco, and Firearms, U.S. Department of the Treasury

COMBUSTIBLE METALS AND DUSTS
Combustible Metals

Combustible metals include virtually all metals since almost all metals will burn under the right conditions. The most commonly encountered combustible metals include magnesium, titanium, sodium, lithium, potassium, hafnium, calcium, zinc, and sodium potassium alloys. Some of these materials will oxidize rapidly enough in the presence of certain gases to generate sufficient heat for them to ignite. How easily they ignite depends upon the physical state of the metal. For example, finely divided particles, molten metal, or fine thin sections tend to ignite much more easily than those in more solid forms. In addition, powdered or flaked metals such as aluminum or magnesium can be explosive under certain conditions.

The various combustible metals react differently when they burn. These reactions require special precautions and extinguishment techniques. Metals such as thorium, uranium, and plutonium emit ionizing radiation, presenting a contamination problem to fire fighting personnel.

Sodium burns vigorously and emits caustic sodium oxide fumes. Titanium burns virtually smoke free, while burning lithium produces large quantities of dense smoke. Zirconium powder, when moistened with water, burns with explosive force; however, when moistened with oil, it burns similarly to an ordinary combustible exposed to a moist atmosphere.

Temperatures reached in fires involving magnesium, zinc, sodium, and other metals are far higher than those reached in flammable liquid fires. This situation is apparent by the high intensity of flame that is present. Some metals that will burn in atmospheres of carbon dioxide, nitrogen, or steam require specialized extinguishing methods. Common extinguishing agents used for ordinary combustible and flammable liquids may cause a violent reaction when applied to burning metals (Figure 9.10 on pg. 230). Water may actually be decomposed into its elements of hydrogen and oxygen, thereby adding to the intensity of the fire.

Combustible Dusts

Combustible dusts (including cocoa, grain, flour, starch, graphite, charcoal, coal, plastic, aluminum, rubber, and magnesium) present a unique hazard because of their ability to form explosive mixtures within an enclosure. Combustible dusts deposited on machinery, structural supports, or other surfaces are highly susceptible to flash fires. If the dusts ignite when suspended in air, the resulting fire may burn with explosive force.

The ignitability of any dust-air mixture is based on several factors, particularly dust versus oxygen concentrations, size of particles present, the impurities in the mixture, and the strength of the ignition source. Just as in the case of flammable liquids and vapors, combustible dusts must be present in air in a certain concentration to create a combustible mixture. The minimum explosive concentration of a given material varies according to the size of the dust particles. As the size of the dust particle decreases, the ease with which the dust will ignite increases. Impurities such as moisture and inert materials reduce the likelihood of ignition occurring. The ability of moisture to absorb heat raises the ignition temperature of the dust. However, moisture in the atmosphere surrounding

Figure 9.10 Combustible metals are used in some automobile engines. These metals can pose a serious problem when firefighters attack fires in these vehicles with water. *Courtesy of Linda Gheen.*

a dust particle has no effect on the fire or explosion once the cloud of dust has been ignited. Hazards such as electric arcs, lights, friction sparks, static sparks, high pressure steam pipes, smoking materials, and open flames are all potential ignition sources and must be strictly controlled or eliminated.

Dust explosions tend to occur as a series, beginning with a small explosion and progressing toward a major one. The first explosion usually involves low concentrations of dust and is relatively harmless. This explosion will loosen more dusts, making it possible for a second, more serious, explosion to occur. This process may repeat itself three or four times until the building is totally destroyed or the dust is consumed. Dust explosions have also been known to progress from building to building. Explosion vents or release devices will greatly reduce the damage sustained by a structure during such an explosion.

Operations generating combustible dusts require several preventive measures. Perhaps the best way to reduce the possibility of a dust explosion is to employ good housekeeping practices. Often, the dust hazard can be reduced if the dust is collected at the point where it is generated. Places such as beams and ledges must be kept clean of dust accumulations. Persons cleaning should be careful to prevent dust from becoming suspended and mixing in the air. Dust should not be blown away; rather, it should be collected with vacuuming devices. Electrical equipment used in the area should conform to the requirements set forth by NFPA 70, *National Electrical Code,* and it should be rated for service in Class II, Division 1, or Class II, Division 2 locations.

NATURAL AND SYNTHETIC FIBERS

The NFPA reports that during the years 1971 through 1978, textile products were the first materials to be ignited in 2,845 fatal fires. From these fires, 4,361 deaths resulted. These statistics indicate that fire inspectors must be aware of the fire problems related to natural and synthetic fibers.

Natural and synthetic fibers used in manufacturing various textiles may be found in every type of occupancy in the form of clothing, furniture upholstering, carpeting, bedding materials, and decorative materials. Natural fibers include those materials composed of 90 percent or more cellulose material such as cotton, hemp, sisal, jute, flax, and other plant and animal fibers. Plant fibers tend to produce carbon monoxide, carbon dioxide, and water when they burn. Protein fibers such as wool and silk are derived from animals. These fibers are generally more difficult to ignite than plant fibers but also emit numerous toxic gases. Burning wool, for example, emits hydrogen cyanide.

Synthetic fibers are woven principally of synthetic or man-made materials. Synthetic fibers are produced by allowing emerging filaments to harden. When exposed to fire conditions, these materials may drip, shrink, or melt in a manner similar to plastics. Many synthetics, however, have been made safer than natural fibers by the addition of fire-retardant materials.

Both natural and synthetic fibers can be made more fire resistant by the addition of fire-retardant chemicals during the manufacturing stage. These chemicals may act to exclude oxygen, absorb heat, or to inhibit the chemical reaction in the flame.

There are a variety of factors that affect the way in which a textile will burn, including chemical composition of the fiber, the fabric finish, the weight of the fabric, the tightness of the weave, and any flame-retardant treatments. In some areas, building and fire prevention codes require that the textiles used in theater scenery as well as the curtains and draperies used in public assembly occupancies be treated with a flame-retardant material. In many cases, flame-retardant fabrics are required in hospitals and hotels where there is high life exposure due to bedridden or sleeping occupants.

The main hazard of storing baled cotton and other fibers is that the surface of the bales is covered with minute fibers. These fibers settle on the floor, shelves, and throughout the warehouse area. Fire tends to travel quickly off the surface of stored bales or carpet and rapidly ignites these fibers. Fire can spread so rapidly that automatic fire doors may not have time to close. Because of this hazard, housekeeping *must* be a high priority where textile products are stored.

Factory Mutual Engineering Corporation (FM) and the National Fire Protection Association (NFPA) provide information about regulations regarding the proper storage and potential hazards associated with textile handling.

10

Storage, Handling, and Use of Hazardous Materials

This chapter provides information that addresses performance objectives described in NFPA 1031, *Standard for Professional Qualifications for Fire Inspector* (1987), particularly those referenced in the following sections:

Fire Inspector I

3-12.2 Storage, Handling, and Use of Flammable and Combustible Liquids.

> **3-12.2.1**
>
> **3-12.2.2**
>
> **3-12.2.3**
>
> **3-12.2.4**
>
> **3-12.2.5**
>
> **3-12.2.6**
>
> **3-12.2.7**

3-12.3 Underground Storage Tanks for Flammable and Combustible Liquids.

> **3-12.3.1**
>
> **3-12.3.2**
>
> **3-12.3.3**

3-12.4 Aboveground Storage Tanks for Flammable and Combustible Liquids.

> **3-12.4.1**
>
> **3-12.4.2**
>
> **3-12.4.3**
>
> **3-12.4.4**

10-90

3-12.5 Inside Storage Tanks for Flammable and Combustible Liquids.

3-12.5.1

3-12.5.2

3-12.5.3

3-12.5.4

3-12.6 Outside Container Storage for Flammable and Combustible Liquids.

3-12.6.1

3-12.6.2

3-12.6.3

3-12.6.4

3-12.7 Inside Container Storage for Flammable and Combustible Liquids.

3-12.7.1

3-12.7.2

3-12.7.3

3-12.7.4

3-12.10 Transportation of Flammable and Combustible Liquids.

3-12.10.1

3-12.10.2

3-12.10.3

3-13.2 Storage, Handling, and Use of Compressed and Liquefied Gases.

3-13.2.1

3-13.2.2

3-13.2.3

3-13.3 Compressed and Liquefied Gas Containers.

10-90

3-13.4 Compressed and Liquefied Gas Transfer Operations.

3-13.5 Compressed and Liquefied Gas Leaks.

3-13.6 Transportation of Compressed and Liquefied Gases.

3-13.6.1

3-13.6.2

3-13.6.3

3-14.1 Properties of Explosives.

3-14.3

3-14.5 Labeling of Explosives, Including Fireworks.

3-14.6

3-14.7 Storage of Explosives, Including Fireworks.

3-14.7.1

3-14.7.2

3-15 Other Hazardous Materials.

3-15.1.2

3-15.2 Specific Hazardous Materials.

3-15.3 Combustible Metals

3-15.3.2

3-15.4 Combustible Dusts

3-15.4.3

Fire Inspector II

4-2 Flammable and Combustible Liquids.

4-3 Compressed and Liquefied Gases.

4-3.2

4-3.3

4-3.4

4-3.5

4-3.7

4-4 Explosives, Including Fireworks.
(See NFPA 495, *Code for the Manufacture, Transportation, Storage, and Use of Explosive Materials,* **and NFPA 1124,** *Code for the Manufacture, Transportation, and Storage of Fireworks.***)**

4-4.1

4-4.2

4-4.5

4-5.2 Combustible Dusts.

4-5.2.1

Chapter 10
Storage, Handling, and Use of Hazardous Materials

For much of their fire prevention efforts, fire inspectors must deal with the storage, handling, and use of hazardous materials. This chapter covers these aspects of flammable and combustible liquids, flammable gases, combustible solids, and explosives. The last section of this chapter discusses local, state, and federal response agencies for hazardous materials incidents.

FLAMMABLE AND COMBUSTIBLE LIQUIDS
Inherent Hazards and Ignition Sources

The hazards of flammable and combustible liquids range from accidental pollution due to leaks to fires to explosions. Fire prevention measures are based on one or more of the following techniques or principles:

- Eliminating or excluding sources of ignition such as electrical, mechanical, or frictional sparks; static electricity; open flames; hot surfaces; and incompatible materials

- Excluding air (oxygen)

- Storing liquids in closed containers or systems

- Ventilating to prevent the accumulation of vapors within the flammable range

- Maintaining an atmosphere of inert gas instead of air (air will support combustion)

The specific regulations regarding the safe transfer, storage, handling, venting, and transportation of flammable and combustible liquids can be found in the following NFPA codes:

- NFPA 30 *Flammable and Combustible Liquids Code*

- NFPA 30A *Automotive and Marine Service Station Code*

- NFPA 31 *Installation of Oil Burning Equipment*

- NFPA 58 *Storage and Handling of Liquefied Petroleum Gases*

- NFPA 325M *Fire Hazard Properties of Flammable Liquids, Gases, and Volatile Solids*

- NFPA 385 *Tank Vehicles for Flammable and Combustible Liquids*

- NFPA 386 *Portable Shipping Tanks for Flammable and Combustible Liquids*

- NFPA 395 *Storage of Flammable and Combustible Liquids on Farms and Isolated Construction Projects*

- NFPA 329 *Handling Underground Leakage of Flammable and Combustible Liquids*

Conditions Conducive to Explosive Atmospheres

Flammable vapors, which are released by flammable liquids, are responsible for flame propagation and explosions. Therefore, storage and handling procedures that prevent vapor release and minimize the amount of liquid exposed to the air are of prime importance in preventing explosive atmospheres.

When flammable vapor-air mixtures do explode, they usually do so in a confined space such as a building, room, or container. The confined space is required to allow a high enough concentration of vapor to collect in order for an explosion to occur. The violence of the explosions will depend upon the quantity of the vapor-air mixture, the concentrations of the vapors, and the type of enclosure containing the mixture. Explosions occurring with the vapor-air mixture near the extreme limits of the flammable range have been found to be less severe than those occurring near the middle of the flammable range.

If flammable and combustible liquids should ignite, extinguishing techniques involve shutting off the fuel supply, excluding the air, cooling the liquid to prevent evaporation, or a combination of these techniques. These extinguishing techniques are basically the same as preventive measures.

Storage of Flammable and Combustible Liquids

Standards and specifications concerning proper storage practices are set forth by such organizations as the American Petroleum Institute (API), the National Fire Protection Association (NFPA 30, *Flammable and Combustible Liquids Code),* Underwriters Laboratories, Inc., and local building or fire prevention codes. When evaluating such requirements, fire inspectors must remember that the size of the tank is relatively unimportant compared to the characteristics of the liquid being stored, the design of the tank, foundations and supports, size and location of vents, and piping and connections used throughout the installation.

There are several different types of storage receptacles used for flammable and combustible liquids. *Containers* have a storage capacity of 60 gallons (227 L) or less. *Portable tanks* are larger than 60 gallons (227 L) and are not intended for a fixed installation. The term *storage tank* refers to a vessel greater than 60 gallons (227 L) and in a fixed location. All portable tanks that have a storage capacity greater than 660 gallons (2 498 L) should be treated according to the same standards as fixed storage tanks.

STORAGE TANKS

Storage tanks can range from smaller tanks holding several hundred gallons (liters) to huge tanks holding millions of gallons (liters) of liquids. There are three basic classifications of storage tanks: aboveground, underground, and tanks located inside buildings. All types of tanks should be installed in accordance with NFPA 30, *Flammable and Combustible Liquids Code.*

Aboveground Storage Tanks

Aboveground storage tanks for flammable and combustible liquids are classified in three categories, depending upon the pressures for which they are designed. Atmospheric tanks are designed for pressures of 0 to 0.5 psi (0 kPa to 3 kPa), low-pressure storage tanks for pressures of 0.5 to 15 psi (3 kPa to 103 kPa), and pressure vessels for pressures greater than 15 psi (103 kPa).

Aboveground tanks should be constructed of steel or concrete unless conditions warrant using some special material. The thickness of the tank walls depends upon the weight and corrosiveness of the liquids the tank will hold. If corrosion is a severe problem, the tank may have a special coating or lining.

The inspector should check that aboveground tanks are placed on sturdy foundations and are provided with adequate supports (Figure 10.1). Exposed supports under the tanks must be protected by fire-resistant materials with ratings of two hours or more. Vertical tanks should be located slightly above the normal ground level in order to keep water from accumulating around them (Figure 10.2). The specified distances from tanks to items such as property lines and public ways depend upon the design of the tanks, their protection from exposure, available fire control resources, the stability of the liquids that the tanks contain, and the internal pressure of the tanks. These distances should be in accordance with the requirements set

Figure 10.1 Stable bases are important for ensuring tank stability. *Courtesy of Illinois Fire Inspectors Association.*

Figure 10.2 Vertical tanks should be installed on a built-up base. *Courtesy of Illinois Fire Inspectors Association.*

forth by the local building codes, fire prevention codes, or the authority having jurisdiction.

Fire inspectors must also consider carefully the spacing between tanks. The spacing between two tanks should be at least 3 feet (1 m). For tanks containing liquids classified as Class I, Class II, or Class IIIA, the distances between the tanks must be at least one-sixth the sum of their diameters. For those tanks containing unstable flammable or combustible liquids, the distances must be at least one-half the sum of their diameters.

Certain tanks should have additional spacing so that inside tanks are accessible for fire fighting operations. These tanks include those that are arranged in an irregular fashion or in rows of three or more, tanks that contain Class I or Class II liquids, and tanks that are in the drainage route of Class I or Class II liquids.

When liquefied petroleum gas containers are located in the same area as flammable or combustible liquid storage tanks, the gas containers and storage tanks must be at least 20 feet (6 m) apart. Fire inspectors should also make sure that there are measures taken to prevent the flammable or combustible liquids from accumulating under the

gas containers. If the liquid storage tanks are enclosed by dikes, any nearby liquefied petroleum gas containers must be on the outside of the dike and at least 10 feet (3 m) from the wall of the dike.

In large storage applications, one of the most important considerations is the properly designed storage vessel. The vessel must be a container that is liquidtight and that is designed to carefully control the release of vapors. In addition to normal venting, emergency venting is required for most aboveground tanks to prevent a Boiling Liquid Expanding Vapor Explosion (BLEVE).

A BLEVE occurs when excessive pressure inside a liquefied container causes the container to explode, breaking into two or more pieces (Figure 10.3). A fire around or under an external tank can cause the liquid in the tank to vaporize, resulting in high internal pressures. If the tank does not

Figure 10.3 A BLEVE is possible whenever liquefied gases are subjected to fire conditions.

have adequate venting to allow the vapor to escape and burn at the vents, it will rupture and explode. It is also possible for a BLEVE to result if the steel in the vapor space is softened by heat and fails.

All tanks must have normal venting to allow air or vapors to flow into or out of the tanks during filling or emptying operations (Figure 10.4). Vents that are clogged, damaged, or otherwise too small may cause the tank to rupture because of internal pressure or to collapse because of an internal vacuum. The inside nominal diameter of the vents must be at least 1¼ inches (32 mm). Containers that have more than one fill or discharge connection must have vents whose sizes are based on maximum possible simultaneous flow. Vent discharges must be arranged to prevent flame impingement on any part of the tank or nearby tanks in case the vapors exiting the vent ignite. Vents located near buildings or public places must be 12 feet (4 m) from the ground and arranged so that vapors are released outside and in a safe area. Vent pipe manifolding should be avoided unless it is required for recovering vapors, conserving vapors, or controlling air pollution. If vent pipe manifolding is used, pipe sizes must be large enough to allow for the maximum possible simultaneous flow from all connected tanks.

Figure 10.4 The inspector should check all vents to be sure they are not damaged or blocked. *Courtesy of Illinois Fire Inspectors Association.*

All aboveground storage tanks must have some device or inherent design that will relieve any excess pressure caused by nearby fires. The only exceptions to this requirement are tanks storing Class IIIB liquids in quantities exceeding 12,000 gallons (45 425 L) that are not in a diked area with, or in the path of, Class I or Class II liquids. Fire inspectors should make sure that the tanks have emergency relief provisions, such as loose manhole covers that will rise under pressure, rupture disks, a weak seam between the roof and shell of the tank, or conventional emergency relief devices designed for the specific application. If a tank has emergency relief provisions that only relieve pressure, it must have normal emergency vents whose total venting capacity is sufficient to prevent the tank from rupturing. All commercial venting devices should be stamped with the opening pressure, the pressure at which the valve reaches the full open position, and the accompanying flow capacity in cubic feet per hour (cubic meters per hour).

Besides emergency relief provisions, all tanks or tank areas storing Class I, II, or IIIA liquids must have dikes or some form of drainage to prevent any leaks or accidental discharges from endangering nearby facilities, property, or waterways (Figure 10.5). If dikes are used, the volume of the diked area should be large enough to contain the entire amount that could be released from the largest tank. The walls of the dike should be liquid-tight and can be constructed of steel, concrete, solid masonry, or earth. Dikes that are made of earth and that are 3 feet (1 m) or more high must have a 2-foot (0.6 m) wide, flat section at the top. Loose combustible materials, empty or full drums, and other items must be kept out of the diked area.

Figure 10.5 Dikes control the flow of spilled liquids. *Courtesy of Illinois Fire Inspectors Association.*

If drainage is used, there should be a slope of at least one percent toward the drainage system, and the basin of the system should be large enough to hold the entire contents of the largest tank. Furthermore, the drainage system must not empty onto nearby property, or into water supplies, public drains, or public sewers.

Fire inspectors must also check the connections for the tanks. All connections through which liquids may flow should have valves located as closely as possible to the tanks. Those connections that are located below the liquid level of the tanks and through which liquids normally do not flow should have a liquidtight closure. For tanks containing Class I liquids, gauging openings that must be made and broken must be located outside buildings and away from any ignition source. The connections should be at least 5 feet (2 m) away from any building opening.

Underground Storage Tanks

Underground tank storage is perhaps the safest form of storage for flammable and combustible liquids if the tanks are designed correctly, installed safely, and maintained properly. This type of storage is most commonly found in bulk and retail fuel storage facilities, car rental agencies, and truck depots.

Underground tanks must be designed for their intended use. The designer must consider how much pressure the surrounding earth or pavement will exert and whether vehicles will cross over the tanks. The tanks may be made of metal, fiber glass, unlined concrete (for liquids with specific gravity of 40 degrees API or greater), or lined concrete (for liquids with a specific gravity lighter than 40 degrees API).

If proper precautions are taken, underground storage tanks may be buried almost anywhere. The tanks must be placed on a firm foundation and surrounded with at least 6 inches (152 mm) of noncorrosive, inert materials, such as clean sand, earth, or gravel, that have been well tamped into place (Figure 10.6). The tank must be placed into the hole in the ground carefully because dropping or rolling the tank can puncture it, break a weld, or scrape off the protective coating if it has one (Figure 10.7). Underground tanks not subject to veh-

Figure 10.6 The base of the hole in which the tank is to be installed should be firmly packed.

Figure 10.7 Appropriate care must be taken when placing the tank into the hole. The tank should not be rolled or dropped into place.

icular traffic over them must be covered with at least 2 feet (0.6 m) of earth, or at least 1 foot (0.3 m) of earth on top of which is a reinforced concrete slab at least 4 inches (102 mm) thick (Figure 10.8 on pg. 244).

If underground tanks are likely to have vehicles passing over them, they must be protected against damage by at least 3 feet (1 m) of earth, or 1½ feet (0.46 m) of well-tamped earth plus 6 inches

(152 mm) of reinforced concrete or 8 inches (203 mm) of asphaltic concrete (Figure 10.9). If concrete is used, it must extend at least 1 foot (0.3 m) horizontally from the edge of the tank in all directions. Any piping that may be damaged must be protected with sleeves, casings, or flexible connections that will absorb any vibrations or impacts.

The vents used in underground storage must be large enough to prevent any liquid or vapor from being blown back toward the opening where the

Figure 10.8 Care should be taken to prevent damaging tanks while they are being covered.

tank is filled. The vents should have at least a 1¼-inch (31 mm) nominal inside diameter. Equipment for recovering vapors must not obstruct or restrict the piping for vents unless the tanks and associated components have protection that prevents back pressure from developing. Vents should also be protected so that dirt, debris, and insect nests will not block them. The vents shall be equipped with flame arresters and must be located so that vapors will not be discharged to an unsafe area.

Vents for Class I liquids must discharge outside buildings, must be higher than the fill pipe opening, and must be at least 12 feet (4 m) above the adjacent ground level (Figure 10.10). Vents for Class II and IIIA liquids also must discharge outside buildings, be higher than the fill pipe opening, and be placed above the normal snow level.

Vent pipings should be protected from physical damage. This can be accomplished by placing them in an inaccessible location or by constructing a barrier around them. The vent pipes must enter the tanks through the top and must be sloped so that they drain toward the tanks. When the vents are manifolded, the pipe sizes must be large enough to allow a discharge within the pressure limitations of the system. Storage facilities for Class I liquids may not have manifolding vents with storage facilities for Class II or Class III liq-

Figure 10.9 Underground tanks that are subject to vehicular traffic must be covered with concrete or asphalt.

Figure 10.10 Vents from underground tanks are often piped to locations away from vehicles and pedestrians.

uids unless provisions are made to keep vapors from Class I liquids from contaminating the Class II or Class III tanks.

Besides vent openings, underground tanks will have openings for fill and discharge lines, gauging, and vapor recovery. All connections for these openings must be vaportight. Any openings to allow the tank to be gauged manually must have liquidtight caps or covers if the openings are separate from the fill pipes. Fill pipes for Class IB and IC liquids must terminate within 6 inches (152 mm) of the tank bottom to minimize the possibility of generating static electricity. Filling, discharge, and vapor recovery connections that are made and broken for Class I, II, and IIIA liquids must be located outside buildings at least 5 feet (1.5 m) from the nearest building openings and in an area free of ignition sources.

Leakage from underground tanks is a major concern due to the possibility that water supplies may be contaminated, the environment damaged, and explosions may occur as a result of accumulated vapors. Leaks as small as one drop per second can result in a loss of 34 gallons (129 L) in a month.

Furthermore, a 1/16-inch (1.6 mm) stream can leak 2,520 gallons (9 539 L) per month, and a ¼-inch (6.4 mm) stream can leak 28,080 gallons (106 300 L) per month.

Perhaps the simplest and most economical method of detecting leaks is to be sure that inventory is stringently controlled, taking into account unavoidable losses from evaporation, product shrinkage, and improper meter calibration.

There are several other tests that may be performed to determine if tanks are leaking. In the air test, the tank is emptied and then pressurized to 10 to 12 psi (69 kPa to 83 kPa). The results of this test are somewhat unreliable, and the method is potentially dangerous since tanks are normally designed to withstand only 5 to 10 psi (34 kPa to 69 kPa). For these reasons, no agency recognizes this test method as acceptable, and fire inspectors should discourage its use. Currently, two types of hydrostatic tests — the pressurization technique and the liquid standpipe or volumetric method — are in use. Other techniques are also used. Fire inspectors should insist that all tests be performed according to the guidelines set forth by NFPA 329, *Handling Underground Leakage of Flammable and Combustible Liquids*.

Under normal circumstances, tanks that are installed correctly may be expected to last approximately 20 years. Tanks that are installed incorrectly or that are buried in corrosive soils, however, may fail in five years or less. Soils containing fill such as cinders, shale, construction debris, or other materials, may be highly corrosive. Tanks installed in these types of soils should have protective coatings or wrappings or cathodic protection, or be made of materials that resist corrosion. Tanks may be dug up and recoated when they reach their projected life expectancy. This may give up to 20 years of additional use.

Problems can occur if different metals are used for tanks and piping. In this case, a chemical reaction known as electrolytic corrosion occurs where the two metals are joined. This corrosion will result in a failure of the connection. Tanks and piping should be constructed of the same materials.

Abandoned tanks can be safeguarded in either of two ways. First, they may be left in the ground permanently. In this case, the fill line, gauge opening, pump suction, and vent lines are disconnected. The underground piping must be capped. The tank is then filled with an inert, solid material.

The second method for safeguarding tanks involves removing them from the ground. All liquid must be removed from the tank lines. All inlets, outlets, and underground piping must be capped. If the tanks are to be disposed of as waste, they must be cleaned and rendered vapor-free. Holes should be made in the tanks so that they will be unfit for later use.

Inside Storage Tanks

The most common types of tanks found inside buildings are those used to store fuel oil. Tanks of less than 660 gallons (2 498 L) require no special fire protection features. If amounts greater than 660 gallons (2 498 L) are to be stored inside the structure, they must be placed in a fire-resistant room or enclosure so they are isolated from the rest of the building. This enclosure must have a 3-hour fire-resistance rating. Such enclosures generally can have a Class A opening that is protected by a self-closing fire door. A raised noncombustible sill or liquidtight ramp is needed at the opening to prevent the liquid from flowing into the structure if a leak should occur.

Tanks containing Class I, Class II, or Class IIIA liquids must have overflow prevention devices such as float valves, fill-line meters, low heat pumps, or liquidtight overflow pipes that are at least one pipe size larger than the fill pipe. The vents and fill pipes must terminate on the outside of the building. Openings provided for vapor recovery equipment must be protected against vapor release by dry break connections or spring-loaded check valves.

Tanks used to store flammable or combustible liquids inside a building have the same wall thickness and design requirements as tanks used for outside storage. Inside tanks do require several additional safeguards, however. They must have automatic-closing, heat-actuated valves for each connection that is used to withdraw liquid and that is located below the liquid level. Such valves prevent liquids from flowing out of the tanks in the event of pipe rupture. Inside tanks must also have ventilation equipment to remove flammable vapors from within the enclosure. Using automatic sprinkler systems to protect the tanks and associated piping should be carefully considered. Manual gauging and other tank openings must have vaportight caps or covers.

CONTAINER AND PORTABLE TANK STORAGE

The most common form of flammable and combustible liquid storage vessels encountered are containers and portable tanks. Virtually every occupancy that the inspector enters will have some type of these vessels in it. This section will take a look at the vessels themselves, as well as the proper method of storage for them.

Containers for storing flammable and combustible liquids come in any one of several forms, including glass containers, metal drums, safety cans, and polyethylene containers. Safety cans should be UL or FM approved. Portable tanks must conform to *Chapter 1, Title 49* of the *Code of Federal Regulations* (DOT Regulations) or NFPA 386, *Portable Shipping Tanks for Flammable and Combustible Liquids.* Table 10.1 highlights the maximum amounts of each classification of liquid that can be stored in each type of container.

Most commonly, the inspector will encounter flammable or combustible liquids stored in containers that are 5 gallons (19 L) or less in size. Although there are other acceptable methods of storing these small amounts, the safest containers are approved safety cans (Figure 10.11). The maximum allowable size for approved safety cans is 5 gallons (19 L). Safety cans are constructed to greatly reduce leakage from container failure. They are also designed to virtually eliminate vapor release from the container under normal conditions. Safety cans use self-closing lids with vapor seals and contain a flame arrester in the dispenser opening. The self-closing lid also acts as a pressure relief device when the can is heated.

Each portable tank must be installed with an emergency venting device to limit internal pressures to 10 psi (69 kPa), or 30 percent of the bursting pressure of the tank, whichever is greater. The total venting capacity of both normal and emergency vents should prevent rupture of the

TABLE 10.1
MAXIMUM ALLOWABLE SIZE OF CONTAINERS AND PORTABLE TANKS

Container Type	Flammable Liquids			Combustible Liquids	
	Class IA	Class IB	Class IC	Class II	Class III
Glass	1 pt	1 qt	1 gal	1 gal	5 gal
Metal (other than DOT drums) or approved plastic	1 gal	5 gal	5 gal	5 gal	5 gal
Safety Cans	2 gal	5 gal	5 gal	5 gal	5 gal
Metal Drum (DOT Spec.)	60 gal	60 gal	60 gal	60 gal	60 gal
Approved Portable Tanks	660 gal	660 gal	360 gal	660 gal	660 gal
Polyethylene DOT Spec. 34, or as authorized by DOT Exemption	1 gal	5 gal	5 gal	60 gal	60 gal

SI Units: 1 pt = 0.473 L; 1 qt = 0.95 L; 1 gal = 3.8 L.

Reprinted with permission for NFPA 30, *Flammable and Combustible Liquids Code,* Copyright©, 1984, National Fire Protection Association, Quincy, MA 02269. This reprinted material is not the complete and official position of the NFPA on the referenced subject which is represented only by the standard in its entirety.

Figure 10.11 Small quantities of flammable and combustible liquids must be stored in approved safety cans.

shell or bottom if the tank is vertical, or the heads or shell if the tank is horizontal. If fusible vents are used, they must be designed to operate at a temperature not exceeding 300°F (149°C).

Inside Storage of Containers and Portable Tanks

The amount of flammable and combustible liquids that can be stored in an occupancy depends upon the occupancy classification. Dwelling occupancies and buildings containing three or fewer dwelling units should not store more than 25 gallons (95 L) of Class I or Class II liquids, or 60 gallons (227 L) of Class IIIA liquids. Assembly occupancies and buildings containing more than three dwelling units may store no more than 10 gallons (38 L) of Class I and Class II liquids, or 60 gallons (227 L) of Class IIIA liquids. All liquids must be stored in approved cabinets and safety cans. Office, educational, and health care occupancies should store only the amount of flammable or combustible liquids needed for operation, maintenance, demonstrations, or treatment. In these facilities, the containers for Class I or Class II liquids must be approved safety cans holding no more than 2 gallons (8 L).

Mercantile occupancies and retail stores storing flammable or combustible liquids in areas accessible to the general public should store no more than the amount needed for displays or operating purposes. If the quantity of liquids exceeds certain limits, the excess must be stored in a room approved for storing flammable liquids. These limits are as follows:

- 60 gallons (227 L) of Class IA liquids
- 120 gallons (454 L) of Class IB liquids
- 180 gallons (681 L) of Class IC liquids
- 240 gallons (908 L) of Class II liquids
- 660 gallons (2 498 L) of combustible liquids
- More than 240 gallons (908 L) of any combination of flammable liquids

Containers in display areas may not be stacked more than 3 feet (1 m) high or two containers high unless they are stored on fixed shelving. The shelving should be sturdily constructed and arranged so that the containers stored on them cannot be easily knocked off.

Warehouses and buildings used strictly for storing flammable liquids and located no more than 50 feet (15 m) from another building or property line should have blank, exposed walls that are constructed of noncombustible materials with a fire-resistance rating of 2 hours or more. Containers that are piled up should be separated by pallets or dunnage so that the piles are stable and there is no excess stress on the container walls. These piles must be at least 3 feet (1 m) from beams, girders, or other obstructions, and at least 3 feet (1 m) below sprinkler deflectors and the discharge openings of fire protection systems. The access aisles must be at least 3 feet (1 m) wide and should be kept clear at all times.

In order to further guard against accidents involving the storage of small quantities of flammable and combustible liquids, it is important that containers and portable tanks be stored in appropriate storage cabinets or rooms. Storage in these places will further protect the liquids from dangerous exposure to fire conditions should they occur.

Storage Cabinets

Small quantities of flammable or combustible liquids in normal operating areas should be stored in approved metal storage cabinets. These cabinets should have a doorsill raised approximately 2 inches (51 mm) or more. In addition, the cabinet must be labeled **"FLAMMABLE — KEEP FIRE AWAY"** in red letters (Figure 10.12). Generally, no more than 120 gallons (454 L) of Class I, Class II, and Class IIIA liquids are permitted to be stored in one cabinet. No more than 60 gallons (227 L) of the 120 gallons (454 L) can be Class I or Class II liq-

Figure 10.12 Flammable liquids storage cabinets must be labeled appropriately. *Courtesy of Plano, Texas Fire Department.*

uids. No more than three metal storage cabinets may be located in the same fire area unless they are 100 feet (30.5 m) or more apart.

Storage cabinets designed to accommodate flammable and combustible liquids can be made of either metal or wood. They must be able to limit the internal temperature at the center of the cabinet, 1 inch (25 mm) from the top, to not more than 325°F (163°C) when subjected to a 10-minute fire test. This fire test employs burners simulating a room fire exposure using the standard time-temperature curve. All joints and seams should remain tight and the door should remain securely closed during the fire test. The cabinet is not required to be vented.

Storage Rooms

To be safe, facilities storing containers of flammable or combustible liquids should have rooms that are specially constructed of fire-resistive materials in order to protect the containers from any nearby fire exposure. Any openings from these rooms into other areas should have a 4-inch (102 mm), raised sill that is liquidtight and composed of noncombustible materials (Figure 10.13). These openings also must have approved, self-closing fire doors. The joint where the walls meet the floor must be liquidtight to prevent any leakage of

spilled liquids into the main facility. Low-pressure steam, hot water, or electrical heating units should be used to heat the facilities. In areas where Class I liquids are stored, electrical installations must be approved for Class I, Division 2 locations as defined by NFPA 70, *National Electrical Code*. The electrical installations in areas containing strictly Class II or Class III liquids may be approved for general use. Inside storage rooms must have some form of ventilation, either gravity or mechanical (Figure 10.14).

Figure 10.14 Shown is a mechanical ventilation shaft inside a flammable and combustible liquids storage room. Note the bonding strap, explosion-proof light, and automatic sprinkler system. *Courtesy of Plano, Texas Fire Department.*

Figure 10.13 Raised doorsills help prevent spilled liquids from seeping into other rooms. *Courtesy of Plano, Texas Fire Department.*

The liquids must be dispensed through approved pumps or self-closing faucets. If the Class I or II storage containers have a capacity of 30 gallons (114 L) or more, they should not be stacked on each other. A clear aisle approximately 3 feet (1 m) wide must be maintained at all times.

If Class I liquids are being dispensed in a room, the room must have mechanical ventilation. This ventilation must provide at least 1 cubic foot per minute of exhaust per square foot of floor area (m^3 per 3 m^2) but not less than 150 cubic feet per minute (4 m^3 per minute). Air must be moved across the room to prevent any accumulation of flammable vapors. Furthermore, the exhaust from the room must exit the building completely, and may not be recirculated. Recirculation is permitted only when the exhaust is continuously monitored by a fail-safe system that automatically sounds an alarm, stops recirculation, and provides full exhaust to the outside of the building if the vapor-air mixture exceeds one-fourth of the lower flammable limit. The dispensing area of a mechanical ventilation system must have an audible alarm to alert personnel if the ventilation system fails. The exhaust installation should follow the guidelines established in NFPA 91, *Installation of Blower and Exhaust Systems for Dust, Stock, and Vapor Removal or Conveying*.

Outside Container Storage

Flammable and combustible liquids are often stored outside in portable containers such as drums (Figure 10.15). Fire inspectors should make sure that the containers are located so that the threat of fire spreading from these containers is minimal. The storage area must have a fence or security guard to protect it against vandalism. The facilities must be kept free of weeds, debris, and other combustible materials that could ignite or spread an outside fire to the storage area. When storage facilities are located next to buildings, the exterior walls of the building must be constructed of noncombustible or approved limited combustible materials having a fire-resistance rating of at least 2 hours. No opening is permitted within 10 feet (3 m) of the storage area. Storage areas must be graded so that spills have at least a 6-inch (152 mm) curb. If the storage area has a curb, there must be some provision for removing rain or spilled liquids that accumulate.

Fire inspectors should allow only containers that have been approved for the liquids that are placed in them. The containers should have one or more devices that will provide sufficient venting capacity to limit the internal pressure of the containers to 10 psi (69 kPa) or 30 percent of the bursting pressure of the container. Each tank should

have a pressure-actuated vent that operates at approximately 5 psi (34 kPa). If fusible venting devices are used, they should be designed to operate at a temperature less than 300°F (149°C). Drains in the storage area must end in a safe location and be accessible during fires. Quantities of more than 1,100 gallons (4 163 L) may not be located adjacent to buildings. Local fire prevention or building codes usually contain tables giving the required spacing distances.

Handling, Transferring, and Transporting Flammable and Combustible Liquids

Assuming that the equipment containing flammable and combustible liquids is properly maintained and that the tanks or containers do not allow the vapors to escape, fire inspectors must next be concerned with the handling, transfer, and transportation of the liquids. It is through these actions that most problems occur. The inspector must always be observant for situations or conditions that pose a hazard. The occupants may demonstrate excellent practices when the inspector is present, but revert to their normal, less safe ways of doing things as soon as the inspector leaves. With a certain amount of experience, the inspector will be able to determine by observing the layout of the facility how normal operations are conducted.

There are several rules to follow to ensure safe handling of flammable and combustible liquids within a facility. Most of them apply to handling small amounts of the liquids, as they are the most common amounts handled. Class I and Class II liquids should be kept in covered containers when not actually in use. When handling any flammable or combustible liquid that is not in a closed container, a way must be provided for safe disposal in case a leak or spill occurs. Class I liquids must not be used in the presence of any possible ignition source, such as open flames, electrical arcs, or heating elements. In addition, Class I liquids cannot be stored in containers that are pressurized with air. In some circumstances, they may be stored in containers that are pressurized with an inert gas.

Finishing operations are one of the most common processes inspectors will assess. Many of the more common finishing processes use materials that can form flammable mixtures if they combine with oxygen. Such materials are usually applied by spraying, flow coating, hand brushing, dipping, or by using electrostatic equipment. Smoking *must* be prohibited in such areas. The electrical wiring equipment must be of the appropriate classification for the particular hazard. Whenever possible, solvents used for cleaning equipment should have flash points greater than 100°F (37°C).

Several additional safeguards are necessary for paint finishing operations. First, adequate ventilation is needed to prevent the residues and vapors of the flammable liquids from accumulating. Fire inspectors should make sure that all potential ignition sources, including open flames and equipment that produce sparks, are removed. Only paints and solvents required for the immediate operation are allowed in the area. Other paints and solvents may be stored in metal cabinets in a special flammable liquids storage room. Because overspray residues are a spontaneous heating hazard, they should be cleaned up regularly.

Operations that use spray booths also require some special safety procedures (Figure 10.16). Each spray booth should have an exhaust duct leading to the outside by the most direct route. These ducts should be constructed of steel and should not pass through floors. Both the booths and ducts should be separated from combustible walls and roof material by clearances that are equal to those required for metal stacks. The booths should

Figure 10.16 Spray booths can present high hazards due to atomized fuels in the air. *Courtesy of Edward Prendergast.*

be used to apply only one type of finish because different finishes may react to cause a fire or explosion. Overspray residues should be routinely removed from the booths and duct systems. All tools that are used for cleaning should be ones that will not produce sparks.

Fire inspectors should consider the inherent hazards associated with dip tanks. All dip tank operations must be located on the ground floor of noncombustible buildings. The covers on the tanks should be kept closed when not in use. The covers should also be the self-closing type that activate if there is a fire (Figure 10.17). In addition, the tanks must have overflow pipes located approximately 6 inches (152 mm) below the top of the tank because

Figure 10.17 When not in use, dip tanks containing flammable or combustible liquids should be kept with their lids closed. *Courtesy of Justrite Manufacturing Co.*

tanks may overflow during a fire. Tanks must also have bottom drains so the tank can be drained both manually and automatically if there is a fire.

Accidents frequently occur during the transfer of liquids from one vessel to another. Though transferring can be hazardous regardless of the amount of liquid being transferred, the inspector should be particularly concerned with loading and unloading in bulk handling operations. Loading and unloading stations for Class I liquids should be located no closer than 25 feet (8 m) from storage tanks, property lines, or adjacent buildings. Loading and unloading stations for Class II and Class III liquids should be located no closer than 15 feet (5 m) from these same objects. These stations should be constructed on level ground. Curbs, drains, natural ground slope, or other means are needed to keep any spills in the original area. Adequate ventilation, natural or mechanical, must be maintained.

Several liquids, including light fuel oils, toluene, gasoline, and jet fuels can develop static charges on their surfaces. (Refer to Chapter 4 for further discussion of bonding and grounding.) If a static discharge is present at the same time that a flammable mixture is near, a fire or explosion can result. To protect against static discharge, tanks must be bonded together with a metal chain or strap. Tanks also need to be grounded to neutralize static charges when Class I liquids are loaded and unloaded. Bonding and grounding are also necessary when Class II or Class III liquids are loaded into containers that have previously contained Class I liquids. All bonding connections must be completed before dome covers, lids, or caps are removed. Bonding is also needed between the dispensing device and the receiving containers.

Perhaps the safest procedure for handling flammable and combustible liquids is to pump them from underground tanks through a piping system to dispensing equipment. The dispensing equipment must be kept in specially designed rooms or out-of-doors. Piping materials must resist any corrosion from the material being transferred. Piping must also be able to withstand the maximum service pressure and temperature as well as thermal shock and physical damage.

All transfer systems must be designed so liquids will not continue to flow by gravity or by siphoning if a pipe breaks. Control valves should be located at easily accessible points along the piping to control the flow. Fire inspectors should be sure that all piping and connections are inspected on a regular basis for signs of deterioration or leakage.

GASOLINE STATION VAPOR RECOVERY SYSTEMS (VRS)

Another common transferring system is the gasoline station vapor recovery system (VRS). Gasoline vapors are released into the atmosphere each time that gasoline is dispensed into motor vehicle fuel tanks or service station bulk storage tanks. These vapors, when mixed in the right proportions with fresh air, create a mixture that can ignite with explosive force. To reduce this fire hazard and to comply with the air pollution standards developed by the Environmental Protection Agency (EPA), many gas stations have begun using vapor recovery systems (VRS).

The gasoline vapors in a gasoline tank occupy whatever space is not filled with liquid. If additional gasoline is put into a tank, it will displace an equal volume of gasoline vapor. The displaced gasoline vapor will be released into the atmosphere.

As gasoline is pumped from a storage tank to a vehicle tank, a vacuum is created. Air then enters the storage tank through the tank vent and occupies the space formerly filled by the gasoline. The gasoline entering the vehicle tank displaces the vapors in the tank and the vapors are released into the surrounding atmosphere (Figure 10.18).

In a VRS, the top of the vehicle tank is connected with the top of the storage tank, creating a closed loop exchange of gasoline and vapor (Figure 10.19 on pg. 254). A similar technique is used when the gasoline transport truck refills the storage tank (Figure 10.20 on pg. 254).

Two types of VRS for gasoline service stations are currently in operation: balance systems, and vacuum assist systems. Both of these systems work on the principle of a closed loop exchange of gasoline and vapors. The difference between the systems is the way in which the closed loop is accomplished.

The balance system has a highly specialized nozzle that provides a tight seal around the fill neck of the vehicles fuel tank (Figure 10.21 on pg.

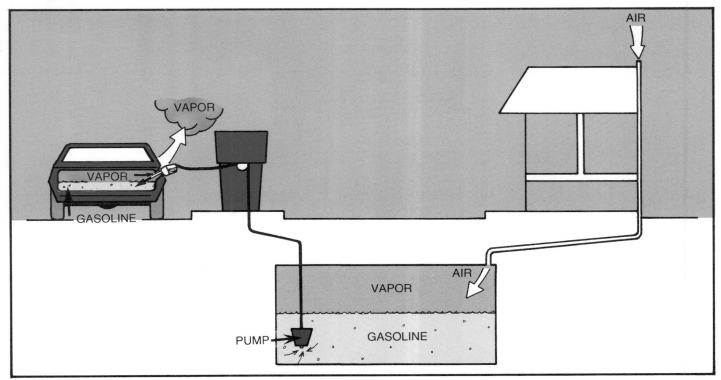

Figure 10.18 Without a vapor recovery system, the release of gasoline vapors to the atmosphere is unavoidable.

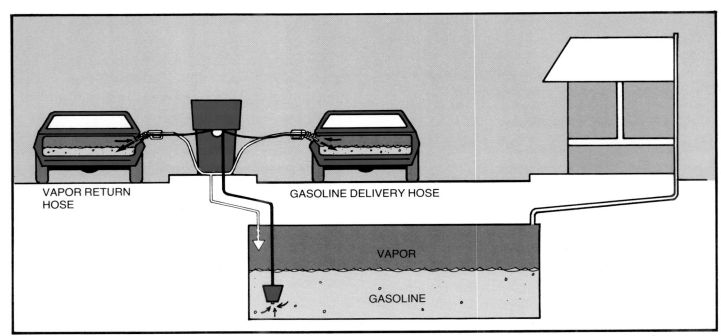

Figure 10.19 When using a vapor recovery system, a closed loop of gasoline and vapors is formed between the storage tank and the vehicle tank.

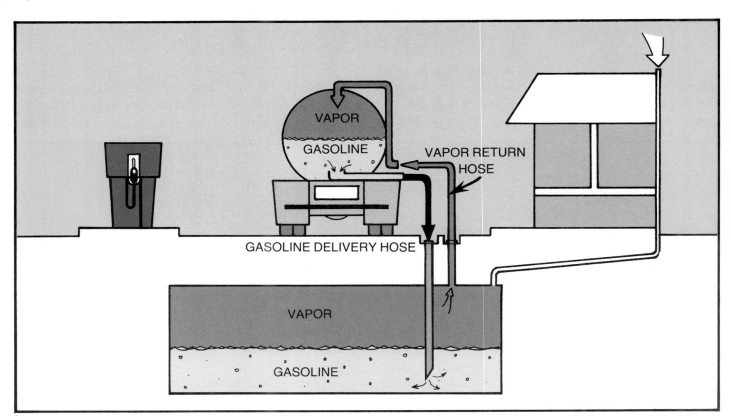

Figure 10.20 The vapor recovery system used to replenish a storage tank operates on the same principle as a vapor recovery system for a dispensing system.

255). This special nozzle, which is larger, heavier, and more complex than an ordinary dispensing nozzle, has an interlock so that the nozzle must be pushed with force into the gasoline tank fill neck before the gasoline will flow. This interlock maintains a tight seal while the gasoline is being dispensed. Because of this tight seal, flammable vapors are not released into the atmosphere.

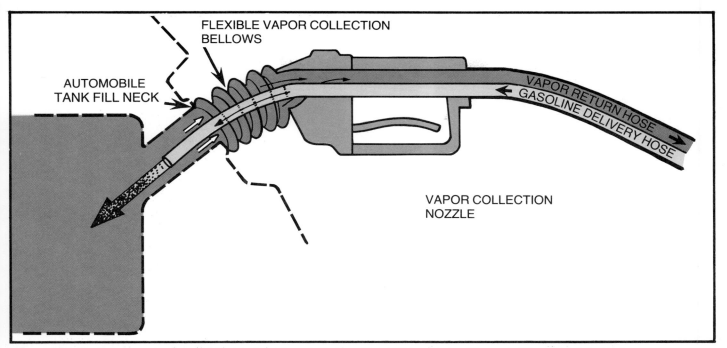

Figure 10.21 The balance system has a highly specialized nozzle that provides a tight seal around the neck of the vehicle's fuel tank.

The vacuum assist system uses a small vacuum generator, which keeps the vapor return hose and piping under a slight negative pressure. If the dispensing nozzle and the gasoline tank fill neck are not tightly sealed, air is drawn into the vapor return hose. This keeps vapors from being released into the atmosphere. This dispensing nozzle used in the vacuum assist system is smaller, lighter, and less complicated than that used in the balance system since the tightness of the seal between the nozzle and the fill neck is not as crucial. Most vacuum assist systems have a small processor that destroys vapors when the amount of vapors being released is greater than the amount of gasoline being exchanged (Figure 10.22).

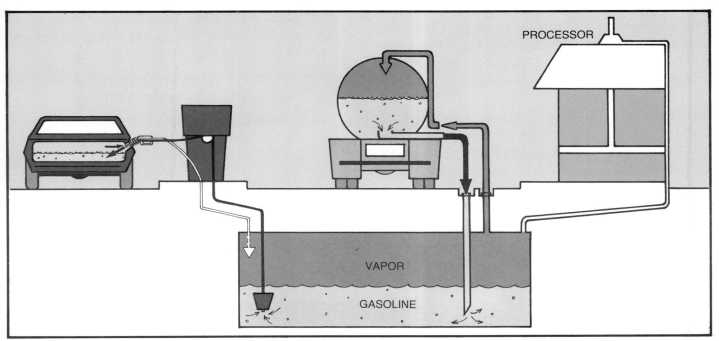

Figure 10.22 Most vacuum assist systems have a small processor that destroys vapors when the amount of vapor being released is greater than the amount of gasoline being exchanged.

VRS Inspections

Inspections made during the installation of a VRS are as important as inspections made while the VRS is in service. Moreover, a complete inspection during installation may save the service station owner time and money that would later be required to correct previously undetected problems.

During any inspection of VRS, special emphasis should be placed on connections, piping, and nozzles and equipment. The dangerous accumulation of flammable vapors can be avoided by ensuring that connections and other fittings are secured and that no sags are present in the vapor return line. Due to the complexity of VRS, the inspector may find it necessary to research the particular systems in the area. Technical data about individual VRS may be obtained from the VRS manufacturer or the appropriate government agency.

Regulation and Transportation of Flammable and Combustible Liquids

The U.S. Department of Transportation (DOT), through a number of its divisions, regulates the interstate (between different states) transportation of flammable and combustible liquids (See Appendix I). Individual states regulate intrastate (within the state) transportation, and local jurisdictions may further regulate transportation. For example, local regulations may designate routes and times of delivery for some products, close some tunnels for certain products, or require that vehicles transporting certain liquids not be left unattended during loading or unloading.

Investigating and Controlling Flammable and Combustible Liquids Incidents

Fire inspectors may be involved if flammable or combustible liquids leak or spill. Some jurisdictions have hazardous materials response teams and most states have such teams if the local jurisdiction does not have one. Local fire inspectors should know the location of the nearest hazardous materials team and the procedures for calling the team.

Leaks may develop because of poor installation of storage tanks, poor maintenance practices, corrosion, mechanical failure, or human error. Whatever the reason, the loss is generally small

and the liquid dissipates in the air without incident. There are times, however, when these liquids flow into conduit, sewers, and drainage areas. Because these flammable and combustible liquids are uncontained, they must be considered an immediate hazard. Although the proper handling of hazardous materials is ultimately the responsibility of the organization in direct contact with the materials, the fire inspector's role is to try to prevent the incident from happening in the first place. The main concern of the inspector should be safety to life and minimal damage to the environment.

Depending upon the location of the leak or suspected leak, inspectors should use common sense and good judgment. They should avoid causing panic by unwarranted evacuation or publicity. Normally, if a hazard has been declared due to an uncontained leak, fire officials assume jurisdiction. If there is a strong potential hazard to life or property, and if conditions warrant evacuation, law enforcement agencies can help with evacuation procedures and assist in crowd control. Planning is essential for these types of emergencies: lack of cooperation and coordination among all parties involved could result in increased risk to life and property.

When flammable and combustible liquid leaks are suspected, a careful and thorough investigation must begin immediately. If the leak is from an underground source, the liquid or its vapors are normally reported in the following areas:

- Inhabited subsurface structures such as basements, subways, and tunnels

- Uninhabited substructures such as sewers, utility conduits, and observation wells near tanks

- Groundwater such as drawn from wells, on or in surface water, or emerging from cuts or slopes in the earth

If the leak or suspected leak is in an inhabited area, the condition should be considered a potential hazard to human life and the area should be evacuated immediately. Law enforcement agencies should be notified to evacuate everyone in the endangered area and to prevent any further traffic from entering the area until it is determined safe.

Some of the local agencies that can offer assistance within the local jurisdiction during an incident are the following:

Local Government Agencies

- Fire Department — Personnel, specialized equipment, and materials.

- Police Department — Traffic and crowd control, evacuation, property protection, and investigation.

- Civil Preparedness — Communications, crowd control, evacuation, and radiological services.

- Public Works — Sand and other containment material, construction equipment, personnel, and maps of natural watershed areas; can also terminate utility service.

- Public Health — Water, air, soil monitoring, and poison control information.

- Building Department — Consultants on manufacturing locations.

State Government Agencies

- National Guard — Communications, crowd control, evacuation, transportation, radiological services, and helicopters.

- State Police and Sheriffs — Crowd and traffic control, accident investigation, evacuation, property protection, and communications.

- Public Health — Monitoring and testing of water, air, and soil for contamination, treatment of injured personnel, access to hospitals, ambulances, and poison control centers.

- Environmental Protection — Access to cleanup contractors and to chemists, biologists, and other scientists with specialized knowledge of hazardous materials.

- Agriculture — Soil and water testing, knowledge of pesticides and fertilizers.

- Fire Marshal, Fire Academy, and Universities — Training and advice in hazardous materials response and suppression, and use of specialized response equipment.

After all precautions have been taken to protect life and property, the next step is to determine the source of the leak and to prevent any further liquid from escaping. Sometimes the leak is in the vicinity of the storage container and discovering its source is not difficult. If the contaminate has entered a sewer, porous rock or soil, pipes or conduit, or a trench filled with porous soil, it may travel a considerable distance before it is detected. Some of the most common places to check for the source of the contaminate are gasoline service stations, automotive garages or agencies, airports, marinas, heating oil distributors, chemical companies, fleet operations (such as taxicab companies), dairies, trucking companies, municipal garages, and any abandoned flammable and combustible liquid tanks. If none of these facilities are within a few hundred feet (meters) of the contaminated area, then organize a general search of the area. Topographical maps may be useful in determining possible flow directions. Any facility where flammable and combustible liquids are stored can be considered a potential location for a hazardous incident.

A leak usually occurs in one of two ways: (1) a liquid spills during transfer and escapes into porous soil or a sewer; (2) a leak develops during storage, transportation, or handling. Equipment and storage devices are prime candidates for being the source of the leak. When inspecting equipment, check for the following:

- Notice the area around the fill pipes where the liquid is transferred. Note if the area is discolored or saturated, concrete is stained, or asphalt is disintegrated. Any of these can indicate repeated spills that may accumulate underground.

- Notice if the area aboveground shows signs of overfilling.

- Check pumping equipment for leaks. Use a combustible gas indicator when checking service station pumps. Open the cover of the unit just far enough to insert the probe. Also check the hose and the nozzle.

- Notice floor and street drain areas. Note if waste liquids have been dumped there.

- Look for signs of dumping waste liquids into any streams or bodies of water (flammable and combustible liquids are lighter than water and will float on the water's surface; furthermore, the contaminate will usually display an oily pattern of colors on the surface of water).

- Check for dead vegetation in the area that might indicate spillage or dumping.

- Use a combustible gas indicator and check underground cavities such as sewers, utility conduit, and telephone manholes.

- Check any nearby steep cuts, excavations, or natural slopes for signs of liquid coming through the soil.

Smoking and other sources of ignition should not be permitted in the suspected area. Any form of electrical connections such as extension cords or switches that could create a spark and ignite flammable vapors should not be touched. Gas and electric service to any buildings in the suspected area should be terminated by qualified personnel. This will remove the fuel from pilot lights and gas burners and the potential spark from every electrical device and outlet throughout the building.

When the area is entered, a combustible gas indicator should be used to detect the presence and concentrations of the vapor. Only a trained operator should use the gas indicator, along with self-contained breathing apparatus in case of toxic vapors or high concentration of combustible vapors.

Venting the danger zone should be the next step in making the area safe. Sometimes natural venting (opening doors and windows) can provide adequate ventilation. Mechanical venting equipment may be required to properly remove vapors from within a hazardous area. If motor driven fans are used, they should be approved for Class I, Group D locations.

For additional information regarding underground leakage of flammable and combustible liquids consult NFPA 329, *Underground Leakage of Flammable and Combustible Liquids*.

Basic Techniques for Extinguishing Fires Involving Flammable and Combustible Liquids

Fire inspectors are expected to know the basic techniques for extinguishing fires involving flammable and combustible liquids. To extinguish these fires, personnel should use a Class B extinguishing agent to either smother or inhibit the chemical chain reaction of the burning liquid vapors. They can use water fog to extinguish combustible liquids.

Fire inspectors will also be expected to teach people to use a hand-held, portable fire extinguisher properly. (Refer to Chapter 7, Fire Protection and Water Supply Systems, for a review of the basic steps for using a portable extinguisher.) Fire inspectors must also know where fire extinguishers should be placed. Places where flammable and combustible liquids are located are considered extra hazardous locations. Because of this classification, NFPA 10, *Portable Fire Extinguishers* dictates that fire extinguishers have a minimum rating of either 20B or 40B, depending on the size of the hazard. Extinguishers with a 20B rating must be located so that the maximum travel distance from the hazard to the extinguisher is not more than 50 feet (15 m). Extinguishers having a 40B rating must be located so that maximum travel distance from the hazard to the extinguisher does not exceed 50 feet (15 m).

COMPRESSED AND LIQUEFIED GASES
Storage of Compressed and Liquefied Gases

Flammable gases are stored in a variety of cylinders and tanks. Flammable gases are also handled through gas pipelines. All these types of containers must be carefully designed, built, and maintained (Figure 10.23). Cylinders for flammable gases must be built according to specifications and regulations set forth by the U.S. Department of Transportation (DOT) in the United States or the Canadian Transport Commission (CTC) in Canada. These specifications cover metal composition, joining methods, wall thickness, heat treatments, marking, proof testing, type of openings required, and cylinder testing. The regulations also govern safety devices, in-service transportation, design pressure, and the gases that the cylinders may contain. The Department of Transportation

Figure 10.23 Gases may be stored in small cylinders or in huge tanks. *Courtesy of Illinois Fire Inspectors Association.*

requirements apply only to those cylinders and tanks involved in interstate travel. Very small containers and those holding nonflammable cryogenics below 40 psi (276 kPa) are exempt.

Tanks used to store relatively small quantities of gas at moderate pressures are designed according to Section VIII (Unfired Pressure Vessels) of the Boiler and Pressure Vessel Code developed by the American Society of Mechanical Engineers (ASME). Tanks designed to hold large quantities of low pressure gas are built according to standards set forth by the American Petroleum Institute (API).

Inspectors should be aware of the techniques for safe installation and storage of compressed and liquefied gases. Careless handling of these materials may result in serious accident or death. Compressed and liquefied gases are most commonly stored in metal cylinders. The contents should be marked in large letters near the top of the cylinder (Figure 10.24).

Cylinders should be stored in a safe, dry, well-ventilated area specifically designed for storing compressed liquefied gases. Liquefied gas cylinders should be stored on a level fireproof floor, with the valve end up. Different types of gases should have a dividing wall between them. Empty cylinders should be stored in a separate area from full cylinders to avoid confusion.

If the cylinders are stored outside, they should be protected from adverse weather conditions, such as accumulations of ice and snow in the winter and continuous direct sunlight in the summer. Gas

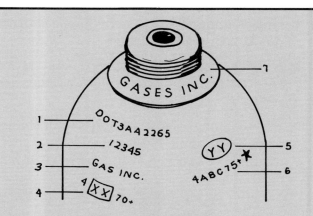

1. Cylinder Specification consisting of three sections:
   (a) DOT — Department of Transportation Regulatory body that governs use of cylinders.
   (b) 3AA — Specification of type and material of cylinder construction.
   (c) 2265 — Service pressure in pounds per square inch.

2. 12345 — Cylinder serial number (See Note A)

3. Gas Inc. — Identifying symbol (See Note A)

4. Manufacturing Data:
   (a) 4-70 — Date of manufacture and original test date
   (b) ⊠⊠ — Inspector's official mark
   (c) + — Cylinder qualifies for 110% filling

5. ⟨Y Y⟩ — Manufacturer's identifying symbol

6. Retest Markings:
   (a) 4-75 — Date of first 5 year hydrostatic retest
   (b) ABC — Retester identifying symbol
   (c) + — Cylinder requalifies for 110% filling
   (d) ★ — Cylinder qualifies for 10 year retest interval

7. Neck ring owner's identification

Notes:
A. Serial number and identifying symbol may be that of purchaser, user, or manufacturer.
B. "Spun" or "Plug" must be stamped near DOT mark when an end closure in the finished cylinder has been welded by the spinning process, or effected by plugging.
C. Markings "5" and "6" are usually shown diametrically opposite other markings on the cylinder neck.

Figure 10.24 Markings on the cylinder should be on the top of the tank, near the valve stem.

cylinders are designed for temperatures no greater than 130° (54°C).

Inspectors should ensure that the following additional safety precautions are observed:

- *Absolutely No Smoking* in the area of the cylinders.
- Adequate ventilation installed to prevent accumulation of flammable gas vapors.
- Electric lights are in a fixed position, enclosed in glass to prevent contact with gas, and the switch to the light is located outside the storage room.
- Glass light enclosures are equipped with a guard to prevent breakage.
- Outside valves on cylinders should be covered with safety caps when cylinders are not in use.
- Stored cylinders are anchored with chains around the tank so that they cannot accidentally be knocked over (Figure 10.25).

Figure 10.25 As a safety precaution, stored cylinders should be anchored with chains so that they cannot inadvertently fall over.

Hazards Associated with Compressed and Liquefied Gas Storage

The primary hazard associated with the storage of compressed and liquefied gases is the high pressure caused by the storage itself. Compressing gas into a container creates a tremendous source of potential energy; this energy is prevented from being released by the container itself. Normally, this stored energy is released in a controlled manner through a valve or other dispensing instrument. The *uncontrolled* release of compressed and liquefied gases may ultimately result in a fire and/or explosion of tremendous proportions. (**NOTE:** A container may be a large storage tank, small storage bottle, pipeline, or any other device designed to handle compressed and liquefied gases.)

There are four basic causes of container failure that lead to uncontrolled release of energy:

- Excessive pressures caused by the expansion of gases when they are heated
- Actual combustion of the gases within the container
- Structural failure of the container caused by flame impingement on the container
- Mechanical damage to the container, such as collision with another object

The uncontrolled release of gas from a container will result in pieces of the tank violently hurtled about and the formation of a vapor cloud. If the gas is flammable or combustible, any ignition source may result in an explosion or fire. If the container fails under fire conditions, this explosion or fire will most likely occur immediately upon failure of the container.

Ignition Sources

Fire inspectors must control ignition sources to prevent fires and explosions. To do so, they must analyze potential ignition sources, including electrical arcs, hot surfaces, open flames, as well as operations involving cutting, welding, grinding, and use of impact tools. All electrical fixtures within the hazard area must be explosion proof. Lift trucks used within the hazard area should be electrically powered. The electrical equipment on the trucks should be designed, constructed, and assembled for use in atmospheres containing flammable vapors. Lift trucks rated for this service will have the symbol EX stamped on the unit.

Static electricity is a potential ignition source where gases are transferred. Flowing gas that is contaminated with metallic oxides, scale, or liquid particles may develop a static electrical charge. If electrically charged gas contacts an ungrounded conductive material, the charge will be transferred to that body. Bonding the two receptacles together with a metal chain or cord and grounding them will prevent the static discharge. If the gas is transferred in a completely closed system, however, no charges can be transferred and the system need not be electrically grounded or bonded.

The combustion process of a mixture of flammable gas and air occurs in a series of steps. First, a flammable gas is released from the equipment, container, or piping holding it. The gas then mixes with the air and diffuses throughout the building or enclosure. How quickly the gas accumulates within the enclosure depends upon the rate of the gas release, the density of the gas, and the ventilation features of the enclosure. Eventually, the proportions of gas and air become such that the mixture falls within the flammable limits for that liquid gas and can be ignited. If the mixture is ignited, the amount of heat increases rapidly and is absorbed primarily by nearby combustibles. If the heated air cannot expand, the pressure will rise in the enclosure, very likely causing structural damage.

Handling Compressed and Liquefied Gases

When inspectors are evaluating handling procedures for gas cylinders, they should ensure that workers are properly trained and are under competent supervision. The following general rules will help control hazards when accepting and handling compressed gas cylinders:

- Accept only cylinders approved for use in interstate commerce for transportation of compressed or liquefied gases.

- Do not allow numbers or marks stamped on cylinders to be removed or changed.

- Do not allow cylinders to be dragged; however, they may be rolled on their bottom edge.

- Protect cylinders from cuts.

- Do not lift cylinders with an electromagnet.

- Do not drop cylinders or let them strike each other violently.

- Do not use cylinders for rollers, supports, or any purpose other than to contain gas.

- Do not tamper with safety devices on valves of cylinders.

- When in doubt about the proper handling of a compressed gas cylinder, or its contents, consult the supplier.

- When cylinders are empty, mark them "empty" or "MT." Replace valve caps if necessary.

GAS TRANSFER OPERATIONS AND SAFETY PRECAUTIONS

One of the major problems associated with transferring a gas from one container to another is that the gas, which may be flammable or toxic, will be released into the atmosphere. Often these gases are released as a result of inadequate or improperly maintained equipment, or operator negligence.

The suppliers of flammable gases should provide information about safe handling of gases. This information should include data about transporting, storing, setting up, and using cylinders safely. Information should also be supplied about using the hardware for the associated pressure system safely.

Bulk gas is usually transferred by replacing storage tank cylinders or by refilling cylinders or tanks. There should be a standard procedure for all disconnect and transfer operations. When deliveries are made, someone should verify that the cylinders are in good condition and are labeled properly. When deliveries are made to hazardous worksites, fire inspectors should require permits.

All cylinders on loading stations must be clearly labeled with the name of the gas being used in order to reduce the possibility that someone will connect the wrong containers (Figure 10.26 on pg. 262). Color coding can be useful, but should not be relied upon totally. The loading hose that is fur-

Figure 10.26 Although color coding is useful, cylinders on loading stations should be clearly labeled to avoid confusion.

nished by the receiver must be properly maintained and should be replaced at least every five years. When not in use, the hose should be stored away from direct sunlight. Bonding is generally not required if the hose has vaportight connections.

The drivers of the trucks should be warned when unloading operations are in progress so they will not attempt to drive away while the hoses are still connected. When the transfer operation is complete, nonreturnable containers must be disposed of properly. This includes having someone remove the plugs, puncture the shells so that they cannot be reused, decontaminate them, and finally dispose of them as nonhazardous waste.

The National Safety Council Research and Development Agency has developed a fact sheet that lists poor handling and storage practices involving gases. This list is reprinted below so that it may be used as a guide for fire inspectors of facilities using compressed gases:

- Failure to check cylinders on receipt for any problems.

- Storage and/or use of cylinders below ground level or in nonventilated or poorly ventilated areas. Leakage has resulted in flammable, toxic, corrosive, or asphyxiating atmospheres.

- "Locking-up" a connected (to cylinder) cutting torch head or instrument in a tool box cabinet. Leakage has allowed an explosive mixture to form.

- Not understanding the properties of the gas or the type of cylinder and associated hardware being used. Forced fits, use of adapters, and incompatible metals have resulted in failure of the container.

- Using transfer arrangements where differences in pressure and lack of check valves were not considered. Failure of the system or cross-contamination of incompatible materials has occurred.

- Not following proper startup, shutdown, and purging procedures, resulting in hazardous mixtures of gases.

- Turning a cylinder valve open very quickly, allowing the gas to expand and cool, and then quickly recompressing the gas, which heats it to a high temperature. Regulators and other devices have failed due to high temperatures.

- The presence of oil or grease in oxygen systems, which can cause an explosion.

- Improperly lifting cylinders by the cap, or by using magnets or slings. Not securing cylinders properly or not using a cap. Cylinders have fallen and been damaged, resulting in failure.

- Modifying a cylinder by welding legs, hooks, and so on. Cylinders have been weakened, resulting in failures (Figure 10.27).

Figure 10.27 Modifying a cylinder weakens it and can lead to failure. *Courtesy of Warren Isman.*

- Storing incompatible gases in the same area.

- Storage or setup of cylinders in the area where poor housekeeping practices have allowed an accumulation of combustible materials. Cylinders have vented because they were involved in fire.

- Neglected and old cylinders have corroded and failed, releasing their contents.

- Using cylinders that have been abused, dented, arc struck, in a fire, have broken or bent valves, or are severely corroded. Weakened cylinders have failed during use.

- Failing to ensure that all pressure system components are rated for the actual service pressure.

- Failing to provide and/or maintain pressure relief devices with appropriate setting and adequate pressure relief capacity.

- Not wearing adequate protective clothing and equipment when transferring liquefied gases. Skin contact has caused severe burns due to cryogenic temperatures.

- Failing to provide a separate relief device or sufficient capacity for each segment of a cryogenic system in which liquefied gas can be trapped or valved off.

- Liquefied gas containers have failed after a fall and a ruptured disk on the inner wall and/or the safety relief on the outer container failed to operate. (NOTE: Liquefied gas containers are multi-walled and are not constructed as strong as high-pressure gas cylinders.)

- Using instruments that have not been calibrated to test for explosive mixture, oxygen contents, and so on. False readings have contributed to various accidents.

Pipelines are generally used for transferring natural gas, LP gas, and some industrial gases, such as hydrogen, oxygen, ethylene, and ammonia. The U.S. Department of Transportation regulates design pressures, pipe materials, valves, meters, service lines, corrosion control, and testing requirements. NFPA 54, *National Fuel Gas Code*, provides additional recommendations and requirements.

When compressed gases are supplied to laboratories in systems using pipe, the pipe may fail and release a large amount of flammable or toxic gases. Installing excess flow devices can reduce the amount of gases that are released, but these devices may stick and remain open. The National Safety Council advocates using a flow-limiting, capillary-tube restrictor in place of a conventional excess flow device. This particular restrictor is safer because it has no moving parts and can fail only if it is plugged.

There are safety precautions that must be taken with capillary-tube restrictors. First, the area in which they are used must have a reliable ventilation system. Next, the maximum gas flow resulting if there is a failure must be small enough to be safely dissipated through dilution and normal room ventilation. Finally, the normal amount of gas used must be less than the safe maximum flow rate. When a capillary-flow restrictor is properly sized and installed, it provides dependable protection against the excess flow of hazardous gases into an area (Figure 10.28).

Transportation of Compressed and Liquefied Gases

Compressed and liquefied gases, like flammable and combustible liquids, are under strict federal regulation. The *Code of Federal Regulations, Title 49 Transportation* details these regulations, including the requirements of the Department of Transportation (DOT) for transporting these gases by air, motor vehicle, rail, and to some extent, vessel. Fire inspectors should have a copy of this code available for reference. Much of the material in

CFR 49 is adopted by reference from nationally recognized organizations (see Appendix J). States also regulate transportation, and local jurisdictions may enact additional regulations. For example, a jurisdiction may prohibit propane cylinders from being transported through tunnels. Fire inspectors must research the local, county, or state regulations that affect their particular jurisdiction.

Procedures for Investigating Known or Suspected Flammable Gas Leaks

When investigating known or suspected flammable gas leaks, immediate precautions should be taken. If investigation has determined that a concentration of gas is inside a building, the following actions should be taken:

- Clear the building or area of all occupants.
- Ventilate the affected portion of the building by opening windows and doors.
- Use every practical means to eliminate ignition sources.
- Terminate gas supply to the area.
- Investigate other buildings in the immediate area to determine if gas is present.

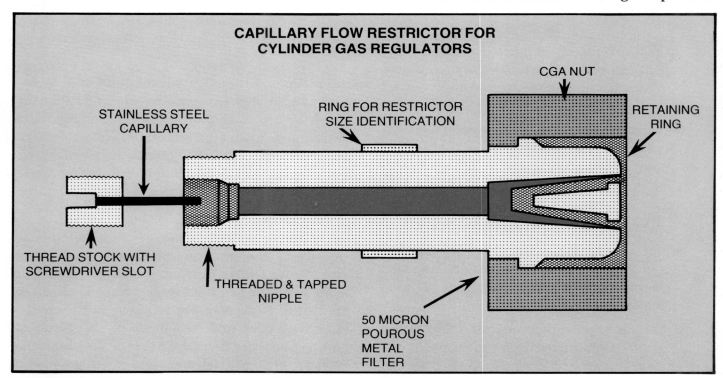

Figure 10.28 A capillary-flow restrictor is recommended to keep excess amounts of gas from flowing into an area.

- Notify all personnel in the area (the danger of explosion and exposure hazards may require evacuation) and the gas supplier (from a telephone away from the gas area).

If artificial illumination is needed to search for the source of a suspected leak, the investigator should use a battery operated flashlight (preferably the safest type) or approved safety lamps. During the search, do not operate electric switches. If the lights are already on, leave them on. If they are not on, do not turn them on.

Check the gas meter, carefully watching the test dial to determine if gas is passing into the meter. If a gas meter is not available, the investigation can be completed in one of two ways:

1. Attach a manometer, calibrated to read in increments of not more than 2 percent, to an appliance orifice. Turn off the gas supply and observe if a pressure drop occurs within three minutes.

2. Insert a pressure gauge between the gas container shutoff valve and the first regulator in the system. Open an appliance valve to drop the pressure between the first regulator and the container shutoff valve 10 psi (69 kPa). Wait for 10 minutes; the pressure gauge reading should remain the same.

Tie-in connections can be tested by using a soap solution. The soap solution is brushed on the exterior of the tie-in connection. When the gas is introduced into the system and the pressure is sufficient to indicate an existing leak, the soap solution will bubble at the location of the leaking gas. As with any testing, safety to employees and the public is a prime consideration.

Colorless and odorless gases are especially hazardous because special meters are needed to detect their presence. Liquefied gases produce a visible fog when released because of the refrigeration effect they have on the moisture in the atmosphere. This fog generally indicates the areas where there is a high concentration of the gas; however, flammable gas-air mixtures may exist at some distances beyond the fogged area.

CONTROL OF LEAKS

The first objective in controlling leaks is to divert, dilute, and disperse the gas so that it does not infiltrate buildings or come into contact with people or ignition sources (Figure 10.29). The next objective is to stop the flow of gas. The techniques used to control the gas leak will depend largely upon the physical and chemical characteristics of the gas and upon the characteristics of the container or pipe.

Figure 10.29 Here, a gas leak is being dispersed to minimize contact with people or ignition sources. *Courtesy of Warren Isman.*

Gas escaping from valve stems can usually be stopped by tightening the valve stem packing nuts. Other leaks can be halted by removing the gas from the container and either cleaning or replacing the defective parts. In other cases, the leak can be stopped by plugging or by clamping a resilient patch over the area where the gas is leaking until the situation can be corrected. If the container is leaking liquefied compressed gas in the liquid state, the container should be turned or tilted until only gas is escaping and then plugged or patched.

Basic Techniques for Extinguishing Fires Involving Compressed and Liquefied Gases

Because fire inspectors know about the hazards and inventories of compressed and liquefied gases in their jurisdictions, they may be called upon to advise fire suppression forces when fires occur involving these materials. Fire inspectors, therefore, should know the proper techniques for extinguishing such fires. Compressed and liquefied gas fires are **NOT EXTINGUISHED** unless the escaping gas or liquid can be shut off because escaping gas presents an explosion hazard that is more serious than the fire hazard. The basic technique is to evacuate the area, deny access, and cool the cylinder or container by applying large volumes of water. It is important to concentrate the streams of water on the upper portion of the cylinder or tank, where the vapor space is. If cylinders cannot be cooled, firefighters should also be evacuated from the area because a rupture (BLEVE) can be expected.

If the fire involves gas escaping from a relief valve, the valve should automatically close when the container is cooled sufficiently, and thus the fire will go out.

Portable Extinguishers for Compressed and Liquefied Gases

Fire extinguisher requirements for compressed and liquefied gases fall into the category of Class B hazards. Extinguishing agents in this category are halogenated agents, carbon dioxide, dry chemicals, foam, and AFFF. Because of the special hazards of fires involving compressed and liquefied gases, fire extinguishers containing other agents are not effective and should not be used. In addition, AFFF and other foam extinguishers, though rated for all Class B fires, are generally not effective on compressed gas fires. This is because foam works by blanketing the surface of the fuel it is extinguishing; usually, this is not possible with compressed gas fires. It is important to reemphasize that even though an extinguisher may be rated for Class B fires, it should not be used unless the flow of fuel can be stopped.

If it is absolutely necessary to extinguish a flowing gas fire, it can be accomplished by directing the agent directly into the flowing stream of gas.

COMBUSTIBLE SOLIDS
Combustible Metals

The requirements for storing combustible metals vary according to the applicable codes and the chemical properties of the metals. In general, storage buildings must be made of noncombustible materials and stored metals should be separated from other combustible materials. Wet scrap and filings should be stored outside because there is always the possibility that they will generate hydrogen gas and ignite spontaneously. Some metals, such as calcium, beryllium, and titanium, must be stored in separate, covered metal containers that are vented to the outside. Furthermore, some metals, such as sodium, require special precautions to keep out moisture. The containers used to store these metals should be kept in a dry, fire-resistive room or building used strictly for storing combustible metals.

The various fire codes of each jurisdiction limit the amount of combustible metals that can be stored. For example, magnesium, one of the most common combustible metals, can be stored in both interior and exterior facilities. If magnesium is stored outside, the pigs, billets, and ingots should be in piles that weigh no more than 1,000,000 pounds (453 600 kg). The piles must be separated by aisles that are at least as wide as half the height of the piles. The metal piles must also be separated from nearby combustibles, buildings, or property lines by a distance that is at least as great as the height of the nearest pile. Magnesium stored inside should be placed on noncombustible floors. The piles of magnesium should weigh no more than 500,000 pounds (268 000 kg) and should be separated by aisles that are at least as wide as half the height of the pile. There is no particular standard regarding the storage of all combustible metals. There are a number of standards that do apply to certain types of combustible metals:

- NFPA 48 *Storage, Handling, and Processing of Magnesium*
- NFPA 65 *Processing and Finishing of Aluminum*
- NFPA 481 *Production, Processing, Handling and Storage of Titanium*
- NFPA 482 *Production, Processing, Handling and Storage of Zirconium*

Combustible Dusts

Using and storing combustible dusts presents a unique hazard because of the possibility of dust explosion. All transfer machinery such as conveyors, augers, hoppers, and spouts must be enclosed in casings or other devices that are dusttight. Magnetic or pneumatic separators must be installed ahead of all crushers, crackers, and the like, to remove any metal that may cause sparks during the operation. Electrical equipment used in dust-laden atmospheres must be of the explosion-proof type as specified by NFPA 70, *National Electrical Code*. Some dusts can produce static electrical charges. Proper grounding or dissipators for static charge must be used for dusts having such properties.

Proper removal of combustible dust is essential to prevent explosions and fires. Vacuum cleaners and dust collection equipment remove dusts from their point of origin and deposit them in a safe location. Suction is applied at locations where dust escapes from processing machinery and the dust is conveyed to safe collection areas. Dust-ignition-proof motors will ensure a safe operation in dust concentrated areas. **CAUTION:** Dust should never be blown from surfaces; this suspends the dust in the air, increasing an explosion hazard. Rather, dust should be removed by vacuuming.

Systems that convey combustible dusts should be designed so that the fans are on the clean side of the collector. The system operates on a suction principle, so dusts do not have to pass through the fan. This reduces the chances of accidental ignition. If this design is unfeasible, the fan and bearing should be of a nonsparking material with ample clearance between the blades and the housing. The fan bearings and motors should be outside the casing unless they are specifically designed and tested for use in a dust-filled atmosphere.

Where processes produce very fine dusts or particles of combustible metals, wet collectors provide efficient dust control. Dust is removed by passing the flow of air through a water curtain and is then collected as a sludge submerged in water.

Good grounding of equipment can help reduce static electricity and reduce potential voltage differences between various parts of the collection system. Static charges can also build in free-falling streams of dust clouds. NFPA 77, *Static Electricity*, addresses the control of static electricity.

Code requirements and regulations relative to combustible dusts explain proper installation and operating procedures for various combustible dusts. The following list of NFPA codes can assist inspectors in determining safe processing and storage of these products.

- NFPA 651 *Manufacture of Aluminum and Magnesium Powder*
- NFPA 61A *Fires and Dust Explosions in Facilities Manufacturing and Handling Starch*
- NFPA 61B *Fires and Dust Explosions in Grain Elevators and Facilities Handling Bulk Raw Agricultural Commodities*
- NFPA 61C *Fires and Dust Explosions in Feed Mills*
- NFPA 61D *Fires and Dust Explosions in the Milling of Agricultural Commodities for Human Consumption*
- NFPA 91 *Installation of Blower and Exhaust Systems for Dust Stock, and Vapor Removal or Conveying*

EXPLOSIVES

When fire inspectors consider storage provided for explosives, they must ensure that the explosives are protected from external sources of energy and that employees are protected in case of explosion. Regulations concerning the manufacture, distribution, and storage of explosives are issued by the Bureau of Alcohol, Tobacco, and Firearms (BATF), which is part of the U.S. Department of the Treasury. Additional regulations are set forth by NFPA 495, *Manufacture, Transportation, Storage, and Use of Explosive Materials,* and may be enforced at the local level if they have been adopted as an ordinance.

General Storage Regulations for Explosives

The BATF regulations contain the requirements for the five types of storage facilities, or magazines. These requirements are as follows:

TYPE 1 MAGAZINE

A Type 1 magazine is a permanent facility for the storage of explosive materials sensitive to initiation by a number 8 test blasting cap and will mass detonate (Figure 10.30). These magazines are fire resistant, weather resistant, theft resistant, and bullet resistant. Materials stored in Type 1 magazines include dynamite or nonelectric blasting caps. Walls must be constructed of 8-inch (203 mm) blocks with hollow spaces filled with well-tamped sand, brick or solid cement block construction 8 inches (203 mm) thick, wood construction covered with 26-gauge metal having ¾-inch (19 mm) plywood or wood sheathing with a 6-inch (152 mm) space between the exterior and interior sheathing and the space filled with dry sand, or 14-gauge metal lined with 4-inches (102 mm) of brick, solid cement block or hardware, or walls filled with 6 inches (152 mm) of sand. Doors must be constructed of steel plate ⅜-inch (9.5 mm) thick and lined with four layers of ¾-inch (19 mm) tongue and groove hardwood flooring, or be constructed of a 14-gauge metal plate lined with 4 inches (102 mm) of hardwood. Roofs must be constructed of 14-gauge metal or ¾-inch (19 mm) wood sheathing covered by not less than 26 gauge metal, or other noncombustible roofing material. All exposed wood must be covered by similar material.

TYPE 2 MAGAZINE

A Type 2 magazine is a portable or mobile magazine used for outdoor or indoor storage of the same materials found in a Type 1 magazine (Fig-

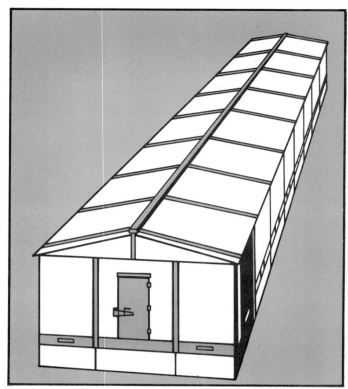

Figure 10.30 A Type 1 magazine is a permanent facility for storing such explosives as dynamite. *Courtesy of TREAD Corporation.*

ure 10.31). These facilities are fire resistant, theft resistant, weather resistant, and bullet resistant. Magazines used strictly for indoor storage need not be bullet resistant. The construction requirements vary based on end use of the magazine. Generally, the top, sides, bottom, and doors must be constructed of metal, lined with at least 4 inches (102 mm) of hardwood or similar bullet-resistive construction. Outdoor magazines must be supported to prevent direct contact between the floor and the ground.

TYPE 3 MAGAZINE

A Type 3 magazine is a portable facility for the temporary storage of explosive materials under constant attendance of a qualified employee. These facilities are bullet resistant, fire resistant, theft resistant, and weather resistant. Type 3 magazines must be secured by a five-tumbler padlock at all times. If the magazine is constructed of wood, the sides, bottom, and doors must be made of 4-inch (102 mm) hardwood and should be braced at all corners. The wood must then be covered with sheet metal of at least 20 gauge. If the magazine is constructed of metal, the sides, bottom, top and

Figure 10.31 A type 2 magazine used for outdoor storage must be supported to prevent direct contact with the ground. *Courtesy of TREAD Corporation.*

doors must be made of 12-gauge metal, lined with nonsparking material.

TYPE 4 MAGAZINE

A Type 4 magazine is a permanent, portable, or mobile magazine for storing explosives that will not detonate when initiated by a number 8 test blasting cap and that will not mass detonate. These facilities are fire resistant, theft resistant, and weather resistant. Materials likely to be found in these magazines include blasting agents, various water gels, black powder, and smokeless powder. This type of magazine must be constructed of masonry, wood covered with metal, or fabricated metal. Doors must be metal or wood covered with metal.

TYPE 5 MAGAZINE

A Type 5 magazine is a permanent, portable, or mobile magazine for storing explosive materials that will detonate when initiated by a number 8 test blasting cap. These facilities include tanks, tank trailers, tank trucks, semitrailers, bulk trucks, and bins. These magazines are theft resistant, and outdoor magazines are also weather resistant. They must be secured with at least one case hardened five-tumbler lock. If the magazine is vehicular, it must be immobilized when unattended.

No detonators, flammables, spark-producing tools, or other metal implements should be stored with other explosives. Piled oxidizers such as ammonium nitrate must be separated from readily combustible fuels. Cases of dynamite must be stored flat, with the top side up. Powder kegs must be stored either on end (bungs down) or on the side (seams down). Containers or explosives must be stored according to corresponding grades and brands. They must be located so that the brand and grade marks are visible. At no time may small arms ammunition be stored in the same magazine with Class A or B explosives.

Older stacks of explosives should be used or shipped first. Packages must be opened or repacked at least 50 feet (15 m) from the magazine. Loose explosives and open packages are not permitted inside the magazine. Where artificial lighting is necessary, only approved safety flashlights or electric lanterns may be used. When magazines need to be heated, hot water radiant heat, forced air directed over hot water, or low pressure steam coils should be used. Smoking must be prohibited in or near the magazine, and employees should not carry matches, lighters, or other flame-producing devices. The interior of the magazine must be kept clean and the surrounding area should be clear of grass, undergrowth, leaves, and other flammable debris. Empty dynamite cases and powder kegs must be removed from the magazine area.

A major consideration in the storage of explosives involves the separation distances between storage facilities and other structures. The Institute of Makers of Explosives has developed a table that gives distances explosives must be from a point of public contact. This table can be seen in Appendix J.

Security Measures for Explosives

The Bureau of Alcohol, Tobacco, and Firearms is responsible for enforcing Federal Explosive Laws under the Organized Crime Act of 1970. Prior to purchasing and using explosives in the United States, individuals must complete a form (4710) from the BATF and have it processed under penalty of perjury. Unfortunately, when manufactured explosives are unavailable, persons can easily obtain the necessary ingredients to compose homemade explosives or can steal explosives. This situation presents a real problem as these explosives may be used in acts of arson (incendiary bombs), terrorism, murder, or illegal entry. Thus, businesses that manufacture, store, or use explosives must have storage security programs.

Magazines should be supervised at all times by competent individuals who will be responsible for enforcing all safety regulations. Someone should open and inspect the magazine at least every three days to make sure that nothing has been damaged or removed from the magazine unless authorized. The security officer should report any theft, loss, or unauthorized removal of explosive materials within 24 hours to the BATF, the police department, and the Bureau of Fire Prevention.

Handling and Use of Explosives

Explosives must never be abandoned or left unattended in the open. The tools used to open explosives packages should be approved nonsparking ones. Empty packing materials such as boxes, paper, or fibers should be disposed of by remote burning.

When blasting is to be performed in congested areas or in close proximity to structures, railways, or highways, the blasting area should first be covered with a mat that will prevent fragments from being thrown. The safety of the general public and workers should be assured by the use of warning precautions which may include sirens or other signals. Before initiating an explosion, the person in charge should give a loud warning signal.

Transportation of Explosives

All operations involving the transportation of explosive materials must be conducted in accordance with the regulations of the Department of Transportation (DOT). Personnel involved in transporting explosives must not drive, load, or unload the vehicle recklessly. Further, individuals performing these duties may not smoke or carry matches or other flame-producing devices.

Before transferring explosives from one vehicle to another within a municipality, the appropriate personnel should notify the fire and police departments. Vehicles used to transport explosives should have a closed body or the load should be covered with a flame retardant and moistureproof tarpaulin. Two fire extinguishers with a combined rating of 2-A-10-B-C should be carried on the vehicle. Explosives must not be transported over or through any prohibited tunnels, subways, bridges, roadways, or elevated highways. At no time should vehicles transporting explosives be left unattended, nor should they be driven within 300 feet (91 m) of each other. The drivers of these vehicles should also avoid congested traffic and densely populated areas when possible. In some localities, specified routes will be named for the carrier.

Upon receipt of explosives at a terminal, the carrier shall notify the Bureau of Fire Prevention of the completed shipment. The carrier shall also notify the consignees of the shipment's arrival. Once the explosives have reached the terminal, the consignee must remove them within 48 hours.

Labeling of Explosives

Requirements for the labeling of packages are included in the *Code of Federal Regulations* (CFR), Title 49. Under these requirements, packaging containing a material meeting the definition of more than one hazard classification must be labeled with both types of labels. For example, an Explosive A material that is also a Poison A must be provided with both types of warning labels. Labels for Class A, Class B, and Class C explosives and blasting agents are orange with a black symbol and printing. These labels are affixed to the shipping container or are a part of it (Figure 10.32).

The following regulations from the National Fire Protection Association may be consulted for further information regarding the storage, use, and transfer of hazardous materials:

- NFPA 30 *Flammable and Combustible Liquids Code*
- NFPA 33 *Spray Application Using Flammable and Combustible Materials*
- NFPA 34 *Dipping and Coating Processes using Flammable or Combustible Liquids*
- NFPA 43C *Storage of Gaseous Oxidizing Materials*
- NFPA 48 *Storage, Handling and Processing of Magnesium*

Figure 10.32 Labels for explosives are orange with black symbols and lettering.

- NFPA 54 *National Fuel Gas Code*
- NFPA 58 *Storage and Handling of Liquefied Petroleum Gases*
- NFPA 59 *Storage and Handling of Liquefied Petroleum Gases at Utility Gas Plants*
- NFPA 61A *Fire and Dust Explosions in Facilities Manufacturing and Handling Starch*
- NFPA 61B *Fires and Explosions in Grain Elevators and Facilities Handling Bulk Raw Agricultural Commodities*
- NFPA 61C *Fire and Dust Explosions in Feed Mills*
- NFPA 61D *Fire and Dust Explosions in the Milling of Agricultural Commodities for Human Consumption*
- NFPA 65 *Processing and Finishing of Aluminum*
- NFPA 70 *National Electrical Code®*
- NFPA 120 *Coal Preparation Plants*
- NFPA 325M *Fire Hazard Properties of Flammable Liquids, Gases, and Volatile Solids*
- NFPA 328 *Control of Flammable and Combustible Liquids and Gases in Manholes, Sewers, and Similar Underground Structures*
- NFPA 395 *Storage of Flammable and Combustible Liquids on Farms and Isolated Construction Projects*
- NFPA 329 *Handling Underground Leakage of Flammable and Combustible Liquids*
- NFPA 385 *Tank Vehicles for Flammable and Combustible Liquids*
- NFPA 386 *Portable Shipping Tanks for Flammable and Combustible Liquids*
- NFPA 481 *Production, Processing, Handling and Storage of Titanium*
- NFPA 490 *Storage of Ammonium Nitrate*
- NFPA 495 *Manufacture, Transportation, Storage, and Use of Explosive Materials*
- NFPA 651 *Manufacture of Aluminum and Magnesium Powder*
- NFPA 1124 *Manufacture, Transportation, and Storage of Fireworks*
- *Fire Prevention Code* — American Insurance Association
- BOCA — *Basic Fire Prevention Code*
- BOCA — *Basic National Building Code*
- *Southern Building Code*
- *Uniform Building Code*
- State and local ordinances

RESPONSE AGENCIES FOR INCIDENTS INVOLVING HAZARDOUS MATERIALS

When a hazardous materials incident occurs, local, state, and federal agencies, as well as hazardous materials response teams, may be called upon to assist in controlling the incident. Local government agencies, especially fire and police departments, are usually the first to arrive at the scene of an incident. These individuals have the initial responsibility for

- Determining what materials are involved
- Determining the approximate amount of material

- Assessing hazards posed by resulting fires or spill
- Controlling access to the area
- Beginning evacuation of the surrounding area
- Rescuing trapped persons, if safely possible
- Implementing fire and spill control procedures
- Preventing further damage to the environment

In incidents such as these, planning and coordination of emergency services is essential. Planning enables hazardous response teams to disperse or contain the material(s) with as little risk to individuals or the environment as possible. Planning also enables those responsible for handling hazardous materials incidents to identify other agencies' particular strengths and limitations so that abatement efforts can be coordinated. The following are a number of state, local, and federal agencies that are typically involved in a hazardous materials incident.

Local Government Agencies

Fire Department — Personnel trained in hazardous materials handling procedures, specialized equipment and materials.

Police Department — Traffic and crowd control, evacuation, property protection, and investigation.

Emergency Civil Preparedness — Communications, crowd control, evacuation, and radiological services.

Public Works — Sand and other confinement material, construction equipment, personnel, and maps of streets, sewers, and natural watershed areas.

Public Health — Water, air, and soil monitoring, and poison control information.

Building Department — Consultants on manufacturing locations and structural damage.

Public Utilities — Gas, electric, water, sewer, and so on.

State Government Agencies

National Guard — Communications, crowd control, evacuation, transportation, radiological services, and helicopters.

State Police and Sheriffs — Crowd and traffic control, accident investigation, evacuation, property protection, and communications. (**NOTE:** In some states, they are in charge of handling the incident.)

Public Health — Monitoring and testing of water, air, and soil for contamination, treatment of injured personnel, access to hospitals, poison control centers, toxicologists, and hygienists.

Environmental Protection — Access to chemists, biologists, and other scientists with specialized knowledge of hazardous materials, and knowledge of cleanup contractors.

Agriculture — Soil and water testing, knowledge of pesticides and fertilizers.

Fire Marshal, Fire Academy, and Universities — Training and advice in hazardous materials response and suppression, and the use of specialized response equipment.

State Sponsored Hazardous Material Response Teams — To bolster the capabilities of local responders.

Federal Government Agencies

Department of Transportation (DOT) — Responsibility for regulating air, rail, ship, highway, and pipeline transportation. Of its various agencies, the following are especially important to emergency response:

- Materials Transportation Bureau (MTB) — Ensures uniform implementation of movement of hazardous materials by highway, rail, water, and air. The MTB designates which substances are hazardous, how they should be packaged and labeled, the use of placarding, the minimum training for and number of personnel required for the handling and transportation of hazardous substances. The MTB also stipulates the types, frequency, and procedures to be used in safety inspections.

- Federal Highway Administration (FHWA) — Highway safety, under the supervision of FHWA, is carried out by the Bureau of Motor Carrier Safety (BMCS). The BMCS conducts inspections of the facilities of interstate commercial carriers and their trucks while on the road. The BMCS serves to enforce proper loading, packaging, labeling, handling, and transportation of hazardous materials.

- Federal Railroad Administration (FRA) — Oversees all railway activities in the United States. Officials from the FRA inspect hazardous materials, track, equipment, signals, train controls, and operations.

National Response Center — Site emergencies involving significant chemical releases should be coordinated with federal response organizations. The federal government has established a National Contingency Plan (NCP) to promote the coordination and direction of federal and state response systems, and to encourage the development of local government and private capabilities to handle emergencies involving chemical releases.

To implement the NCP, a national organization was established, including a National Response Team (NRT), a network of Regional Response Teams (RRTs), a group of On-Scene Coordinators (OSCs), and a National Response Center (NRC). The NRC is the national terminal point for receipt of notification of significant chemical releases, and the OSCs are the interface between the onsite personnel and the federal response organizations. The OSC is the federal official responsible for ensuring that necessary response actions are taken to protect the public and the environment from the effects of a chemical release. Many federal agencies have specific technical expertise available to assist the OSC.

If a significant chemical release occurs at a hazardous waste site, the National Response Center in Washington, D.C. should be contacted (Telephone: 1-800-424-8802). The NRC will activate federal response under the National Contingency Plan.

United States Coast Guard (USCG) — All waterways, ports, and shipping by water are the responsibility of the USCG. Anyone loading, discharging, storing, or transporting hazardous materials must obtain the appropriate permit from the USCG. The USCG has National Strike Forces that combine skilled personnel and high seas equipment for the containment, cleanup, and disposal of hazardous material spills on water. The USCG also provides information on the chemical and physical properties of hazardous materials through its Chemical Hazards Response Information System (CHRIS). The system includes a four-volume manual, a regional contingency plan, the Hazard Assessment Computer System (HACS), and a response team. The four-volume set is as follows:

Volume 1: *A Condensed Guide to Chemical Hazards*
Volume 2: *The Hazardous Chemical Data Manual*
Volume 3: *The Hazard Assessment Handbook*
Volume 4: *Response Methods Handbook*

The Hazard Assessment Computer System (HACS) is Volume 3 on a computer. The computer can provide faster and more explicit assessment of an incident that can be done without such specialized equipment. The National Response Center (NRC) is operated 24 hours a day, 7 days a week, and can provide advice over the telephone. The number to call for assistance is 1-800-424-8802.

National Transportation Safety Board (NTSB) — The sole function of the NTSB is to conduct investigations and safety studies on accidents involving all forms of transportation. The NTSB is independent of the DOT and reports directly to the Executive Branch.

Environmental Protection Agency (EPA) — Water and air quality, solid waste, and pesticides are the primary concerns of the EPA (Figure 10.33 on pg. 274). The EPA has the Oil and Hazardous Materials Technical Assistance Data System (OHMTADS) which includes chemical, biological, toxicological, and commercial data on more than 1,000 chemicals. The EPA can be contacted through the National Response Center at 1-800-424-8802.

Department of Energy (DOE) — The DOE, under the Federal Radiological Monitoring and Assistance Plan (FRMAP), assists with planning and responses involving radioactive materials. When notified, a regional coordinator evaluates the problem, provides advice, and, when appropriate, dispatches a technical response team and notifies the Nuclear Regulatory Commission.

Nuclear Regulatory Commission (NRC) — The primary purpose of the NRC is to license and regulate radioactive materials. NRC allows DOT to regulate the transportation of radioactive materials, but the DOT and the NRC together have established regulations controlling packaging, labeling, handling, loading, and storing of radioactive materials. As general practice, the NRC regulates the routes used to transport radioactive materials.

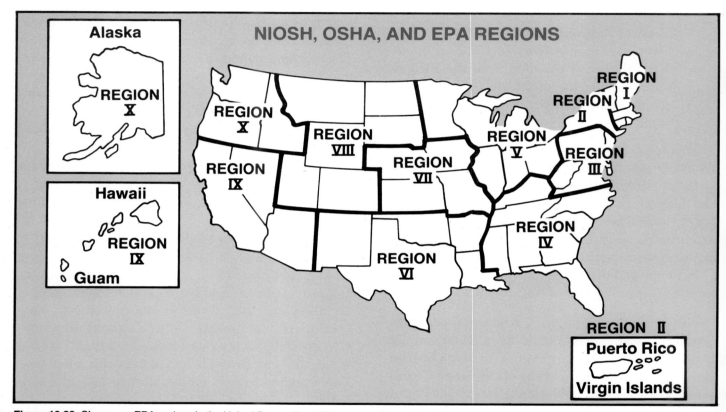

Figure 10.33 Shown are EPA regions in the United States. The EPA can provide technical data on many chemicals.

APPENDIX A
Excerpts of Procedural Codes

A-1: Duty to Inspect

Uniform Fire Code, 1985
Responsibility for Enforcement
Sec. 2.101 The chief shall be responsible for the administration and enforcement of this code. Under his direction, the fire department shall enforce all ordinances of the jurisdiction pertaining to:

(a) The prevention of fires.

(b) The suppression or extinguishing of dangerous or hazardous fires.

(c) The storage, use and handling of explosive, flammable, combustible, toxic, corrosive and other hazardous gaseous, solid and liquid materials.

(d) The installation and maintenance of automatic, manual and other private fire alarm systems and fire extinguishing equipment.

(e) The maintenance and regulation of fire escapes.

(f) The maintenance of fire protection and the elimination of fire hazards on land in buildings, structures and other property, including those under construction.

(g) The means and adequacy of each exit in the event of fire, from factories, schools, hotels, lodging houses, asylums, hospitals, churches, halls, theaters, amphitheaters and all other places in which people work, live or congregate from time to time for any purpose.

(h) The investigation of the cause, origin and circumstances of fire.

F-102.1 Enforcement officer: It shall be the duty and responsibility of the chief of the fire department or of the fire prevention bureau, or if there is not a jurisdiction fire department or fire prevention bureau, such officer as shall be designated by the appointing authority of the jurisdiction, or his duly authorized representative, to enforce the provisions of the Fire Prevention Code as herein set forth. The designated enforcement officer of this code is herein referred to as the fire official.

F-102.2 Inspections: The fire official shall inspect all structures and premises except single-family dwellings, and dwelling units in two-family and multifamily dwellings as often as may be necessary for the purpose of ascertaining and causing to be corrected any conditions liable to cause fire, contribute to the spread of fire, interfere with fire fighting operations, endanger life or any violations of the provisions or intent of this code or any other ordinance affecting fire safety.

A-2: Liability

F-102.5 Administrative liability: The fire official, officer or employee charged with the enforcement of this code, while acting for the jurisdiction, shall not thereby be rendered liable personally, and is hereby relieved from all personal liability for any damage that may accrue to persons or property as a result of any act required or permitted in the discharge of official duties. Any suit instituted against any officer or employee because of an act performed in the lawful discharge of duties and under the provisions of this code shall be defended by the legal representative of the jurisdiction until the final termination of the proceedings. The fire official or any subordinates of the fire official shall not be liable for costs in any action, suit or proceeding that may be instituted in pursuance of the provisions of this code; and any official, officer or employee, acting in good faith and without malice, shall be free from liability for acts performed under any of its provisions or by reason of any act or omission in the performance of the official duties in connection herewith.

F-102.6 Jurisdictional liability: The jurisdiction shall not be liable under this code for any damage to persons or property, by reason of the inspection or reinspection of buildings, structures or equipment authorized herein, or failure to inspect or reinspect such buildings, structures or equipment or by reason of the approval or disapproval of any building, structure or equipment authorized herein.

A-3: Right of Entry

1-3.4 Right of Entry.

1-3.4.1 To the full extent permitted by law, any fire official engaged in fire prevention and inspection work is authorized at all reasonable times to enter and examine any building, structure, marine vessel, vehicle or premises for the purpose of making fire safety inspections. Before entering a private building or dwelling, the fire official shall obtain the consent of the occupant thereof or obtain a court warrant authorizing entry for the purpose of inspection except in those instances where an emergency exists. As used in this section, "emergency" means circumstances which the fire official knows, or has reason to believe, exist, and which reasonably may constitute immediate danger to life and property.

1-3.4.2 Persons authorized to enter and inspect buildings, structures, marine vessels, vehicles, and premises as herein set forth shall be identified by proper credentials issued by this jurisdiction.

1-3.4.3 It shall be unlawful for any person to interfere with a fire official carrying out any duties or functions prescribed by this Code.

1-3.4.4 It shall be unlawful for any unauthorized person to use an official badge, uniform, or other credentials so as to impersonate a fire official for the purpose of gaining access to any building, structure, marine vessel, vehicle, or premises in this jurisdiction.

A-4: Alternatives

1-3.3 Alternatives.

1-3.3.1 Whenever this Code requires a particular system, condition, arrangement, material, equipment, or any other particular provision, the Fire Marshal may accept alternatives provided that such alternatives shall afford a substantially equivalent level of safety.

1-3.3.2 Application for Alternatives. Each application for an alternative shall be filed with the Fire Marshal and shall be accompanied by such evidence, letters, statements, results of tests or other supporting information as may be required to justify the request. The Fire Marshal shall keep a record of actions on such applications and a signed copy of the Fire Marshal's decision shall be provided for the applicant.

A-5: Revision of Codes

Section 15. Any person who shall violate any of the provisions of the Code hereby adopted; or shall fail to comply therewith; or shall violate or fail to comply with any order made thereunder; or shall build in violation of any details, statements, specifications, or plans submitted or approved thereunder; or shall operate not in accordance with the provisions of any certificate, permit, or approval issued thereunder, and from which no appeal has been taken; or who shall fail to comply with such an order as affirmed or modified by the Fire Marshal or by a court of competent jurisdiction within the time fixed herein shall severally for each and every violation and noncompliance, respectively, be guilty of a misdemeanor punishable by a fine of not less than *(Dollar Amount)* nor more than *(Dollar Amount)* or by imprisonment for not less than *(Number)* days nor more than *(Number)* days or by both such fines and imprisonment. The imposition of a penalty for any violation shall not excuse the violation nor shall the violation be permitted to continue. All such persons shall be required to correct or remedy such violations or defects within a reasonable time, and when not otherwise specified, the application of the above penalty shall not be held to prevent the enforced removal of prohibited conditions.

A-6: Board of Appeals

1-5 Board of Appeals. (See Appendix D-1-5 and Section 16 of Appendix C.)

1-5.1 Appointment. A Board of Appeals shall be appointed consisting of members, who by education and experience, are qualified to pass upon the application of this Code as it affects the interests of the general public. Board members shall not be officers, agents, or employees of this jurisdiction. All members and any alternate members shall be appointed and serve in accordance with the terms and conditions of the authority having jurisdiction. The Board shall establish rules and regulations for conducting its business and shall render all decisions and findings in writing to the Fire Marshal, with a copy to the appellant.

No more than one of said members or their alternates shall be engaged in the same business, profession or line of endeavor. No member of the Board of Appeals shall sit in judgment on any case in which the member, personally, is directly interested.

1-5.2 Purpose. The Board of Appeals shall provide for reasonable interpretation of the provisions of this Code and rule on appeals from decisions of the Fire Marshal.

1-5.3 Duties. The Board of Appeals shall meet whenever directed by the appointing authority for the purpose of interpreting the provisions of this Code and to consider and rule on any properly filed appeal from a decision of the Fire Marshal, giving at least five days notice of hearing, but in no case shall it fail to meet on an appeal within 30 days of the filing of notice of appeal. All of the meetings of the Board shall be open to the public.

APPENDIX B
Summaries of Recent Court Decisions Regarding Liability

Adams v. State, 555 P. 2nd 235 (1976)

   A suit against the state was filed following a fire in a motel in which five persons died. The State of Alaska had inspected the motel eight months before the fire and had failed to issue a letter to the owner citing the violations of the State Fire Safety Code, despite the fact the inspector had indicated to his superior that the motel presented an "extreme life hazard." The Supreme Court of Alaska reversed and remanded for trial a lower court's granting of the state's motion for judgment on the pleading. The court ruled that the statute that immunizes the state from tort claims arising out of failure to perform discretionary function did not immunize the state from negligent failure to alleviate known fire hazards.

Coffey v. City of Milwaukee, 74 Wis. 2d 526, 247 N.W. 2d 132 (1976)

   A tenant in an office building brought suit against the City of Milwaukee following a fire which damaged his tenant space. Despite arrival of the fire department in time to control and extinguish the fire, a defective standpipe was unable to furnish the necessary water to fight the fire. The building had been inspected, but the inspector had failed to detect and/or order replacement of the defective standpipe.

   The Supreme Court of Wisconsin affirmed a lower court ruling that overruled the city's and inspector's demurrers. It stated that building inspections do not involve a "quasijudicial" function within the meaning of governmental tort immunity statute and that the city could not claim it was merely performing a "public duty," since there was no distinction drawn between "public duty" and "special duty" owed to the tenant under the circumstances.

Grogan v. Kentucky, 577 S.W., 2d 4 (1979)

   Following the Beverly Hills Supper Club fire (May 28, 1977), suit was filed against the City of Southgate and the Commonwealth of Kentucky for failure to enforce laws dealing with fire safety. The Supreme Court of Kentucky held the city and commonwealth not liable, and that government ought to be free to enact laws without exposing its supporting taxpayers to liability for failures of omission in its attempt to enforce them.

Halvorsen v. Dahl, 89 Wash., 2d 673, 574 P. 2d 1190 (1978)

   The Supreme Court for the State of Washington revised and remanded for trial a superior court dismissal charging liability against the City of Seattle following a hotel fire in which a man was killed. The court ruled that the Seattle housing code did impose special duty to those individuals who reside in buildings covered by the code, and that the city had long-term knowledge of violations of that code by the building and had undertaken to force compliance on several occasions but had not followed through.

Gordon v. Holt. 412, N.Y.S., 2d., 534 (1979)

   The City of Utica, New York was defendant in a suit filed on behalf of tenants and owners of an apartment house following a fire. They claimed the city was liable because the building department issued a certificate of occupancy when there were known major violations of the building code. The Supreme Court, Appellate Division, Fourth Department, held that the city building code did not create a special duty flowing from the city to specific tenants of a particular building so as to permit those tenants to file suit for breach of duty owed to the public at large to provide adequate police and fire protection.

Cracraft v. City of St. Louis Park, 279 N.W., 2d 801 (1979)

Suit was brought against the City of St. Louis Park, Minnesota following an explosion and fire in a high school involving a 55-gallon (208 L) drum of duplicating fluid. Two students were killed and a third seriously injured. The courts held that the city was not liable as no special circumstances existed so as to create a special duty on the part of the city toward individual members of the public injured in the explosion.

Wilson v. Nepstad, 282 N.W., 2d, 664 (1979)

Following an apartment fire in Des Moines, Iowa which involved deaths and injuries, a district court dismissed a municipal court tort claim action. The Supreme Court of Iowa reversed and remanded the case on the basis that certain statutes were intended to impose municipal tort liability for negligence based on breach of statutory duty, in this case inspecting the property in a negligent manner and issuing an "inspection certificate" which by implication warranted the premises to be safe for the purpose of human habitation.

Modlin v. City of Miami Beach, Fla., 201 So. 2d 70 (1967)

The City of Miami Beach was the defendant in a case filed in behalf of a woman when a store mezzanine collapsed. The case against the city was based on negligent inspection during construction some five years previous which failed to discover the defect which led to the collapse. The Florida Supreme Court ruled that the building inspector was owed no special duty to the woman killed; and therefore, the city was not liable for the negligent inspection by its employee.

APPENDIX C
Model Code Organizations

This section explains model code organizations and the ways in which they provide a national forum for presenting and discussing proposed code revisions. It is important that fire inspectors understand this process so they can actively participate in code reform and revision.

These nonprofit organizations use a democratic committee system of active members to update and produce code revisions. Any individual, organization, or professional group can become a member and participate in the code revision process; however, designated individuals vote for their assigned jurisdiction. State and local officials are encouraged to participate in model code organizations because they are responsible for code enforcement in their respective jurisdictions.

NATIONAL FIRE PROTECTION ASSOCIATION, INC. (NFPA)
Batterymarch Park
Quincy, Massachusetts 02269
(617) 770-3000

Publishes the *National Fire Codes,* a compilation of standards, recommended practices, manuals, guides, and model laws. The NFPA also publishes the *Fire Protection Handbook,* various fire-related textbooks, instructional aids, and produces audiovisual materials. The basis for NFPA is the dissemination of timely standards developed by more than 170 technical committees. The NFPA consists of over 31,000 members from fire departments, business, industry, health care, architecture, engineering, manufacturing, local and state government, and other professional trade areas and organizations. Three of their most widely adopted standards are NFPA 54, *National Gas Code;* NFPA 70, *National Electrical Code;* and NFPA 101, *Life Safety Code.*

Services Provided
- Code maintenance through changes
- Code and standards revision or confirmation every three to five years
- Formal interpretation procedures
- Engineering advisory service

Code Development and Amendment Revision Process
Step 1: Public notice is posted in *Fire News* and in the *Federal Register* requesting proposals to amend an existing document or develop a new document. A technical committee may also develop its own proposal.

Step 2: Technical committees are assigned to review all proposals and prepare a report.

Step 3: A *Technical Committee Report* (TCR) is published and 60 days are allowed for public review and comments.

Step 4: All public comments and technical committee's reaction to the comments are published in *Technical Committee Documentation* (TCD).

Step 5: During the NFPA Annual (May) and Fall (November) meetings, TCR and TCD proposals are submitted for approval to the members.

Step 6: The NFPA Standards Council issues a revised document based on the evidence provided.

Step 7: The NFPA Board of Directors heads the appeals process, and the Standards Council heads the complaint process.

BUILDING OFFICIALS & CODE ADMINISTRATORS INTERNATIONAL, INC. (BOCA)
4051 West Flossmoor Road
Country Club Hills, Illinois 60477
(312) 799-2300

Publishes the *BOCA Plumbing Code, BOCA Building Code, BOCA Mechanical Code, BOCA National Fire Prevention Code, BOCA Property Maintenance Code,* and *BOCA Energy Conservation Code, The Building Official and Code Administration, BOCA Bulletin,* research reports, and Professional Development Series. (NOTE: These codes were formerly known as the *Basic Plumbing Code, Basic Building Code, Basic Mechanical Code,* and so forth.)

Services Provided
- Code maintenance through annual changes
- Examination planning
- Consultation
- Administrative studies of local government code enforcement agencies

Building Code Revision Process
Step 1: Submit proposed changes (additions, deletions, or modifications) to the executive office on BOCA code change forms.

Step 2: Proposed changes are published.

Step 3: Appropriate committees review, evaluate, and document the proposed changes in preparation for the public hearing during the BOCA midwinter conference.

Step 4: Approval, denial, and modification recommendations are discussed and noted at the midwinter conference.

Step 5: Recommendations from the midwinter conference are published and open to challenges. Time is allowed for parties to challenge the recommendations.

Step 6: During the July BOCA conference, challenged proposals receive a second hearing.

Step 7: Block votes are cast on unchallenged proposals and adopted according to the recommendation.

Step 8: BOCA's active members vote on the proposed changes on recommendations.

Step 9: Actions and recommendations are published.

Step 10: Code changes are published as either a supplement to the current code book or as a new edition.

INTERNATIONAL CONFERENCE OF BUILDING OFFICIALS (ICBO)
5360 South Workman Mill Road
Whittier, California 90601
(213) 699-0541

Publishes the *Uniform Building Code,* the *Uniform Fire Code, Building Standards* magazine, and related construction codes.

Services Provided

- Code maintenance through annual charges
- Examination planning
- Code interpretation
- Consultation
- Voluntary certification of members

Building Code Revision Process

Step 1: Submit proposed changes to ICBO code coordinator, including an explanation for the proposed change and supporting data. An ICBO code change form is available for submission but is not mandatory.

Step 2: Proposed changes are examined for duplication and acceptable in-house standards.

Step 3: The proposed changes are analyzed by ICBO staff developers. Possible effects the proposed changes might have on the codes are emphasized and noted.

Step 4: The proposed changes, supporting data, and analysis are sent to an appropriate technical committee for review.

Step 5: Technical committees conduct a semiannual open hearing on the proposed changes. Interested parties are given an opportunity to comment at this time. The committees then approves or disapproves the proposal, approves a revised proposal, or tables the proposal for further study.

Step 6: The status of the proposals, based on the results of the semiannual hearing, are published in the annual report of the code development committees. These results are subject to challenge by interested parties. If the results are unchallenged, they are assumed approved and published as an annual supplement. Every three years a new edition of the code book is published.

Step 7: Challenged results from the technical committee conference are reviewed during the ICBO annual meeting (usually held in October). Representatives of member jurisdictions vote on the challenged results. The results of this vote conclude the proposals.

Step 8: Code changes are published as an annual supplement or every three years as a new edition.

SOUTHERN BUILDING CODE CONGRESS INTERNATIONAL, INC. (SBCCI)

900 Montclair Rd.
Birmingham, Alabama 35213-1206
(205) 591-1853

Publishes the *Standard Building Code, Standard Fire Prevention Code,* and related construction codes.

Services Provided

- Code maintenance through annual charges
- Code interpretations
- Consultation

Building Code Revision Process

Step 1: Submit proposed changes to SBCCI by April 1.

Step 2: Proposed changes are published in *The Blue Book* and sent to members 90 days prior to the annual conference (usually held in October or November).

Step 3: Interested parties are given an opportunity to comment during the month of July.

Step 4: Committees review the suggested changes and publish their recommendations *The Red Book*, during the month of October.

Step 5: Recommendations are submitted to the SBCCI Annual Conference and discussed. The members approve, deny, or table the recommendations for further study. A two-thirds majority vote is required to veto a committee's recommendation.

Step 6: A letter ballot is sent to active members. They cast their vote and return their ballot before February.

Step 7: Code changes are published as a pamphlet, an annual supplement, or every three years in a new edition of the code book.

OTHER GROUPS AND ASSOCIATIONS INVOLVED IN CODE FORMULATION AND ADMINISTRATION

AMERICAN PLANNING ASSOCIATION
1776 Massachusetts Avenue N.W.
Washington, D.C. 20005
(202) 872-0611

Major Publications
- *Journal of the American Planning Association*
- *Planning*
- *APA News: Planning Advisory Service Reports*
- *Land Use Law and Zoning Digest*

AMERICA INSTITUTE OF ARCHITECTS
1735 New York Avenue N.W.
Washington, D.C. 20006
(202) 785-7300

Major Publications
- *AIA Journal*
- *AIA J Memo*

ASSOCIATION OF MAJOR CITY BUILDING OFFICIALS (AMCBO)
c/o NCSBCS
481 Carlisle Drive
Herndon, Virginia 22070

Membership limited to cities of 500,000 plus.

BUILDING RESEARCH ADVISORY BOARD
2101 Constitution Avenue N.W.
Washington, D.C. 20418
(202) 389-6348

Major Publications
- Technical Reports
- Survey of Practices
- Newsletter

COUNCIL OF AMERICAN BUILDING OFFICIALS (CABO)
5201 Leesburg Pike
Falls Church, Virginia 22041
(703) 931-4533

Major Publication
- *CABO One and Two Family Dwelling Guide*

INTERNATIONAL ASSOCIATION OF ELECTRICAL INSPECTORS (IAEI)
802 Busse Highway
Park Ridge, Illinois 60068
(312) 696-1455

INTERNATIONAL ASSOCIATION OF FIRE CHIEFS (IAFC)
1329 18th Street N.W.
Washington, D.C. 20036
(202) 833-3420

Major Publications
- *Fire Chief*
- *Washington Scene*
- *Labor Management Report*

INTERNATIONAL ASSOCIATION OF PLUMBING AND MECHANICAL OFFICIALS (IAPMO)
5032 Alhambra Avenue
Los Angeles, California 90032
(213) 223-1471

Major Publications
- *Uniform Plumbing Code*
- *Uniform Plumbing Code Training Manual*
- *Uniform Mechanical Code*
- *Uniform Solar Energy Code*
- *Uniform Swimming Pool Code*

NATIONAL ASSOCIATION OF HOUSING AND REDEVELOPMENT OFFICIALS (NAHRO)
2600 Virginia Avenue N.W.
Washington, D.C. 20037
(202) 333-2020

Major Publications
- *Journal of Housing*
- *NAHRO Monitor*
- Rehabilitation and Neighborhood Conservation Series
- Directory of Local Agencies
- *Housing for the Handicapped and Disabled: A Guide for Local Action*
- Handbook of Commissioners

APPENDIX D
Authoritative Reference Sources

FIRE PREVENTION CODES

NFPA 1, *National Fire Codes®* — An 11-volume set of fire codes and standards containing authoritative guidance on almost every phase of fire protection. At least 25 percent of the codes are updated annually. Although these fire codes are minimum standards and are purely advisory, they are widely used as a basis of good practice for laws or ordinances. It is possible to adopt them by reference rather than incorporate them by law. This procedure permits the easy adoption of revised editions as they are published. (**NOTE:** *National Fire Code®* is a Registered Trademark of the National Fire Protection Association, Inc., Quincy, MA.)

The BOCA *National Fire Prevention Code*
Building Officials & Code Administrators International, Inc.
4051 West Flossmoor Road
Country Club Hills, Illinois 60477

Uniform Fire Code
International Conference of Building Officials
5360 South Workman Mill Road
Whittier, California 90601

Standard Fire Prevention Code
Southern Building Code Congress International, Inc.
900 Montclair Road
Birmingham, Alabama 35213-1206

National Fire Code of Canada
National Research Council Canada
Ottawa, Ontario
Canada K1A OR6

BUILDING CODES

Building codes are designed to control construction, alteration, repair, moving, and demolition of buildings. They provide rules for public safety that can be applied as law under the authority of police power. In addition to state building codes, there are five recognized model building codes currently used in the United States and Canada. The codes may differ in form and arrangement.

Uniform Building Code
International Conference of Building Officials
5360 South Workman Mill Road
Whittier, California 90601

The BOCA *National Building Code*
Building Officials & Code Administrators International, Inc.
4051 W. Flossmoor Road
Country Club Hills, Illinois 60477

Standard Building Code
Southern Building Code Congress International, Inc.
900 Montclair Road
Birmingham, Alabama 35213-1206

National Building Code of Canada
National Research Council Canada
Ottawa, Ontario
Canada K1A OR6

ADDITIONAL REFERENCE SOURCES
National Fire Protection Association (NFPA)

Fire Protection Handbook — A comprehensive reference that covers almost every aspect of fire protection engineering. It is often referred to as the "Bible" of fire protection.

NFPA Inspection Manual — A pocket-size publication written primarily for private fire inspection; however, most of the information contained in this guide is helpful to fire inspectors. This publication explains various types of hazards encountered during inspections.

Conducting Fire Inspections. A Guide For Field Use — A pocket-size publication that identifies individual classes of occupancies. This guide highlights special areas to inspect and includes checklists for maintaining accurate inspections.

Life Safety Code Handbook — A publication that explains and illustrates NFPA 101, *Life Safety Code* in greater detail.

National Electrical Code Handbook — A publication that explains and illustrates NFPA 70, *National Electrical Code* in greater detail.

SPECIAL INTEREST BULLETINS

A series of over 300 short bulletins published by the *American Insurance Services Group, Inc.* (AIA) on various fire service related subjects. Many of the topics covered in the bulletins are of interest to fire inspectors.

In addition, many organizations and agencies actively promote various phases of fire protection and prevention:

American Insurance Association (AIA)
85 John Street
New York, New York 10038

Alliance of American Insurance
20 North Wacker Drive
Chicago, Illinois 60606

Bureau of Explosives
Association of American Railroads
1920 L Street, N.W.
Washington, D.C. 20036

Canadian Underwriters Association
Fire Protection Engineering Division
460 St. John Street
Montreal 1, Quebec
Canada

Factory Mutual System
1151 Boston-Providence Turnpike
Norwood, Massachusetts 02062

Industrial Risk Insurance (IRI)
85 Woodland Street
Hartford, Connecticut 06102

International Association of Fire Chiefs (IAFC)
1329 18th Street N.W.
Washington, D.C. 20036

International Association of Fire Fighters (IAFF)
1750 New York Avenue N.W.
Washington, D.C. 20006

National Bureau of Standards (NBS)
United States Department of Commerce
Washington, D.C. 20234

National Fire Protection Association (NFPA)
Batterymarch Park
Quincy, Massachusetts 02269

National Safety Council
Chicago, Illinois 60611

Southwest Research Institute (SWRI)
6220 Culebra Road
San Antonio, Texas 78284

Underwriters Laboratories, Inc.
333 Pfingsten Road
Northbrook, Illinois 60062

United States Department of Agriculture
Information Division
Washington, D.C. 20250

United States Department of Interior
Safety Division
Bureau of Mines
Washington, D.C. 20240

United States Department of Treasury
Bureau of Alcohol, Tobacco, and Firearms
1200 Pennsylvania Avenue, N.W.
Washington, D.C. 20226

United States Fire Administration
Federal Emergency Management Administration
Washington, D.C. 20036

APPENDIX E
Standard Map Symbols

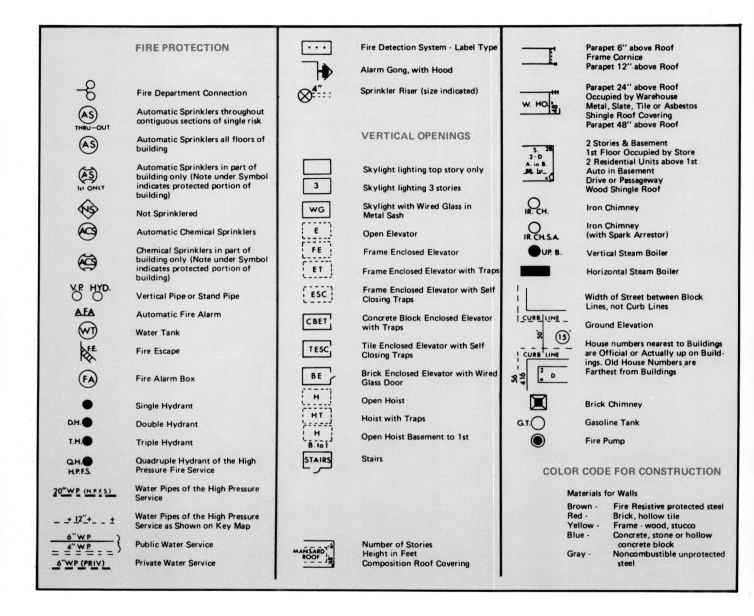

FIRE PROTECTION

	Fire Department Connection
AS (THRU—OUT)	Automatic Sprinklers throughout contiguous sections of single risk
AS	Automatic Sprinklers all floors of building
AS (1st ONLY)	Automatic Sprinklers in part of building only (Note under Symbol indicates protected portion of building)
NS	Not Sprinklered
ACS	Automatic Chemical Sprinklers
ACS	Chemical Sprinklers in part of building only (Note under Symbol indicates protected portion of building)
V.P HYD.	Vertical Pipe or Stand Pipe
AFA	Automatic Fire Alarm
WT	Water Tank
F.E.	Fire Escape
FA	Fire Alarm Box
●	Single Hydrant
D.H. ●	Double Hydrant
T.H. ●	Triple Hydrant
Q.H. ● H.P.F.S.	Quadruple Hydrant of the High Pressure Fire Service
20"W.P (H.P.F.S.)	Water Pipes of the High Pressure Service
_ -+ 12"+_ - _ ±	Water Pipes of the High Pressure Service as Shown on Key Map
6" W.P / 4" W.P	Public Water Service
6"W.P (PRIV)	Private Water Service

VERTICAL OPENINGS

• • •	Fire Detection System - Label Type
	Alarm Gong, with Hood
⊗ 4"	Sprinkler Riser (size indicated)
	Skylight lighting top story only
3	Skylight lighting 3 stories
WG	Skylight with Wired Glass in Metal Sash
E	Open Elevator
FE	Frame Enclosed Elevator
ET	Frame Enclosed Elevator with Traps
ESC	Frame Enclosed Elevator with Self Closing Traps
CBET	Concrete Block Enclosed Elevator with Traps
TESC	Tile Enclosed Elevator with Self Closing Traps
BE	Brick Enclosed Elevator with Wired Glass Door
H	Open Hoist
HT	Hoist with Traps
H B. to 1	Open Hoist Basement to 1st
STAIRS	Stairs
MANSARD ROOF	Number of Stories / Height in Feet / Composition Roof Covering

Parapet 6" above Roof
Frame Cornice
Parapet 12" above Roof

W. HO. — Parapet 24" above Roof
Occupied by Warehouse
Metal, Slate, Tile or Asbestos
Shingle Roof Covering
Parapet 48" above Roof

S. 2-D A. in B. 1st — 2 Stories & Basement
1st Floor Occupied by Store
2 Residential Units above 1st
Auto in Basement
Drive or Passageway
Wood Shingle Roof

IR. CH. — Iron Chimney

IR. CH. S.A. — Iron Chimney (with Spark Arrestor)

UP. B. — Vertical Steam Boiler

— Horizontal Steam Boiler

CURB LINE — Width of Street between Block Lines, not Curb Lines

(15) — Ground Elevation

CURB LINE — House numbers nearest to Buildings are Official or Actually up on Buildings. Old House Numbers are Farthest from Buildings

— Brick Chimney

G.T. — Gasoline Tank

— Fire Pump

COLOR CODE FOR CONSTRUCTION

Materials for Walls

Brown -	Fire Resistive protected steel
Red -	Brick, hollow tile
Yellow -	Frame - wood, stucco
Blue -	Concrete, stone or hollow concrete block
Gray -	Noncombustible unprotected steel

APPENDIX F
Example of a Citation Program

CODE CITATION POLICY AND PROCEDURES

Section I. Purpose

1.1   To gain compliance with the *Uniform Fire Code,* California Administrative Code and Title 19 when all reasonable efforts have been unsuccessful.

1.2   A course of legal action to be taken when a condition exists that causes a threat to life or property from fire or explosion.

Section II. Background

2.1   During the year (1977), the Fire Department wrote 2,426 violations, achieved 1,698 corrections and had 728 outstanding violations.

2.2   Our present process of enforcement (City Attorney, District Attorney, Office Hearings, filing of complaint, etc.) does not lead itself to providing uniformity of compliance within the community. The majority of fire violations written are characteristic of the following three conditions:

2.2.1   *Transient* problems such as overcrowding of public entertainment facilities, illegal parking in fire lanes, mischievous fire setting.

2.2.2   *Changeable* or portable situations such as illegal locking devices on public exit doors and obstructions to aisles or exitways.

2.2.3   *Maintenance* of fire extinguishing and alarm systems, portable situations such as electrical violations, housekeeping including outdoor fire hazards.

2.3   Transient violations are specific occurrences which should be acted upon immediately through a citation process. Changeable violations are corrected by the person responsible (in most cases) on a temporary basis, but are changed back after the inspector leaves the premises. The same situations are encountered yearly and are not being permanently abated.

2.4   Citizen awareness of the Fire Department's ability to cite for violations would create an effective deterrent in maintaining corrective abatements on a more permanent basis. The citation process would be used on a discretionary basis and would be very cost effective from the standpoint of available manpower utilization and steady increasing workload demand.

Section III. Policy

3.1   Members of the Fire Prevention Division who are authorized by the Division chief in charge shall have the authority to issue citations for fire and life safety violations of the Uniform Fire Code and the California Administrative Code, Title 19, Ordinance No. 1088, Resolution No. 1296.7.

3.2   It is the intent of the Fire Prevention Division to achieve compliance of the majority of code violations by traditional means of inspection, notification, the granting of reasonable time limits to comply and reinspection.

3.3   Citations shall be issued by Fire Prevention Personnel when the following conditions exist:

3.3.1   Failure to gain reasonable compliance for Uniform Fire Code Violations.

3.3.2   Deliberate or mischievous fire setting not involving property loss.

3.3.3   Justification is evident that the violation was restored after inspection (Inspection records must verify the facts of violation).

3.3.4   Obstruction of fire lanes.

3.3.5   Upon direction of the City Attorney's office.

Section IV. Definitions

4.1   Reasonable time to comply.

4.1.1   Generally means after the third visit or second reinspection, with proper time intervals between visits for responsible party to make corrections of conditions.

4.2   Justification is evident violation was restored after inspection.

4.2.1   When the same violation has been written two times during any one year period of a facility inspected two times or more annually.

4.2.2   When the same violation has been written two times during any two year period of a facility inspected one time per year.

Section V. Procedure

5.1   Citations issued under the conditions as set forth in policy should have previous notification history as set forth below.

5.1.1   Exiting or overcrowding of public assembly facilities requires two previous notices of violation.

5.1.2   Engine company referrals to Prevention Division require engine company survey and reinspection, notice of violation written by Inspector.

5.1.3   Prevention inspection (originated by Fire Prevention) requires survey report and notice of violation.

5.1.4   Repeated violation which is corrected during inspection process and restored to violation condition afterwards, requires previous history of specific abatement process within the previous two years with the same responsible party.

5.1.5   Illegal controlled burning requires a warning notice of violation issued. Second offense by same party requires citation to be issued.

5.1.6   Deliberate mischievous fire setting by persons of responsible age requires a citation to be issued to person or persons committing act.

5.2   Citing for misdemeanor violations. The citation is a release stating the defendant will appear in court or post bail in lieu of physical arrest. If not sure if you have the owner, check the liquor license and cite whomever is listed. (It may be a corporation). If no liquor license, cite whomever is listed on the business license. If this is not available, check City Business License Division for correct owner.

5.2.1   Try to give citation directly to the owner. You may have to return at a later time to catch him or her in the place of business. If the owner is not available, mail the citation (unsigned & certified mail) to the owner's address listed on the liquor or business license.

5.2.2   The citation must be signed by the responsible party for the premises or the person committing the act in cases of transient fire setting or illegal burning.

5.2.3   All misdemeanors by the bail schedule require a court appearance by the violator. At that time the defendant appears and either (a) pleads guilty, in which case we hear nothing more about it; (b) has matter continued, which is not our concern; or (c) pleads not guilty, in which case the court sets a date for trial and for pretrial conference, and notifies the City Attorney's office or the acting agency. It is the latter hearing date, set by the court, that requires the citing officer's appearance.

5.3   Citing for Infractions. The procedure for writing the citation for an infraction is the same as for a misdemeanor violation.

5.3.1   All infractions can be paid at the court clerk's office or by mail (after phoning clerk to find out what their final assessment will be). Do not indicate to the recipient of the citation what you think the fine will be.

5.4   Failure to Appear in Court. When the violator fails to appear in court on the appointed date, the citation is forwarded to the District Attorney's office, which will then notify the agency issuing the citation. A declaration must be filled out by the issuing officer at which time a warrant will be issued for the violator. A maximum of three weeks from the due date on the citation will be allowed for the officer to complete the paper work and the District Attorney to act.

5.5   Failure to Abate Violation. After the date of appearance has expired a reinspection of the violation is warranted. Failure to abate violation requires that a second citation be issued.

5.6   Attach copies of previous notices issued with your copy of citation. Submit citation copy and notices to the Fire Prevention Officer, who will review them and forward to the proper authorities.

5.7   The Department citation logs shall be filled out and kept current. Log will be posted on wall behind Inspector's desk. The Prevention Officer and/or Division Chief shall be notified the following day of any citations issued by Fire Prevention personnel.

5.8   The Division Chief shall maintain a citation book record indicating which citation book numbers have been assigned to officers.

5.9   Attached is a schedule of recommended bails set for violations. Bail costs are required if violator is physically arrested and may be used by judge to determine first offense fines. (**NOTE:** The bail schedule is lengthy and is not printed here.)

Section VI. Exceptions
6.1   All exceptions or deviations shall be discussed with the Division Chief and City Attorney prior to writing of a citation.

Courtesy of Palm Springs, California Fire Department.

APPENDIX G
Standpipe Testing and Inspection Procedure

OBJECTIVE:

To ensure that standpipe systems will perform adequately for fire department use, all standpipes will be tested on a regular basis. Property owners are required to maintain fire protection systems in working order, and biannual testing of standpipe systems should assure that these systems are serviceable.

ALL SYSTEMS:

- Begin by completing the General Information section of the Standpipe Test Form.

- Complete the applicable Dry Systems or Wet Systems section of the Test Form during the test.

- Note defects in the Remarks section of the Test Form. Notify the responsible party that repairs must be performed within a reasonable time (generally two weeks). If more time is necessary, notify the Fire Prevention Bureau.

- If the standpipe is found to be inoperable, notify the appropriate office. Repairs must be made by a licensed contractor. When repairs are completed, the owner is to notify the fire department, which will schedule a retest.

- When the test is completed, return the form so it can be filed appropriately.

  Tests shall include each riser in the building. At any time, if a defect is noted in the system, notify a responsible party representing the building at once. Do not proceed with the test if there is any danger of damage resulting without verbal permission by the responsible party.

DRY SYSTEMS TEST PROCEDURE

Air Test

Prior to filling with air, dry standpipes shall be tested for continuity with air pressure. The steps for conducting an air pressure test are as follows:

Step 1: Locate all fire department connections and verify that they are accessible and clear of debris inside if uncapped. Survey all parts of the system to verify that there are no weaknesses, open valves, or missing caps. *Verify that all connection threads are compatible with local fire department threads.* (Blindcaps may be used for this purpose.)

Step 2: Attach a 2½-inch (65 mm) gate valve, if one is not present, and an air pressure gauge cap to the uppermost outlet of each riser. If there are more than two risers, repeat the air test until all risers have been tested.

Step 3: Connect the air compressor to the fire department connection and pressurize the system. Personnel at top will advise if pressure is rising and when 25 psi (175 kPa) is reached. At 25 psi (175 kPa), close the valve, allowing air to enter the system, and shut down the compressor. The system should hold 25 psi (175 kPa) for five minutes.

Step 4: If pressure cannot be held, survey the system to locate leaks and correct them if possible. If corrections cannot be made, indicate on the Test Form and notify a responsible party representing the building that repairs are necessary, allowing a set reasonable time for corrections to be made.

Step 5: To release pressure:
- (a) close the 2½-inch (65 mm) outlet valve at the air pressure gauge.
- (b) bleed pressure at the bleeder valve on the air pressure gauge cap.
- (c) remove the gauge cap.
- (d) open the 2½-inch (65 mm) valve slowly.
(USE CAUTION AS DEBRIS MAY BE EJECTED).

Water Test

A water test is performed only after successful completion of the air pressure test. Laying a supply line will most likely be necessary.

Step 1: Send a crew to the roof and connect an in-line gauge/valve and the flow diffuser to the roof manifold. If there is no roof manifold, stretch a 2½-inch (65 mm) hoseline to the roof from the uppermost outlet.

Step 2: Connect a 2½-inch (65mm) gate valve to the fire department connection. Connect one 3-inch (77 mm) line from the fire department pumper to the gate valve and allow water to enter into the system at hydrant pressure.

Step 3: Increase pump discharge pressure slowly until approximately 10 psi (70 kPa) is indicated on the gauge. This is done to flush air and debris from the system.

Step 4: Shut off the flow from the diffuser at the in-line gauge/valve. Pressurize the system until the pump discharge gauge reads 50 psi (345 kPa) greater than the pump discharge pressure required for a working stream at the roof. Pressure should be 100 psi (690 kPa), the standard desired nozzle pressure, or 200 psi (1 380 kPa), whichever is greater. Close the gate valve at the fire department connection and shut down the pump. The pressure at the roof gauge should remain unchanged for 10 minutes. If it does not, survey the entire system for leaks. Note any problems on the Test Form and notify a responsible party representing the building of repairs that are necessary, allowing a reasonable time for corrections to be made.

Step 5: Use the diffuser to flush every other riser, if possible, either directly from the roof manifold or with a 2½-inch (65 mm) handline from the uppermost outlet.

Step 6: Bleed pressure and drain the system at the lowest available outlet or drain. Do not hammer against clapper valves in the Fire Department connection to drain. When the system is drained, the test is complete.

WET SYSTEMS TEST PROCEDURE

Systems that are normally kept wet do not require an air pressure test.

Step 1: Locate all fire department connections and verify that they are accessible and clear of debris inside if uncapped. Survey all parts of the system to verify there are no weaknesses, open valves, or missing caps. If present, inspect occupant use hose cabinets for proper equipment and hose condition, noting all deficiencies on the Test Form. *Verify that connection threads are compatible with local fire department threads.* (A blindcap may be used for this purpose.)

Step 2: If the system appears in good condition, begin the test. If not, have a responsible party representing the building inspect the potential defects and either give verbal permission to complete the test or stop the test and set a reasonable time period for corrections to be made. Note problems on the test form.

Step 3: If there is no fire pump, locate the indicating control valve for supplying the system and ensure that it is fully open. Proceed with Step 7.

Step 4: Begin in the fire pump room. Verify that the OS&Y control valve for city water supplying the system is fully open. There should be pressure gauges at the fire pump intake and discharge.

Step 5: Inspect to see if the fire pump is monitored for "Power Loss" and "Pump Running" conditions on a fire alarm control panel, and if not, indicate under Remarks on the Test Form.

Step 6: If there is a fire pump, leave one person at the pump to record rated flow, rated pressure, and pressures developed during the test.

Step 7: Send a crew to the roof and connect the in-line gauge/valve and the flow diffuser to the roof manifold or connect a hose to the roof from the uppermost outlet.

Step 8: Open the valve fully, and record the flow and roof residual pressure.

Step 9: If a fire pump is present, record the pump intake pressure, and the pump discharge pressure. (**NOTE:** The pump should operate automatically and shut off within seven minutes after the water flow is shut down. Internal combustion engine driven pumps and some electric pumps must be stopped manually using the "Stop" button. DO NOT PLACE THE CONTROLLER IN THE "OFF" POSITION.)

Step 10: Shut off the flow from the diffuser and connect a 2½-inch (65 mm) gate valve to the fire department connection. Connect one 3-inch (77 mm) hoseline from the fire department pumper to the gate valve and allow water to enter the system at hydrant pressure.

Step 11: Pressurize the system until the pump discharge gauge reads 50 psi (345 kPa) greater than the fire pump discharge pressure noted from Step 7 or 200 psi (1 380 kPa), whichever is greater. Close the gate valve at the fire department connection and shut down the pump. The pressure at the roof gauge should remain unchanged for 10 minutes. If it does not, survey the entire system for leaks. Note any problems on the Test Form and notify a responsible party representing the building of repairs that are necessary, allowing a reasonable time for corrections to be made.

Step 12: Repeat Steps 7 and 8 for every other riser, if possible, directly from the roof manifold or with a 2½-inch (65 mm) handline from the uppermost outlet using the fire department pumper. When all risers are tested and flushed, the test is complete.

Courtesy of Fort Worth, Texas Fire Department.

APPENDIX H
Plano Fire Department
Approving Authority Check List
For Fire Sprinkler Systems

1. Name of Facility: _____
2. Facility Physical Add.: _____
3. Municipality & Zip _____
4. Sprinkler Contractor: _____
5. Contractor: Physical Add.: _____
6. Cont. Municipality & Zip.: _____
7. Cont. Registration No.: _____
8. Contractor Cont. No: _____
9. Name of Responsible Layout Tech.: _____

10. **Authority Hazard Classification:**
10.1 Total Area: _____ sq. ft.
10.2 Light Hazard Area: _____ sq. ft.
10.3 Ordinary Hazard, Gp. 1 Area: _____ sq. ft.
10.4 Ordinary Hazard, Gp. 2 Area: _____ sq. ft.
10.5 Ordinary Hazard, Gp. 3 Area: _____ sq. ft.
10.6 Extra Hazard, Gp. 1 Area: _____ sq. ft.
10.7 Extra Hazard, Gp. 2 Area: _____ sq. ft.
10.8 High Piled Storage Area: ($\cdot$12') _____ sq. ft.

11.0 **Environmental Treatment:**
11.1 Area Wet: _____ sq. ft.
11.2 Area Dry: _____ sq. ft.
11.3 Area Dry Pendent: _____ sq. ft.
11.4 Area Pre-Action: _____ sq. ft.
11.5 Area Deluge: _____ sq. ft.
11.6 Area Anti-freeze: _____ sq. ft.

12.0 **Minimum Number of Risers:** (Calculate for largest floor in multi-floor structures. Round up to the next highest whole number. See NFPA 13, 3-3.1.)

12.1 General:

(Sum of areas (_____) $$\frac{\text{“10.2” + “10.3” + “10.4” + “10.5”}}{52{,}000} = \boxed{}$$
(max. area)

12.2 Extra Hazard:

$$\frac{\text{“10.6” + “10.7”}}{25{,}000} = \boxed{}$$

12.3 High Piled Storage

$$\frac{\text{“10.8”}}{40{,}000} = \boxed{}$$

12.4 Total $\boxed{}$

NOTE: Single systems serving high piled storage and ordinary hazard areas shall not exceed a total of 52,000 square feet for combined area and not more than 40,000 square feet for high piled storage (NFPA 13, 3-3.1)

			Y	N	
12.5	Are there enough risers?		☐	☐	

			Y	N	
13.0	**The sprinkler system is:**		Y	N	
	Hydraulically calculated		☐	☐	
	Pipe Schedule		☐	☐	

14.0 Sprinkler Spacing:
Are sprinkler locations compatible with the per-sprinkler maximum area limitations?

			Y	N	NA
14.1	**Light Hazard:** (NFPA 13, 4-2.2.1)		Y	N	NA
14.1.1	Smooth Ceiling and Beam and Girder.				
14.1.1.1	Calculated	—225 sq. ft.	☐	☐	☐
14.1.1.2	Schedule	—200 sq. ft.	☐	☐	☐
14.1.2	Open Wood Joists				
	Schedule or Calc.	—130 sq. ft.	☐	☐	☐
14.1.3	Other Construction				
	Schedule or Calc.	—130 sq. ft.	☐	☐	☐
14.2	**Ordinary Hazard** (NFPA 13, 4-2.2.2				
14.2.1	No high piled storage.				
	All configurations	—130 sq. ft.	☐	☐	☐
14.2.2	High piled storage. (NFPA 13, 4-2.2.4)				
14.2.2.1	Schedule, all const.	—100 sq. ft.	☐	☐	☐
14.2.2.2	Calc., all const.	—130 sq. ft.	☐	☐	☐
14.3	**Extra Hazard:** (NFPA 13, 4-2.2.3)				
14.3.1	Schedule, all const.	—90 sq. ft.	☐	☐	☐
14.3.2	Calc., all const.	—100 sq. ft.	☐	☐	☐
14.4	**Spacing Between Upright and Pendent Heads:**				
14.4.1	Maximum of 7½' from any wall.				
	(Small room rule is not valid in Delaware.)		☐	☐	☐
14.4.2	Maximum of 15' between sprinklers (NFPA 13, 4-2.1.5.1)		☐	☐	☐
14.4.3	Minimum of 4" from a wall. (NFPA 13, 4-2.1.5.2)		☐	☐	☐
14.4.4	Minimum of 6' from each other?				
	(Residential minimum is 8')		☐	☐	☐

NOTE: On line items 14.4.1 and 14.4.2, some residential sprinklers have been listed for greater spacings. Check the listings.

		Y	N	NA
14.5	**Sidewall Sprinkler Spacing:**	Y	N	NA
14.5.1	Are the specified sprinklers listed for this hazard?	☐	☐	☐

NOTE: Sidewall sprinklers are special application devices. Most are listed only for light hazard. Check the listing.

14.5.2	Are the specified sprinklers within their listed coverage?	☐	☐	☐

15.0 **Obstructions;** Have obstructions such as columns, ☐ ☐ ☐
firdowns, beams offsets, etc., been considered so that every
square foot of floor area is covered by sprinklers?

16.0 **Temperature Ratings** (Degrees Fahrenheit)
(NFPA 13, Table 3-16.6.1)

	Classification	Max. Clg. Temp.	Quantity
16.1	Ordinary (135-170)	100	_____
16.2	Intermediate (175-225)	150	_____
16.3	High (250-300)	225	_____
16.4	Extra High (400-475)	300	_____
16.5	Very Extra High (400-475)	375	_____
16.6	Ultra High I (500-575)	475	_____
16.7	Ultra High II (650	625)	_____

		Y	N	
16.8	Do these fit the expected ceiling temperatures?	☐	☐	
16.9	Are these compatible with NFPA 13, Figure 3-16.6.3 (a) Heater and Danger Zones at Unit Heaters, and Table 3-16.6.3 (A), Distance of Sprinklers from Heat Sources?	☐	☐	

		Y	N	NA
17.0	**For areas not involving high piled storage:**			
17.1	Are the contractor-selected hazard classifications correct?	☐	☐	☐
17.2	Does the hydraulically most remote area size-and-density-point fall on or to the right of the required density curve of NFPA 13 Figure 2-2.1 (B).	☐	☐	☐
17.3	Is the selected remote area truly the hydraulically most remote?	☐	☐	☐
17.4	Review the hydraulic computations?	☐	☐	☐
17.4.1	Has the minimum hose allowance per NFPA 13, Table 2-2.1 (B) been included at the base of the riser?	☐	☐	☐
17.4.2	Is the total water demand greater than the remote area times density (A x d) + the minimum hose allowance of NFPA 13, Table 2-2.1 (B)? (It should be a minimum of 5% to 10% greater)	☐	☐	☐
17.4.3	Have all the fittings, valves and special devices been included?	☐	☐	☐
17.4.4	Does the Supply-Demand Curve show adequate safety factor?	☐	☐	☐
17.4.5	Has the static head loss been included in the computations?	☐	☐	☐
17.4.6	Is the plotted water supply consistent with municipal records?	☐	☐	☐

		Y	N	NA
18.0	**For Schedule Pipe Systems:**			
18.1	Are the applied sizes of pipe in accordance with the hazard classification and schedules of NFPA 13, paragraphs 3-5 through 3-7?	☐	☐	☐
18.2	Calcuate the static head loss for the highest sprinkler head as P = h x .434 _____			
18.3	Is this value + 15 psi less than the pressure available at the base of the riser at the flows required for a schedule system per NFPA 13, Table 2-2.1 (A).	☐	☐	☐
19.0	**Dry Systems:**			
19.1	Is (Are) the systems divided into volumes not more than 500 gallons for gridded systems or 750 gallons for non-gridded systems? (NFPA 13, 5-2.3.1)	☐	☐	☐
19.2	Is (Are) systems with check valve subdivision divided into branches with not more than 400 gallons for gridded system and 600 gallons for non-gridded systems? (See NFPA 13, 5-2.3.1)	☐	☐	☐
19.3	Is the dry pipe valve located in a heated and lighted room?	☐	☐	☐
19.4	If the system capacity exceeds 350 gallons for gridded systems or 500 gallons for non-gridded systems, have quick opening devices been provided?	☐	☐	☐
20.0	**Anti-freeze Systems:**			
20.1	Is the system size less than 40 gallons? (See NFPA 13, A-5-5-2)	☐	☐	☐
20.2	Is the propr anti-freeze solution specified on the drawing? (NFPA 13, 5-5.3)	☐	☐	☐
21.0	**Pre-Action Systems:**			
21.1	Are there less than 1000 closed sprinklers on a system?	☐	☐	☐
22.0	**High Piled Storage:** Have the proper standards been applied			
22.1	Indoor General Storage — NFPA 231	☐	☐	☐
22.2	Rack Storage of Materials — NFPA 231C	☐	☐	☐
22.3	Storage of Rubber Tires — NFPA 231D	☐	☐	☐
22.4	Storage of Baled Cotton — NFPA 231E	☐	☐	☐
22.5	Storage of Rolled Paper — NFPA 231F	☐	☐	☐
23.0	**Additional Provisions:** Does the design detail the following:			
23.1	Alarm provisions: Water motor gong or electric bell?	☐	☐	☐
23.2	Fire Department Connection?	☐	☐	☐
23.3	Inspector's Test?	☐	☐	☐
23.4	Supervisory Provisions?			
23.4.1	Valves? (NFPA 13, 3-14.2.3)	☐	☐	☐
23.4.2	Pre-Action Systems (NFPA 13, 5-3.5.3)	☐	☐	☐

		Y	N	NA
23.4.3	Deluge Systems (NFPA 13, 5-3.6)	☐	☐	☐
23.5	Are hanger details correctly specified?	☐	☐	☐
23.6	Are elevations of piping indicated?	☐	☐	☐
23.7	Is the drawing to a standard scale?	☐	☐	☐
23.8	Is a riser diagram provided?	☐	☐	☐
23.9	Are the canopies sprinklered?	☐	☐	☐

24.0   **Design**   Accepted ☐

                   Rejected ☐   With the following comments:

1. _____

2. _____

3. _____

4. _____

5. _____

Name of Reviewer: _____

Date: _____ Signature: _____

APPENDIX I
Federal Standards Regarding Transportation of Hazardous Materials

NOTE: These standards have been adopted as part of *Code of Federal Regulations (CFR 49.)*

American Society of Mechanical Engineers (ASME)
- "ASME Boiler and Pressure Vessel Code." Section VIII (Division I) and IX of the 1977 edition and addenda thereto through December 31, 1978; except paragraph UW-11 (a)(7) of the code does not apply.

Association of American Railroads (AAR)
- "AAR Specifications for Tank Cars." 1970 edition.

Compressed Gas Association (CGA)
- CGA Pamphlet C-3, "Standard for Welding and Brazing on Thin Walled Containers," 1975 edition.
- CGA Pamphlet C-6, "Standards for Visual Inspection of Compressed Gas Cylinders," 1975 edition.
- CGA Pamphlet C-7, Appendix A, "A Guide for the Precautionary Markings for Compressed Gas Cylinders," dated May 15, 1971, Addenda issued January 1976.
- CGA Pamphlet C-8, "Standard for Requalification of DOT-3HT Cylinders," 1972 edition.
- CGA Pamphlet S-1.2, "Safety Relief Device Standards Part 2 — Cargo and Portable Tanks for Compressed Gases," 1966 edition.

American National Standards Institute (ANSI)
- ANSI Standard B9.1, "Safety Code for Mechanical Refrigeration," 1964 edition.
- ANSI Standard B.16.5, "Steel Pipe Flanges and Fittings," 1968 edition.
- ANSI Standard N.14.1, "Packaging of Uranium Hexafluoride for Transport," 1971 edition.

American Society for Testing and Materials (ASTM)
- ASTM Standard D1310, "Standard Method of Test for Flash Point of Volatile Flammable Materials by Tag Open-Cup Apparatus," 1967 edition.
- ASTM Standard D323, "Test for Vapor Pressure of Petroleum Products (Reid Method)," 1958(68) edition.
- ASTM Standard D1056, "Sponge and Expanded Cellular Rubber Products, Spec. and Tests for," 1968 edition.
- ASTM Standard G 23-69, "Standard Recommended Practice for Operating Light- and Water-Exposure Apparatus (Carbon-Arc Type) for Exposure of Nonmetallic Materials," 1969 edition (reapproved 1975).
- ASTM Standard D-638, "Test for Tensile Strength of Plastics," 1976 edition.

- ASTM Standard D-1505, "Test for Density of Plastics by the Density Gradient Technique," 1968 edition.

- ASTM Standard C148-77, "Standard Methods of Polariscopic Examination of Glass Containers," 1977 edition.

- ASTM Standard E487-74, "Standard Test Method for Constant-Temperature Stability of Chemical Materials," 1974 edition.

National Fire Protection Association (NFPA)
- NFPA 58, "Standard for the Storage and Handling of Liquefied Petroleum Gases," 1972 edition.

Bureau of Explosives, Association of American Railroads
- Bureau of Explosives Pamphlet No. 6, "Illustrating Methods for Loading and Bracing Carload and Less Than Carload Shipments of Explosives and Other Dangerous Articles," 1962 edition.

- Bureau of Explosives Pamphlet No. 6A, "Illustrating Methods for Loading and Bracing Carloads and Less Than Carload Shipments of Loaded Projectiles, Loaded Bombs, Etc.," 1943 edition.

- Bureau of Explosives Pamphlet No. 6C, "Illustrating Methods for Loading and Bracing Trailers and Less Than Trailer Shipments of Explosives and Other Dangerous Articles via Trailer-on-Flat-Car (TOFC) or Container-on-Flat-Car (COFC)," September, 1968.

- Bureau of Explosives Pamphlets 1 and 2, "Emergency Handling of Hazardous Materials in Surface Transportation," June, 1973.

National Association of Corrosion Engineers (NACE)
- NACE Standard TM-01-69, "Test Method Laboratory Corrosion Testing of Metals for the Process Industries," 1969 edition.

Institute of Makers of Explosives (IME)
- IME Standard 22, "IME Standard for the Safe Transportation of Class C Detonators (Blasting Caps) in a Vehicle With Certain Other Explosives," Revised March 21, 1979, (IME Safety Library Publication No. 22).

International Atomic Energy Association (IAEA)
- IAEA "Regulations for the Safe Transport of Radioactive Materials," 1967 edition and 1973 revised edition. Safety Series No. 6.

United States Atomic Energy Commission (USAEC)
- Title 10, CFR, Part 71, "Packaging of Radioactive materials for Transport and Transportation of Radioactive Materials Under Certain Conditions."

United States Department of Commerce (USDC)
- USDC, National Bureau of Standards Handbook H28 (1957) — Part II, "Screw-Thread Standards for Federal Services 1957," December, 1966 edition.

- USDC, CAPE-1962, one of the series of "Civilian Applications Program Engineering Drawings" which is a package of information including drawings and bills of materials, describing phenolic-foam insulated, protective overpacks.

- USDC, USDOE Material and Equipment Specifications No. SP-9, Rev. 1, and Supplement, "Fire Resistant Phenolic Foam."
- USDC, ORO-651, "Uranium Hexafluoride Handling Procedures and Container Criteria," Revision 3, 1972 edition.

American Water Works Association (AWWA)
- AWWA Standard C207-55, "AWWA Standard for Steel Pipe Flanges," 1955 edition.

American Welding Society (AWS)
- AWS Code B-3.0), "Standard Qualification Procedure," 1972 edition.
- AWS Code D-1.0, "Code for Welding in Building Construction," 1966 edition.

International Maritime Dangerous Goods Code (IMCO Code)
- IMCO Code Volumes I, II, III, and IV, 1977 edition, and Amendments 14-76, 15-77, and 16-78 thereto.

General Services Administration (GSA)
- GSA, Federal Specification RR-C-901b, "Cylinders, Compressed Gas: With Valve or Plug and Cap; ICC-3AA," August 1, 1977.

National Institute for Occupational Safety and Health (NIOSH)
- "Registry of Toxic Effects of Chemical Substances," 1978 edition.

APPENDIX J
American Table of Distances for Storage of Explosive Materials

As Revised and Approved by The Institute of Makers of Explosives— February 1986

QUANTITY OF EXPLOSIVE MATERIALS[1,2,3,4]		DISTANCES IN FEET							
		Inhabited Buildings [9]		Public Highways Class A to D [11]		Passenger Railways— Public Highways with Traffic Volume of more than 3,000 Vehicles/Day [10,11]		Separation of Magazines [12]	
Pounds Over	Pounds Not Over	Barri-caded[6,7,8]	Unbarri-caded	Barri-caded[6,7,8]	Unbarri-caded	Barri-caded[6,7,8]	Unbarri-caded	Barri-caded[6,7,8]	Unbarri-caded
2	5	70	140	30	60	51	102	6	12
5	10	90	180	35	70	64	128	8	16
10	20	110	220	45	90	81	162	10	20
20	30	125	250	50	100	93	186	11	22
30	40	140	280	55	110	103	206	12	24
40	50	150	300	60	120	110	220	14	28
50	75	170	340	70	140	127	254	15	30
75	100	190	380	75	150	139	278	16	32
100	125	200	400	80	160	150	300	18	36
125	150	215	430	85	170	159	318	19	38
150	200	235	470	95	190	175	350	21	42
200	250	255	510	105	210	189	378	23	46
250	300	270	540	110	220	201	402	24	48
300	400	295	590	120	240	221	442	27	54
400	500	320	640	130	260	238	476	29	58
500	600	340	680	135	270	253	506	31	62
600	700	355	710	145	290	266	532	32	64
700	800	375	750	150	300	278	556	33	66
800	900	390	780	155	310	289	578	35	70
900	1,000	400	800	160	320	300	600	36	72
1,000	1,200	425	850	165	330	318	636	39	78
1,200	1,400	450	900	170	340	336	672	41	82
1,400	1,600	470	940	175	350	351	702	43	86
1,600	1,800	490	980	180	360	366	732	44	88
1,800	2,000	505	1,010	185	370	378	756	45	90
2,000	2,500	545	1,090	190	380	408	816	49	98
2,500	3,000	580	1,160	195	390	432	864	52	104
3,000	4,000	635	1,270	210	420	474	948	58	116
4,000	5,000	685	1,370	225	450	513	1,026	61	122
5,000	6,000	730	1,460	235	470	546	1,092	65	130

QUANTITY OF EXPLOSIVE MATERIALS		DISTANCES IN FEET							
6,000	7,000	770	1,540	245	490	573	1,146	68	136
7,000	8,000	800	1,600	250	500	600	1,200	72	144
8,000	9,000	835	1,670	255	510	624	1,248	75	150
9,000	10,000	865	1,730	260	520	645	1,290	78	156
10,000	12,000	875	1,750	270	540	687	1,374	82	164
12,000	14,000	885	1,770	275	550	723	1,446	87	174
14,000	16,000	900	1,800	280	560	756	1,512	90	180
16,000	18,000	940	1,880	285	570	786	1,572	94	188
18,000	20,000	975	1,950	290	580	813	1,626	98	196
20,000	25,000	1,055	2,000	315	630	876	1,752	105	210
25,000	30,000	1,130	2,000	340	680	933	1,866	112	224
30,000	35,000	1,205	2,000	360	720	981	1,962	119	238
35,000	40,000	1,275	2,000	380	760	1,026	2,000	124	248
40,000	45,000	1,340	2,000	400	800	1,068	2,000	129	258
45,000	50,000	1,400	2,000	420	840	1,104	2,000	135	270
50,000	55,000	1,460	2,000	440	880	1,140	2,000	140	280
55,000	60,000	1,515	2,000	455	910	1,173	2,000	145	290
60,000	65,000	1,565	2,000	470	940	1,206	2,000	150	300
65,000	70,000	1,610	2,000	485	970	1,236	2,000	155	310
70,000	75,000	1,655	2,000	500	1,000	1,263	2,000	160	320
75,000	80,000	1,695	2,000	510	1,020	1,293	2,000	165	330
80,000	85,000	1,730	2,000	520	1,040	1,317	2,000	170	340
85,000	90,000	1,760	2,000	530	1,060	1,344	2,000	175	350
90,000	95,000	1,790	2,000	540	1,080	1,368	2,000	180	360
95,000	100,000	1,815	2,000	545	1,090	1,392	2,000	185	370
100,000	110,000	1,835	2,000	550	1,100	1,437	2,000	195	390
110,000	120,000	1,855	2,000	555	1,110	1,479	2,000	205	410
120,000	130,000	1,875	2,000	560	1,120	1,521	2,000	215	430
130,000	140,000	1,890	2,000	565	1,130	1,557	2,000	225	450
140,000	150,000	1,900	2,000	570	1,140	1,593	2,000	235	470
150,000	160,000	1,935	2,000	580	1,160	1,629	2,000	245	490
160,000	170,000	1,965	2,000	590	1,180	1,662	2,000	255	510
170,000	180,000	1,990	2,000	600	1,200	1,695	2,000	265	530
180,000	190,000	2,010	2,010	605	1,210	1,725	2,000	275	550
190,000	200,000	2,030	2,030	610	1,220	1,755	2,000	285	570
200,000	210,000	2,055	2,055	620	1,240	1,782	2,000	295	590
210,000	230,000	2,100	2,100	635	1,270	1,836	2,000	315	630
230,000	250,000	2,155	2,155	650	1,300	1,890	2,000	335	670
250,000	275,000	2,215	2,215	670	1,340	1,950	2,000	360	720
275,000	300,000	2,275	2,275	690	1,380	2,000	2,000	385	770

Numbers in ( ) refer to explanatory notes.

EXPLANATORY NOTES ESSENTIAL TO THE APPLICATION OF THE AMERICAN TABLE OF DISTANCES FOR STORAGE OF EXPLOSIVE MATERIALS

NOTE — "Explosive materials" means explosives, blasting agents and detonators.

NOTE — "Explosives" means any chemical compound, mixture, or device, the primary or common purpose of which is to function by explosion. A list of explosives determined to be within the coverage of "18 U.S.C. Chapter 40, Importation, Manufacturer, Distribution and Storage of Explosive Materials" is issued at least annually by the Director of the Bureau of Alcohol, Tobacco and Firearms of the Department of the Treasury. For quantity and distance purposes, detonating cord of 50 grains per foot should be calculated as equivalent to 8 lbs. of high explosives per 1,000 feet. Heavier or lighter core loads should be rated proportionately.

NOTE — "Blasting agents" means any material or mixture, consisting of fuel and oxidizer, intended for blasting, not otherwise defined as an explosive: Provided, that the finished product, as mixed for use or shipment, cannot be detonated by means of a No. 8 test blasting cap when unconfined.

NOTE — "Detonator" means any device containing any initiating or primary explosive that is used for initiating detonation. A detonator may not contain more than 10 grams of total explosives by weight, excluding ignition or delay charges. The term includes, but is not limited to, electric blasting caps of instataneous and delay types, blasting caps for use with safety fuses, detonating cord delay connectors, and nonelectric instantaneous and delay blasting caps which use detonating cord, shock tube, or any other replacement for electric leg wires. All types of detonators in strengths through No. 8 cap should be rated at 1½ lbs. of explosives per 1,000 caps. For strengths higher than No. 8 cap, consult the manufacturer.

NOTE — "Magazine" means any building, structure, or container, other than an explosives manufacturing building, approved for the storage of explosive materials.

NOTE — "Natural Barricade" means natural features of the ground, such as hills, or timber of sufficient density that the surrounding exposures which require protection cannot be seen from the magazine when the trees are bare of leaves.

NOTE — "Artificial Barricade" means an artificial mound or revetted wall of earth of a minimum thickness of three feet.

NOTE — "Barricaded" means the effective screening of a building containing explosive materials from the magazine or other building, railway, or highway by a natural or an aritificial barrier. A straight line from the top of any sidewall of the building containing explosive materials to the eave line of any magazine or other building or to a point twelve feet above the center of a railway or highway shall pass through such barrier.

NOTE — "Inhabited Building" means a building regularly occupied in whole or part as a habitation for human beings, or any church, schoolhouse, railroad station, store, or other structure where people are accustomed to assemble, except any building or structure occupied in connection with the manufacture, transportation, storage or use of explosive materials.

NOTE — "Railway" means any steam, electric, or other railroad or railway which carries passengers for hire.

NOTE — "Highway" means any public street, public alley, or public road. "Public Highways Class A to D" are highways with average traffic volume of 3,000 or less vehicles per day as specified in "American Civil Engineering Practice" (Abbett, Vol. 1, Table 46, Sec. 3-74, 1956 Edition, John Wiley and Sons).

NOTE — When two or more storage magazines are located on the same property, each magazine must comply with the minimum distances specified from inhabited buildings, railways, and highways, and, in addition, they should be separated from each other by not less than the distances shown for "Separation of Magazines," except that the quantity of explosive materials contained in detonator magazines shall govern in regard to the spacing of said detonator magazines from magazines containing other explosive materials. If any two or more magazines are separated from each other by less than the specified "Separation of Magazines" distances, then such two or more magazines, as a group, must be considered as one magazine, and the total quantity of explosive materials stored in such group must be treated as if stored in a single magazine located on the site of any magazine of the group, and must comply with the minimum of distances specified from other magazines, inhabited buildings, railways, and highways.

NOTE — Storage in excess of 300,000 lbs. of explosive materials, in one magazine is generally not required for commercial enterprises.

NOTE — This Table applies only to the manufacture and permanent storage of commercial explosive materials. It is not applicable to transportation of explosives or any handling or temporary storage necessary or incident thereto. It is not intended to apply to bombs, projectiles, or other heavily encased explosives.

Courtesy of Institute of Makers of Explosives.

APPENDIX K
Characteristics of Extinguishers (Metrics)

Extinguishing Agent	Method of Operation	Capacity	UL or ULC Classification
Water	Stored Pressure	9.5 L	2-A
Water	Pump Tank	5.7L	1-A
	Pump Tank	9.5L	2-A
	Pump Tank	15.1L	3-A
	Pump Tank	18.9L	4-A
Water (Antifreeze Calcium Chloride)	Cartridge or Stored Pressure	4.73, 5.7 L	1-A
	Cartridge or Stored Pressure	9.5 L	2-A
	Cylinder	125 L	20 A
Water (Wetting Agent)	Stored Pressure	5.7 L	2-A
	Carbon Dioxide Cylinder (wheeled)	94.6 L	10-A
	Carbon Dioxide Cylinder (wheeled)	170.3 L	30-A
	Carbon Dioxide Cylinder (wheeled)	227 L	40-A
Water (Soda Acid)	Chemically Generated Expellant	4.73, 5.7 L	1-A
	Chemically Generated Expellant	9.5 L	2-A
	Chemically Generated Expellant (wheeled)	64.3 L	10-A
	Chemically Generated Expellant (wheeled)	125 L	20-A
Water (Loaded Stream)	Stored Pressure	9.5 L	2 to 3-A:1-B
	Cartridge or Stored Pressure (wheeled)	125 L	20-A
AFFF	Stored Pressure	9.5 L	3-A 20 B
	Nitrogen Cylinder (wheeled)	125 L	20-A 160 B
Carbon Dioxide	Self-Expellant	.9 to 2.26 kg	1 to 5-B:C
	Self-Expellant	4.5 to 6.8 kg	2 to 10-B:C
	Self-Expellant	9 kg	10-B:C
	Self-Expellant (wheeled)	22.6 to 45.3 kg	10 to 20-B:C
Dry Chemical (Sodium Bicarbonate)	Stored Pressure	.45 kg	1 to 2-B:C
	Stored Pressure	.68 to 1.13 kg	2 to 10-B:C
	Cartridge or Stored Pressure	1.24 to 2.26 kg	5 to 20-B:C
	Cartridge or Stored Pressure	2.7 to 13.6 kg	10 to 160-B:C
	Nitrogen Cylinder or Stored Pressure (wheeled)	34 to 158.7 kg	40 to 320-B:C
Dry Chemical (Potassium Bicarbonate)	Stored Pressure	.45 to .9 kg	1 to 5-B:C
	Cartridge or Stored Pressure	1.02 to 2.26 kg	5 to 20-B:C
	Cartridge or Stored Pressure	2.4 to 4.5 kg	10 to 80-B:C
	Cartridge or Stored Pressure	7.2 to 13.6 kg	40 to 120-B:C
	Cartridge	21.7 kg	120-B:C
	Nitrogen Cylinder or Stored Pressure (wheeled)	56.7 to 142.8 kg	80 to 640-B:C

Extinguishing Agent	Method of Operation	Capacity	UL or ULC Classification
Dry Chemical (Potassium Chloride)	Stored Pressure	.9 to 1.1 kg	5 to 10-B:C
	Stored Pressure	2.26 to 4 kg	20 to 40-B:C
	Stored Pressure	4.5 to 9 kg	40 to 60-B:C
	Stored Pressure	61.2 kg	160-B:C
Dry Chemical (Ammonium Phosphate)	Stored Pressure	.45 to 2.26 kg	1 to 2-A and 2 to 10-B:C
	Stored Pressure or Cartridge	1.13 to 3.8 kg	1 to 4-A and 10 to 40-B:C
	Stored Pressure or Cartridge	4 to 7.7 kg	2 to 20-A and 10 to 80-B:C
	Stored Pressure or Cartridge	7.7 to 13.6 kg	3 to 20-A and 30 to 120-B:C
	Cartridge	20.4 kg	20-A and 80-B:C
	Nitrogen Cylinder or Stored Pressure (wheeled)	49.9 to 142.8 kg	20 to 40-A and 60 to 320-B:C
Dry Chemical (Foam Compatible)	Cartridge or Stored Pressure	2.1 to 4 kg	10 to 20-B:C
	Cartridge or Stored Pressure	4 to 12.2 kg	20 to 30-B:C
	Cartridge or Stored Pressure	8.2 to 13.6 kg	40 to 60-B:C
	Nitrogen Cylinder or Stored Pressure (wheeled)	68 to 158.7 kg	80 to 240-B:C
Dry Chemical (Potassium Chloride)	Cartridge or Stored Pressure	1.13 to 2.26 kg	10 to 20-B:C
	Cartridge or Stored Pressure	4.3 to 9 kg	40 to 60-B:C
	Cartridge or Stored Pressure	8.8 to 13.6 kg	60 to 80-B:C
	Stored Pressure (wheeled)	56.7 to 90.7 kg	160-B:C
Dry Chemical (Potassium Bicarbonate Urea Base)	Stored Pressure	2.26 to 4.98 kg	40 to 80-B:C
	Stored Pressure	4 to 10.4 kg	60 to 160-B:C
	Stored Pressure	79.3 kg	480-B:C
Bromotrifluoromethane	Stored Pressure	1.3 kg	2-B:C
Bromochlorodifluoromethane	Stored Pressure	.9 to 1.8 kg	2 to 5-B:C
	Stored Pressure	2.49 to 4 kg	1-A and 10-B:C
	Stored Pressure	5.90 to 9.97 kg	1 to 4-A and 20 to 80-B:C

APPENDIX L
Elevator Testing Procedure

OBJECTIVE

All high-rise elevators must be inspected and tested on a regular basis. This is necessary to ensure that building occupants have an adequate level of fire safety if a fire occurs while an elevator is in use. Testing is also necessary to ensure an elevator's readiness for fire fighting personnel as a personnel and equipment deployment device during a high-rise fire, rescue, or EMS incident.

The American Society of Mechanical Engineers (ASME) Standard A17.1-1981, Rule 211.3, addresses the requirements of elevator recall during a fire or other emergency condition. In addition, Chapter 51 of the 1982 *Uniform Building Code* and Article 620 of the *National Electrical Code* define requirements for the installation and operation of elevators. These requirements are defined as follows:

Phase I — Emergency Recall Operation

Phase II — Emergency In-Car Operation

TESTING EQUIPMENT
- One canister of aerosol smoke detector tester
- Elevator operation keys
- Two portable hand-held radios

TIME ALLOTMENT AND PERSONNEL REQUIREMENTS

For system acceptance tests, the minimum personnel requirements are one suppression company and a staff member representing the Fire Protection Division. The minimum time for correctly performing the acceptance test is one hour for the first elevator and 30 minutes for each additional elevator.

For biannual performance tests, the minimum personnel requirement is one suppression company. The minimum time needed to correctly perform the biannual acceptance test is one hour for the first elevator and 30 minutes for each additional elevator.

GENERAL — ALL TESTS

Begin the test by meeting with the building manager to obtain the preliminary information necessary to complete the inspection form. Try to schedule tests during periods of light use in order to minimize the inconvenience to building occupants. Advise the building management that the elevator(s) will be controlled by the fire department during the test and that building occupants will not be allowed to use the elevator being tested. In addition, advise the building occupants that an alarm condition will be initiated within the protective signaling system; thus, the proprietary or central station operator should be notified that an alarm condition will be transmitted.

Begin the test by performing the pretest inspection. This is done to make sure that the elevators meet mechanical and electrical codes. The assistance of the building engineer is needed during this initial portion of the test. Never use fire department personnel to check electrical circuits or components; the building engineer will have a sound understanding of these systems and can provide the inspection team with information that will help to quickly complete the test. If the elevator or its subsystems do not meet the criteria established, stop the test and advise the building engineer that repairs need to be made.

PHASE I — EMERGENCY RECALL OPERATION TEST

Step 1: Notify fire dispatch that inspection companies will be performing an elevator operations test, giving the location of the test and the mode of communication that will be used by personnel performing the test.

Step 2: Assign one firefighter to the elevator car with a portable radio. This firefighter will be responsible for reporting the elevator operations described in Step 4.

Step 3: Locate the three-position key-operated switch. This switch should be near the lobby or the main floor for a particular elevator bank. The key to operate the switch can be obtained from either the building engineer or, if provided, the fire department key box installation.

Step 4: Insert the key and place the switch in the "ON" position. Once in the "ON" mode, the following actions should occur automatically:

- All elevators controlled by this switch will return nonstop to the main floor and the doors will open and remain open.

- All elevators traveling away from the main floor will reverse at the next available floor without opening their doors and proceed to the main floor without stopping.

- Elevators standing at a floor other than the main floor will close their doors and proceed to the main floor without stopping.

- All hall buttons will cease operating.

- Once the elevator has started to return, the emergency stop switch will be deactivated.

Step 5: If the elevators do not perform as described in Step 4, stop the test and notify building management. If the elevator performs adequately, continue with the test.

Step 6: Place the three-way switch in the "OFF" position. Following the instructions listed on the container of aerosol smoke detector tester, apply the test agent to the elevator lobby corridor smoke detector. This should cause an alarm to be sent to the fire alarm control panel. The elevators should again operate as described in Step 4, except that they should not return to the floor on which the detector was activated. If an alarm condition is not transmitted, stop the test and advise the building management that repairs will need to be made.

Step 7: With the protective signaling system still in an alarm condition and the alarm signaling devices silenced, place the three-way switch in the "BYPASS" or "BYP" position. This allows for testing of the elevators emergency in-car operating features. The key should not be removable in this position; if it is, stop the test and advise the building management to make repairs. If the test is stopped, have the protective signaling system reset.

PHASE II — EMERGENCY IN-CAR OPERATION TEST

Step 1: Locate the two-way key switch on the car operating panel. It should be located within or adjacent to the control panel. Insert the key and activate the emergency in-car controller. Upon activation, the door light beam should become inoperative. If it does not, stop the test and advise building management to make repairs.

Step 2: Contact the firefighter assigned to monitor the hall control panel via portable radio. Instruct the firefighter to randomly press both the designated car call switches ("UP" and "DOWN"). The elevator should not respond to these calls.

Step 3: Choose any floor for the elevator to travel to. Upon activating the floor control switch, press the "DOOR CLOSE" button but do not allow the door to close. Release of pressure on the button prior to full closure should result in the doors reopening immediately. If this does not occur, stop the test and notify building management.

Step 4: Press the "DOOR CLOSE" switch and allow the door to close completely. The elevator call should now respond to the floor that was randomly chosen in Step 3.

Step 5: When the elevator arrives at the called floor, the elevator door should not open automatically. Press the "DOOR OPEN" switch momentarily; if pressure to the switch is released prior to full open, the doors should immediately reclose. If not, stop the test and notify building management.

Step 6: Depress the "DOOR OPEN" switch and allow the elevator door to fully open; once this action is complete, the doors should not close.

Step 7: Notify the firefighter assigned to the hall elevator control panel to place the three-way switch in the "OFF" position and have the protective signaling system reset. Close the elevator doors, reset the car two-way switch to the "OFF" position, and return to the lobby.

Step 8: Return the elevator control keys to either the building management or the fire department key box installation. Once this task is completed, advise the property representative that the test is complete.

Courtesy of Plano, Texas Fire Department.

INDEX

IFSTA MANUALS

FIRE SERVICE ORIENTATION & INDOCTRINATION

History, traditions, and organization of the fire service; operation of the fire department and responsibilities and duties of firefighters; fire department companies and their functions; glossary of fire service terms.

FIRE SERVICE FIRST RESPONDER

Covers all objectives for U.S. DOT First Responder Training Courses as well as NFPA 1001 Emergency Medical Care sections. The special emphasis on maintenance of the ABC's features updated CPR techniques. Also included are scene assessment and safety, patient assessment, shock, bleeding, control spinal injuries, burns, heat and cold emergencies, medical emergencies, poisons, behavioral emergencies, emergency childbirth, short-distance transfer, and emergency vehicles and their equipment.

ESSENTIALS OF FIRE FIGHTING

This manual was prepared to meet the objectives set forth in levels I and II of NFPA 1001, *Fire Fighter Professional Qualifications*. Included in the manual are the basics of fire behavior, extinguishers, ropes and knots, self-contained breathing apparatus, ladders, forcible entry, rescue, water supply, fire streams, hose, ventilation, salvage and overhaul, fire cause determination, fire suppression techniques, communications, sprinkler systems, and fire inspection.

SELF-INSTRUCTION FOR ESSENTIALS

Over 260 pages of structured questions and answers for studying the *Essentials of Fire Fighting*. Each unit begins with the NFPA Standard No. 1001 required performance objectives. This self-instruction book will help you to learn many of the important topics and review the basic text.

IFSTA'S 500 COMPETENCIES FOR FIREFIGHTER CERTIFICATION

This manual identifies the competencies that must be achieved for certification as a firefighter for levels I and II. The text also identifies what the instructor needs to give the student, NFPA standards, and has space to record the student's score, local standards, and the instructor's initials.

FIRE SERVICE GROUND LADDER PRACTICES

Various terms applied to ladders; types, construction, maintenance, and testing of fire service ground ladders; detailed information on handling ground ladders and special tasks related to them.

HOSE PRACTICES

This new edition has been updated to reflect the latest information on modern fire hose and couplings, including large diameter hose. Details basic methods for handling hose and coupling construction; care, maintenance, and testing; hose appliances and tools; basic methods of handling hose; supply and attack methods; special hose operations.

SALVAGE AND OVERHAUL PRACTICES

Planning and preparing for salvage operations, care and preparations of equipment, methods of spreading and folding salvage covers, most effective way to handle water runoff, value of proper overhaul and equipment needed, and recognizing and preserving arson evidence.

FORCIBLE ENTRY

This comprehensive manual contains technical information about forcible entry tactics; tools; and door, window, and wall construction. Forcible entry methods are described for door, window, and wall entry. A new section on locks and through-the-lock entry makes this the most up-to-date manual available for forcible entry training.

SELF-CONTAINED BREATHING APPARATUS

Beginning with the history of breathing apparatus and the reasons they are needed, to how to use them, including maintenance and care, the firefighter is taken step by step with the aid of programmed-learning questions and answers throughout to complete knowledge of the subject. The donning, operation, and care of all types of breathing apparatus are covered in depth, as are training in SCBA use breathing-air purification and recharging cylinders. There are also special chapters on emergency escape procedures and interior search and rescue.

FIRE VENTILATION PRACTICES

Objectives and advantages of ventilation; requirements for burning, flammable liquid characteristics and products of combustion; phases of burning, backdrafts, and the transmission of heat; construction features to be considered; the ventilation process including evaluating and size-up is discussed in length.

FIRE SERVICE RESCUE PRACTICES

Sections include water and ice rescue, trenching, cave rescue, rigging, search-and-rescue techniques for inside structures and outside, and taking command at an incident. Also included are vehicle extrication and a complete section on rescue tools. The book covers all the information called for by the rescue sections of NFPA 1001 for Fire Fighter I, II, and III, and is profusely illustrated.

HAZARDOUS MATERIALS FOR FIRST RESPONDERS

Designed to assist first-arriving companies in identification of hazardous materials and scene assessment. Covers initial scene control and operations to maintain safety for all responders. Includes characteristics of hazardous materials, identifying hazardous materials, pre-incident planning, personal protective equipment, command and control of incidents, operations at hazardous materials incidents, and control agents.

STUDY GUIDE FOR HAZARDOUS MATERIALS FOR FIRST RESPONDERS

Written to complement the *Hazardous Materials for First Responders* manual, this study guide allows students to review procedures and ensure they understand characteristics of hazardous materials, pre-incident planning, command, and on-scene tactics. Through the use of questions and answers, this study guide promotes retention of essential information. Included are case studies that simulate hazardous material situations and incidents.

THE FIRE DEPARTMENT COMPANY OFFICER

This manual focuses on the basic principles of fire department organization, working relationships, and personnel management. For the firefighter aspiring to become a company officer and the company officer who wishes to improve management skills, this manual will be invaluable. This manual will help individuals develop and improve the necessary traits to effectively manage the fire company.

FIRE CAUSE DETERMINATION

Covers need for determination, finding origin and cause, documenting evidence, interviewing witnesses, courtroom demeanor; and more. Ideal text for company officers, firefighters, inspectors, investigators, insurance and industrial personnel.

PRIVATE FIRE PROTECTION & DETECTION

Automatic sprinkler systems, special extinguishing systems, standpipes, detection and alarm systems. Includes how to test sprinkler systems for the firefighter to meet NFPA 1001.

INDUSTRIAL FIRE PROTECTION

Devasting fires in industrial plants occur at a rate of 145 fires every day. *Industrial Fire Protection* is the single source document designed for training and managing industrial fire brigades. A must for all industrial sites, large and small, to meet the requirements of the Occupational Safety and Health Administration's (OSHA) regulation 29 CFR part 1910, Subpart L, concerning incipient industrial fire fighting.

CHIEF OFFICER

The role of the fire service has expanded from solely fire suppression to include public education, emergency medical services, and hazardous materials control. This manual provides an overview of the skills needed by today's chief officer. Included are coordination of emergency medical services, master planning, disaster planning, budgeting, information management, labor relations, and the political process. Referenced where appropriate to NFPA 1021, *Fire Fighter Professional Qualifications*, for levels V and VI.

SI CHIEF OFFICER

This manual assists those reading the text in retaining the principles and concepts necessary to be a skilled chief officer. It includes structured questions and answers, exercises, and a new supplemental section on Resolution by Objectives. It will serve as a review of the text and assist in promotional exam preparation.

PUBLIC FIRE EDUCATION

Public fire education planning, target audiences, seasonal fire problems, smoke detectors, working with the media, burn injuries, and resource exchange.

FIRE INSPECTION AND CODE ENFORCEMENT

This revised edition is designed to serve as a reference and training manual for fire department inspection personnel. Includes authority and responsibility; inspection procedures; principles of fire protection and fire cause determination; building construction for fire and life safety; means of egress; extinguishing equipment and fire protection systems; plans review, storage, handling, and use of hazardous materials.

STUDY GUIDE FOR FIRE INSPECTION AND CODE ENFORCEMENT

The *Fire Inspection and Code Enforcement Study Guide* is designed to supplement the *Fire Inspection and Code Enforcement* manual by providing questions and answers to key areas addressed within the manual. This study guide can help the student obtain a thorough understanding of NFPA 1031, levels I and II. Included are case studies that simulate inspection responsibilities.

WATER SUPPLIES FOR FIRE PROTECTION

Designed to help improve understanding of the principles, requirements, and standards used to provide water for fire fighting. Revised (1987) to include information about rural water supplies. Includes water supply management, water system fundamentals, fire hydrants, fire flow testing, static sources, relay operations, and shuttle operations.

FIRE DEPARTMENT PUMPING APPARATUS

The driver/operator's encyclopedia on operating fire department pumps and pumping apparatus. Includes the driver/operator; operating emergency vehicles; types of pumping apparatus; positioning apparatus; fire pump theory; operating fire pumps; apparatus maintenance; apparatus testing; apparatus purchase and specifications. Also included are detailed appendices covering the operation of all major manufacturers' pumps.

STUDY GUIDE FOR FIRE DEPARTMENT PUMPING APPARATUS

This Study Guide is designed to supplement *Fire Department Pumping Apparatus*. It identifies important information and concepts by providing questions and answers developed directly from the manual. When properly used, it ensures a better understanding of driver/operator responsibilities and proficiency.

FIRE STREAM PRACTICES

Characteristics, requirements, and principles of fire streams; developing, computing, and applying various types of streams to operational situations; formulas for application of hydraulics; actions and reactions created by applying streams under different circumstances.

FIRE PROTECTION ADMINISTRATION

A reprint of the Illinois Department of Commerce and Community Affairs publication. A manual for trustees, municipal officials, and fire chiefs of fire districts and small communities. Subjects covered include officials' duties and responsibilities, organization and management, personnel management and training, budgeting and finance, annexation and disconnection.

BUILDING CONSTRUCTION

This 170-page manual covers building construction features vital to developing fire fighting tactics in a structure. Subjects include construction principles, assemblies and their fire resistance, building services, door and window assemblies and special types of structures.

FIREFIGHTER SAFETY

Basic concepts and philosophy of accident prevention; essentials of a safety program and training for safety; station house facility safety; hazards en route and at the emergency scene; personal protective equipment; special hazards, including chemicals, electricity, and radioactive materials; inspection safety; health considerations.

AIRCRAFT FIRE PROTECTION AND RESCUE PROCEDURES

Aircraft types, engines, and systems; conventional and specialized fire fighting apparatus, tools, clothing, extinguishing agents; dangerous materials; communications; pre-fire planning; and airfield operations.

GROUND COVER FIRE FIGHTING PRACTICES

Ground cover fire apparatus, equipment, extinguishing agents, and fireground safety; organization and planning for ground cover fire; authority, jurisdiction, and mutual aid; techniques and procedures used for combating ground cover fire.

INCIDENT COMMAND SYSTEM

Developed by a multi-agency task force, this manual is designed to be used by fire, police, and other government groups during an emergency. ICS is the approved basic command system as taught at the National Fire Academy. Includes components of ICS, major incident organization, and strike team kind/types and minimum standards.

LEADERSHIP FOR THE COMPANY OFFICER

A 12- to 15-hour course designed for the new company officer or the firefighter who anticipates promotion to company officer. Includes introduction to leadership, leadership techniques, theories of human motivation, determining leadership style, leadership styles, and demanding leadership situations.

FIRE SERVICE PRACTICES FOR VOLUNTEER AND SMALL COMMUNITY FIRE DEPARTMENTS

A general overview of material covered in detail in Forcible Entry, Ladders, Hose, Salvage and Overhaul, Fire Streams, Apparatus, Ventilation, Rescue, Inspection, Self-Contained Breathing Apparatus, and Public Fire Education.

PRINCIPLES OF EXTRICATION

This manual is designed for use by firefighters and emergency medical service personnel who respond to extrication emergencies. The text includes information on the organization and administration of a rescue company, taking charge and scene assessment, site management and incident command, rescue vehicles and equipment, personal protective equipment for rescuers, extrication from automobiles, extrication from buses, extrication and evacuation of passenger trains, agricultural extrication and rescue, and industrial extrication.

HAZ MAT LEAK & SPILL GUIDE

A brief, practical treatise that reviews operations at spills and leaks. Sample S.O.P. and command recommendations along with a decontamination guide.

TRANSPARENCIES

Multicolored overhead transparencies to augment *Essentials of Fire Fighting* and other texts. Since costs and availability vary with different chapters, contact IFSTA Headquarters for details.

SLIDES

2-inch by 2-inch slides that can be used in any 35 mm slide projector. Subjects include:

Sprinklers (4 modules)
Smoke Detectors Can Save Your Life
Matches Aren't For Children
Public Relations for the Fire Service
Public Fire Education Specialist (Slide/Tape)

FIREFIGHTER VIDEOTAPE SERIES

Designed to reinforce basic skills and increase knowledge on a variety of fire fighting topics. Excellent for use with *Essentials* or *Volunteer* to review and emphasize different topics. Available for Firefighter levels I, II, and III.

IFSTA BINDERS

Heavy-duty three-ring binders for organizing and protecting your IFSTA manuals. Available in two sizes: 1 1/2 inch and 3 inch.

WATER FLOW SUMMARY SHEETS

50 summary sheets and instructions for use; logarithmic scale to simplify the process of determining the available water in an area.

PERSONNEL RECORD FOLDERS

Personnel record folders should be used by the training officer for each member of the department. Such data as training, seminars, and college courses can be recorded, along with other valuable information. Letter or legal size.

ADDITIONAL PUBLICATIONS AND TRAINING MATERIALS ARE AVAILABLE, CALL FOR A FREE CATALOG.
Call Toll Free 1-800-654-4055 or FAX # 1-405-744-8204

SHIP TO: DATE: _____

NAME _____

CUSTOMER ACCOUNT NO: _____ PHONE _____

STREET ADDRESS (Shipped UPS) _____

CITY _____ STATE _____ ZIP _____

SIGNATURE _____

SOCIAL SECURITY NO. OR FEDERAL ID NO. _____

(NOTE: Order cannot be processed without signature.)

☐ VISA ☐ MASTERCARD ☐ AMERICAN EXPRESS

CARD # _____ EXP. DATE _____

Payment Enclosed ☐   Bill Me Later ☐

ifsta® ORDER FORM

10/90

Send to:
Fire Protection Publications
Oklahoma State University
Stillwater, Oklahoma 74078-0118
1-800-654-4055
Or Contact Your Local Distributor

Allow 4 to 6 weeks for delivery.

FILL IN THE ITEMS AND QUANTITIES DESIRED

QUANTITY	TITLE	LIST PRICE	TOTAL

Orders **outside** United States, contact Customer Services for shipping and handling charges.

Obtain postage and prices from current IFSTA Catalog or they will be inserted by Customer Services.

NOTE: Payment with your order saves you postage and handling charges when ordering from Fire Protection Publications.

SUBTOTAL _____

Postage and Handling
if applicable _____

TOTAL _____

TOLL-FREE NUMBER
1-800-654-4055

FAX YOUR ORDER
1-405-744-8204

SHIP TO: DATE: _____

NAME _____

CUSTOMER ACCOUNT NO: _____ PHONE _____

STREET ADDRESS (Shipped UPS) _____

CITY _____ STATE _____ ZIP _____

SIGNATURE _____

SOCIAL SECURITY NO. OR FEDERAL ID NO.

(NOTE: Order cannot be processed without signature.)

☐ VISA ☐ MASTERCARD ☐ AMERICAN EXPRESS

CARD # _____ EXP. DATE _____

Payment Enclosed ☐   Bill Me Later ☐

ifsta ® **ORDER FORM**

10/90

Send to:
Fire Protection Publications
Oklahoma State University
Stillwater, Oklahoma 74078-0118
1-800-654-4055
Or Contact Your Local Distributor

Allow 4 to 6 weeks for delivery.

FILL IN THE ITEMS AND QUANTITIES DESIRED

QUANTITY	TITLE	LIST PRICE	TOTAL

Orders **outside** United States, contact Customer Services for shipping and handling charges.

Obtain postage and prices from current IFSTA Catalog or they will be inserted by Customer Services.

NOTE: Payment with your order saves you postage and handling charges when ordering from Fire Protection Publications.

SUBTOTAL _____

Postage and Handling
if applicable _____

TOTAL _____

TOLL-FREE NUMBER
1-800-654-4055

FAX YOUR ORDER
1-405-744-8204

COMMENT SHEET

DATE _____ NAME _____

ADDRESS _____

ORGANIZATION REPRESENTED _____

CHAPTER TITLE _____ NUMBER _____

SECTION/PARAGRAPH/FIGURE _____ PAGE _____

1. Proposal (include proposed wording, or identification of wording to be deleted),
    OR PROPOSED FIGURE:

2. Statement of Problem and Substantiation for Proposal:

RETURN TO: IFSTA Editor   SIGNATURE _____
            Fire Protection Publications
            Oklahoma State University
            Stillwater, OK 74078

Use this sheet to make any suggestions, recommendations, or comments. We need your input to make the manuals
the most up to date as possible. Your help is appreciated. Use additional pages if necessary.